Smart Antennas and Electromagnetic Signal Processing for Advanced Wireless Technology

with Artificial Intelligence Applications and Coding

RIVER PUBLISHERS SERIES IN COMMUNICATIONS

Indexing: All books published in this series are submitted to the Web of Science Book Citation Index (BkCI), to SCOPUS, to CrossRef and to Google Scholar for evaluation and indexing.

The "River Publishers Series in Communications" is a series of comprehensive academic and professional books which focus on communication and network systems. Topics range from the theory and use of systems involving all terminals, computers, and information processors to wired and wireless networks and network layouts, protocols, architectures, and implementations. Also covered are developments stemming from new market demands in systems, products, and technologies such as personal communications services, multimedia systems, enterprise networks, and optical communications.

The series includes research monographs, edited volumes, handbooks and textbooks, providing professionals, researchers, educators, and advanced students in the field with an invaluable insight into the latest research and developments.

For a list of other books in this series, visit www.riverpublishers.com

Smart Antennas and Electromagnetic Signal Processing for Advanced Wireless Technology

with Artificial Intelligence Applications and Coding

Paul R. P. Hoole

D.Phil. Eng. Oxford University Wessex

Institute of Technology

United Kingdom

LONDON AND NEW YORK

Published 2020 by River Publishers
River Publishers
Alsbjergvej 10, 9260 Gistrup, Denmark
www.riverpublishers.com

Distributed exclusively by Routledge
4 Park Square, Milton Park, Abingdon, Oxon OX14 4RN
605 Third Avenue, New York, NY 10158

First published in paperback 2024

Smart Antennas and Electromagnetic Signal Processing for Advanced Wireless Technology with Artificial Intelligence Applications and Coding / by Paul R. P. Hoole.

Routledge is an imprint of the Taylor & Francis Group, an informa business

Publisher's Note
The publisher has gone to great lengths to ensure the quality of this reprint but points out that some imperfections in the original copies may be apparent.

While every effort is made to provide dependable information, the publisher, authors, and editors cannot be held responsible for any errors or omissions.

ISBN: 978-87-7022-206-8 (hbk)
ISBN: 978-87-7004-264-2 (pbk)
ISBN: 978-1-003-33956-4 (ebk)

DOI: 10.1201/9781003339564

Dedication

Chrishanthy Hoole, MBBS
Esther Hoole, LLB, LLM
Ezekiel Hoole, B.Sc.
Elisabeth Hoole, BA Econ. reg.

The four most treasured people that God generously blessed me with.
Soli Deo Gloria

Contents

Preface

This book seeks to provide an in-depth knowledge of both antenna theory and related signal processing techniques that make smart antennas a powerful and an integral tool of 5G wireless systems, future 6G system for Internet of Things (IoT), and in imaging using electromagnetic signals. Most books either focus on pure antenna theory and techniques or on signal processing techniques relevant to antennas. But in the current state-of-the-art technology, when we speak of antennas and antenna applications, the software that is tagged to the antenna doing the signal processing is an integral part of the performance and applicability of the antenna as a whole. Although viewing, understanding, analyzing, and designing the antenna system as a whole, which includes hardware and signal processing software, is very demanding, there is currently no way of avoiding such complexity in order to be a good researcher or engineer in this field. Although the book makes demanding reading, since it combines in-depth science of antenna electromagnetics with electromagnetic signal processing necessary for intelligent or smart antennas, it will prove to be very rewarding. The book is also scattered with details related to smart antennas and signal processing and artificial intelligence (AI) techniques which are often overlooked but a necessary part of good design, implementation, and operation of wireless systems. Applications of the smart antennas, including the various electromagnetic signal processing techniques, presented in this book are found in communication engineering, electric power grid communication, control and command, medical signal and image processing, monitoring and diagnostics, radar systems, safety and security systems, vehicle navigation and collision avoidance, remote sensing, IoT, and space technology, to name a few.

A review of the predecessor to this book, *Smart Antennas and Signal Processing*, published 20 years ago, had the following to say: *The book is refreshingly positioned by concentrating on applied antenna engineering...The style attempts, with some success to make a complex and wide-ranging subject accessible...a useful addition to the bookshelf of many engineers who are responsible for using antennas as part of a complete system. There are a number of up-to-date considerations which usefully read across to many systems* (IEE Electromagnetics Professional Network Online). This book builds on the previous book, with a substantial amount of new material, including the important application of AI in antenna beamforming and the new developments in 5G wireless technology.

The first four chapters of the book cover the essential and fundamental theory of all types of antennas at a senior undergraduate level. The fifth chapter presents the details of generating a steerable antenna beam without the use of reflectors or other mechanical structures which are commonly used. It also presents three-element and two-element array antennas for mobile wireless systems. The sixth chapter deals with one powerful application of electromagnetic two-dimensional signal or image processing technique used to make antennas able to image systems in applications in aerospace and military industries. The seventh to ninth chapters address advanced techniques used in smart antennas for beam forming, tracking a moving antenna in mobile stations and the uses of AI in smart antenna beam forming. The technological aspects of the wireless systems, of which the smart antennas are an essential part, are addressed in chapters 10 and 11. In the last chapter, we present coding techniques for smart antennas. The book grew out of many years of teaching and collaborative research on smart antennas and electromagnetic signal processing.

To the best of our knowledge, this is the only book that covers in reasonable depth traditional antenna theory, antennas in imaging systems, and the most recent developments in smart antenna signal processing and advanced wireless technology. The eyes, ears, and brain of intelligent wireless systems including the 5G and future 6G wireless systems are the smart antennas. Antennas are the eyes and ears, and the signal processing module is the brain of smart antennas. Antennas are no longer thought of as stand-alone devices but have incorporated into them signal processing, tracking, and beamforming algorithms that make them very powerful, active devices used in communication, diagnostic, military, safety, digital power, IoT, and imaging systems. The book seeks to capture these overall developments and exciting features of modern wireless and antenna technology through a theoretical and practical approach. A common thread that unifies the book is the concept of antenna as a temporal–spatial filter, a device that is capable of operating on signals in the time and three-dimensional spatial domains.

Some useful computer code listings are given so that the students may actually see how antennas work, learn how to develop signal processing modules for antennas, and understand and visualize the design and implementation of smart antennas and complex routines using the concept of encoded antenna signals and synthetic antennas. It is our view that when an engineering student sees numbers associated with equations, he or she becomes more confident about the underlying theory as well as its use in practical situations. The computer code listings given in the book are meant only for beginners and are not intended for professional or commercial use.

With the rapid increase in satellite, mobile, and remote sensing systems, properly designed antennas would greatly increase the effectiveness of the systems, in addition to increasing the efficiency of the use of the frequency spectrum. Smaller antennas are used as probes in biomedical applications and other diagnostic systems, in addition to their wide use in mobile and military technology. This book aims to provide a clear, in-depth, and concise description of antenna analysis, smart antenna, and signal processing techniques applied. Although the emphasis is on the theoretical understanding and design of smart antennas, the book contains practical insight into antenna troubleshooting, selection, and installation. The book will be of value to the undergraduate student, the graduate student, and the practicing engineer.

The motivation for writing this book arose from the need that the authors felt for a text on antennas which would cover all the important aspects related to antennas in an attractive, comprehensive, in-depth, and practical manner. The audiences addressed are undergraduate students and graduate students just getting into the demands of problem formulation and solution in antennas and related subjects. There are many textbooks on electromagnetic fields that give a brief introduction to antennas. There are other books that go into great details and analysis of antennas. Even those books that give a detailed analysis of antennas often fail to interest the student by missing out the important technical, signal/image processing, computer coding details related to antennas. Very few books cover the signal processing aspects of antenna design and system engineering. Furthermore, books that are generally addressed to practicing engineers tend to play down the importance of the theoretical foundation that is necessary for better design and troubleshooting of antenna associated systems. This book makes an attempt to fill in the gap with a small useful volume.

I would like to acknowledge the volumes of affection and strength the families of my wife and mine have been to me in all my endeavors, and in particular my wife Chrishanthy and our three children Esther, Ezekiel, and Elisabeth.

Great are the works of the LORD, studied by all who delight in them (Psalm 111: 2).

The author would appreciate feedback from readers that make use of the book, for any further improvements that could be made in future editions.

Paul Hoole, Wessex Institute of Technology, December 2020.

List of Contributors

U.R. Abeyratne, *School of Electrical and Electronic Engineering, Nanyang Technological University, Singapore*

Kho Lee Chin, *Faculty of Electrical and Electronic Engineering, Universiti Malaysia Sarawak, Malaysia*

Ng Kim Chong, *School of Electrical and Electronic Engineering, Nanyang Technological University, Singapore*

Dennis Goh, *School of Electrical and Electronic Engineering, Nanyang Technological University, Singapore*

E. Gunawan, *School of Electrical and Electronic Engineering, Nanyang Technological University, Singapore*

Stetson Oh Kok Leong, *School of Electrical and Electronic Engineering, Nanyang Technological University, Singapore*

H.M.C.J. Herath, *Department of Electrical and Electronic Engineering, University of Peradeniya, Peradeniya, Sri Lanka*

H.M.G.G.J.G. Herath, *Department of Electrical and Electronic Engineering, University of Peradeniya, Peradeniya, Sri Lanka*

P.R.P. Hoole, *Wessex Institute of Technology, Chilworth, Southampton, United Kingdom*

S.R.H. Hoole, *FIEEE as Professor of Electrical Engineering (Retired), Michigan State University, USA*

Tan Pek Hua, *School of Electrical and Electronic Engineering, Nanyang Technological University, Singapore*

H. Kunsei, *Papua New Guinea University of Technology, Lae, Papua New Guinea*

Ade Syaheda Wani Marzuki, *Faculty of Electrical and Electronic Engineering, Universiti Malaysia Sarawak, Malaysia*

Dayang Azra Awang Mat, *Faculty of Electrical and Electronic Engineering, Universiti Malaysia Sarawak, Malaysia*

K. Pirapaharan, *Faculty of Engineering, University of Jaffna, Jaffna, Sri Lanka*

K.M.U.I. Ranaweera, *Department of Electrical and Electronic Engineering, University of Peradeniya, Peradeniya, Sri Lanka*

K.S. Senthilkumar, *Department of Computers and Technology, St. George's University, Grenada, West Indies*

D.N. Uduwawala, *Department of Electrical and Electronic Engineering, University of Peradeniya, Peradeniya, Sri Lanka*

Dayang Nurkhairunnisa Abang Zaidel, *Faculty of Electrical and Electronic Engineering, Universiti Malaysia Sarawak, Malaysia*

List of Figures

List of Tables

List of Abbreviations

Physical Constants

Permittivity of free space, $\varepsilon_\circ = 8.854 \times 10^{-12}$ F/m
Permeability of free space, $\mu_\circ = 4\pi \times 10^{-7}$ H/m
Boltzmann's constant, $\mathrm{k} = 1.38 \times 10^{-23}$ J/K
Planck's constant h = 6.63×10^{-34} J-s
Electric charge of an electron, $\mathrm{e} = -1.602 \times 10^{-19}\mathrm{C}$
Mass of an electron $\mathrm{m_e} = 9.11 \times 10^{-31}$ kg
Mass of a proton $\mathrm{m_p} = 1.67 \times 10^{-27}$ kg
Speed of electromagnetic waves and light in free space, $\mathrm{c} = 3 \times 10^8\ \mathrm{m/s}$
Impedance of free space, $\mathrm{z_\circ}$ or $\eta_\circ = 376.7\ \Omega$

Conductivity, σ (in Siemen/meter), at room temperature

Conductors:

Silver (6.17×10^7)
Copper (5.8×10^7)
Aluminium (3.82×10^7)
Brass (2.56×10^7)
Tungsten (1.83×10^7)
Nickle (1.45×10^7)
Iron (1.03×10^7)
Michrome (0.1×10^7)
Mercury (1.0×10^6)
Graphite (3.0×10^4)
Sea water (4.0)
Intrinsic Germanium (2.2)
Ferrite (1.0×10^{-2})

Intrinsic semiconductors:

Intrinsic Silicon (0.44×10^{-3})

Insulators:

Distilled water (1.0×10^{-4})
Bakellite (1.0×10^{-9})
Glass (1.0×10^{-12})

Mica (6.17 $\times 10^{-15}$)
Fused quartz (6.17 $\times 10^{7}$)

Relative Permittivity, $\varepsilon_{\mathbf{r}}$, at low frequencies
Air 1.0006
Bakelite 4.8
Glass 6.0
Lucite 3.2
Nylon 3.6
Plexiglas 3.45
Polythylene 2.26
Polystyrene 2.5
Quartz 3.8
Dry Soil 3.0
Teflon 2.1
Water 80

Relative Permeability $\mu_{\mathbf{r}}$
Diamagnetic materials:
Bismuth 0.9999834
Silver 0.99998
Copper 0.999991
Vacuum 1.0 (Nonmagnetic)
Paramagnetic materials:
Aluminium 1.00002
Nickle chloride 1.00004
Ferromagnetic materials:
Cobalt 250
Nickel 600
Mild steel 2000
Iron 5000
Mumetal 100,000
Supermalloy 800,000

Frequency Bands: General
Very Low Frequency VLF 3-30 kHz
Low Frequency (Long waves) LF 30-300 kHz
Medium Frequency (medium wave) MF 0.3-3 MHz
High Frequency (Short Waves) HF 3-30 MHz
Very High Frequency VHF 30-300 MHz
Ultra High Frequency UHF 0.3-3 GHz

Super High Frequency (centimeter wave) SHF 3-30 GHz
Extra High Frequency (millimeter wave) EHF 30-300 GHz

Frequency Bands: Microwaves
L Band 1-2 GHz
S Band 3-4 GHz
C Band 4-8 GHz
X Band 8-12 GHz
Ku Band 12-18 GHz
K Band 18-26 GHz
Ka Band 26-40 GHz
V Band 40-75 GHz
W Band 75-111 GHz

Basic Mathematical Rclations
$e^{j\theta} = \cos\theta + j\ \sin\theta$
$e^{-j\theta} = \cos\theta - j\ \sin\theta$
$e^{j\theta} + e^{-j\theta} = 2\cos\theta$
$e^{j\theta} - e^{-j\theta} = -j\ 2\sin\theta$
$\cos(-\theta) = \cos\theta$
$\sin(-\theta) = -\sin\theta$
$\cos(A+B) = \cos A \cos B - \sin A \sin B$
$\cos(A-B) = \cos A \cos B + \sin A \sin B$
$\sin(A+B) = \sin A \cos B + \cos A \sin B$
$\sin(A-B) = \sin A \cos B - \cos A \sin B$
$x \cos A - y \sin B = (x^2 + y^2)^{1/2} \cos(A + \tan^{-1}(y/x))$
$x \cos A + y \sin B = (x^2 + y^2)^{1/2} \cos(A - \tan^{-1}(y/x))$
$\sin(2A) = 2 \sin A \cdot \cos B$
$\cos(2A) = \cos^2 A - \sin^2 A = 1 - 2\sin^2 A = 2\cos^2 A - 1$
$2\sin^2 A = 1 - \cos(2A)$
$2\cos^2 A = 1 + 2\cos(2A)$
$\mathbf{A} \cdot \mathbf{B} = |\mathbf{A}|\,|\mathbf{B}| \cos\theta$
$\mathbf{A} \times \mathbf{B} = |\mathbf{A}|\,|\mathbf{B}| \sin\theta\, \mathbf{u}_n$

$$\mathbf{grad} f(n_1, n_2, n_3) = \nabla f(n_1, n_2, n_3) = \left(\frac{1}{h_1}\frac{\partial f}{\partial n_1}\mathbf{u}_{n1} + \frac{1}{h_2}\frac{\partial f}{\partial n_2}\mathbf{u}_{n2} + \frac{1}{h_3}\frac{\partial f}{\partial n_3}\mathbf{u}_{n3}\right)$$

$$\nabla \cdot \mathbf{F}(x, y, z) = \frac{\partial F_x}{\partial x} + \frac{\partial F_y}{\partial y} + \frac{\partial F_z}{\partial z}$$

$$\nabla \cdot \mathbf{F}(r, \phi, z) = \frac{1}{r}\frac{\partial(rF_r)}{\partial r} + \frac{1}{r}\frac{\partial F_\phi}{\partial \phi} + \frac{\partial F_z}{\partial z}$$

$$\nabla \cdot \mathbf{F}(r, \theta, \phi) = \frac{1}{r^2}\frac{\partial(r^2 F_r)}{\partial r} + \frac{1}{r\sin\theta}\frac{\partial(F_\theta \sin\theta)}{\partial \theta} + \frac{1}{r\sin\theta}\frac{\partial F_\phi}{\partial \phi}$$

$$\oint_S \mathbf{F}\cdot d\mathbf{s} = \int_v \nabla\cdot\mathbf{F}\, dv$$

$$\nabla\times\mathbf{F}(x,y,z) = \left\{\frac{\partial F_z}{\partial y} - \frac{\partial F_y}{\partial z}\right\}\mathbf{u}_x + \left\{\frac{\partial F_x}{\partial z} - \frac{\partial F_z}{\partial x}\right\}\mathbf{u}_y + \left\{\frac{\partial F_y}{\partial x} - \frac{\partial F_x}{\partial y}\right\}\mathbf{u}_z$$

$$\nabla\times\mathbf{F}(r,\phi,z) = \left\{\frac{1}{r}\frac{\partial F_z}{\partial \phi} - \frac{\partial F_\phi}{\partial z}\right\}\mathbf{u}_r + \left\{\frac{\partial F_r}{\partial z} - \frac{\partial F_z}{\partial r}\right\}\mathbf{u}_\phi + \frac{1}{r}\left\{\frac{\partial(rF_\phi)}{\partial r} - \frac{\partial F_r}{\partial \phi}\right\}\mathbf{u}_z$$

$$\nabla\times\mathbf{F}(r,\theta,\phi) = \frac{1}{r\sin\theta}\left\{\frac{\partial(F_\phi\sin\theta)}{\partial\theta} - \frac{\partial F_\theta}{\partial\phi}\right\}\mathbf{u}_r + \frac{1}{r}\left\{\frac{1}{\sin\theta}\frac{\partial F_r}{\partial\phi} - \frac{\partial(rF_\phi)}{\partial r}\right\}\mathbf{u}_\theta + \frac{1}{r}\left\{\frac{\partial(rF_\theta)}{\partial r} - \frac{\partial F_r}{\partial\theta}\right\}\mathbf{u}_\phi$$

$$\oint_C \mathbf{F}\cdot d\mathbf{l} == \int_S (\nabla\times\mathbf{F})\cdot d\mathbf{s}$$

$$\nabla\times\nabla\phi = 0$$

$$\nabla\cdot\nabla\times\mathbf{A} = 0$$

$$\nabla(\phi\psi) = \phi(\nabla\psi) + \psi(\nabla\phi)$$

$$\nabla\cdot(\phi\mathbf{A}) = \mathbf{A}\cdot(\nabla\phi) + \phi\ (\nabla\cdot\mathbf{A})$$

$$\nabla\times(\phi\mathbf{A}) = (\nabla\phi)\times\mathbf{A} + \phi\ (\nabla\times\mathbf{A})$$

$$\mathbf{A}\cdot(\mathbf{B}\times\mathbf{C}) = \mathbf{B}\cdot(\mathbf{C}\times\mathbf{A}) = \mathbf{C}\cdot(\mathbf{A}\times\mathbf{B})$$

$$\mathbf{A}\times\mathbf{B}\times\mathbf{C} = \mathbf{B}(\mathbf{A}\cdot\mathbf{C}) - \mathbf{C}(\mathbf{A}\cdot\mathbf{B})$$

Symbols used in Text

$\mathbf{A}$	Vector magnetic potential (Weber/metre)
A_e	Antenna aperture area (m^2)
AF	Array Factor
$\mathbf{B}$	Magnetic flux density (Tesla)
B	Bandwidth
C	Capacitance (Farad)
c	Velocity of Light (meter/second^2)
D	Antenna Directivity
$\mathbf{D}$	Electric Flux Density (Coulomb/m^2)
$\mathbf{E}$	Electric field Intensity (Volt/metre)
f	Frequency (Hertz)
$\mathbf{F}$	Force (Newton)
G	Conductance (Siemens)
G	Antenna Gain
$\mathbf{H}$	Magnetic field Intensity (Ampere/metre)
I	Current (Amperes)
$\mathbf{J}$	Current density (Ampere/m^2)
J	Jacobian Matrix
l	Length (metre)

$\boldsymbol{L}$	Inductance (Henry)
m	Magnetic dipole Moment (Ampere m^2)
M	Mutual Inductance (Henry)
$\mathbf{M}$	Magnetic polarization vector (Ampere/metre)
N	Number of Turns
p	Probability Density Function
P	Power (Watt)
P_i	Power Density
$\mathbf{P}$	Poynting Vector
Q, q	Electric Charge (Coulombs)
r	Distance (metre)
R	Ohmic Resistance (Ohm)
R_r	Radiation Resistance (Ohm)
$\mathbf{R}$	Distance vector (Metre)
s	Surface area (metre^2)
S	Radiation Intensity
S	Standing Wave Ratio
t	time (seconds)
$\boldsymbol{T}$	Temperature (Kelvin)
T	Period (second)
T	Pulse Duration
u	speed
v	Volume (metre^3)
v	Velocity (Metre/second)
V	Potential difference (Volts)
w	Neural Network weights
W	Work, Energy (Joule)
x	Neural Network input signals
y	Neural Network output signals
y	Normalized admittance
Y	Admittance (Siemens)
z	Normlized impedance
Z	Impedance (Ohm)
Z_0,	Characterisitc Impedance of transmission line (Ohm)
Z_0, η_0	Characterisitc Impedance of free space (Ohm)
$\mathbf{u_x}$, $\mathbf{u_y}$, $\mathbf{u_z}$	Unit vectors in Cartesian coordinates
$\mathbf{u_r}$, $\mathbf{u_\phi}$, $\mathbf{u_z}$	Unit vectors in cylindrical coordinates
$\mathbf{u_r}$, $\mathbf{u_\theta}$, $\mathbf{u_\phi}$	Unit vectors in spherical coordinates
α	Angle (Radian)

α	Attenuation constant (Neper/metre)
β	Angle (Radian)
β	Phase constant (Radian/metre)
γ	Propagation constant (1/metre)
δ	Angle (Radian)
δ	Skin depth (Metre)
Γ	Reflection coefficient
θ	Elevation Angle (Radian)
ϕ	Azimuth Angle (Radian)
η	Neural Network kearning rate
τ	Transmission coefficient
τ	Time Delay
ε	Permittivity (Farad/meter)
ε_r	Relative permittivity
μ	Permeability (Henry/metre)
μ_r	Relative Permeability
ρ_l	Line Electric charge density (Coulomb/metre)
ρ_s	Surface Electric charge density (Coulomb/metre^2)
ρ_v	Volume Electric charge density (Coulomb/metre^3)
σ	Conductivity (Siemens/metre)
σ	Radar cross section (Metre^2)
ϕ	Angle (Radian)
ω	Radian frequency (Radian/second)
Ω	Beam solid angle (Steradian)

List of prefixes (multipliers)

tetra (10^{12})
giga (10^9)
mega (10^6)
kilo (10^3)
hecto (10^2)
deka (10)
deci (10^{-1})
centi (10^{-2})
milli (10^{-3})
micro (10^{-6})
nano (10^{-9})
pico (10^{-12})
femto (10^{-15})
atto (10^{-18})

1

Introduction

P.R.P. Hoole

Abstract

In this chapter, we review the fundamentals of antennas and antenna system equations including the Friis equation for antennas in communication systems and the radar equation for antennas used in radar systems. The chapter presents the basic operation principle of antennas, the basic antenna characteristics, and the antenna as an electronic filter in the frequency domain and as a spatial filter in the space domain. The basic antenna parameters are presented, including antenna gain, antenna directivity, antenna impedance, effective aperture area, and antenna temperature. The basic electromagnetic equations including energy transferred by electromagnetic waves radiated by antennas are presented. The basic factors to be considered when an antenna is designed or selected for an application are presented using an aircraft–satellite communication antenna requirement. The smart antenna is defined and described.

1.1 Elementary Principle

An antenna is a device that transmits and receives electromagnetic waves or signals. When transmitting, it has to send the signals in a particular direction or in all directions. Radiation must be maximized in a given direction or in some cases omni-directionally. The transmitting and receiving patterns of any antenna are identical. Figure 1.1 shows the transmitting and receiving radiation patterns of an antenna.

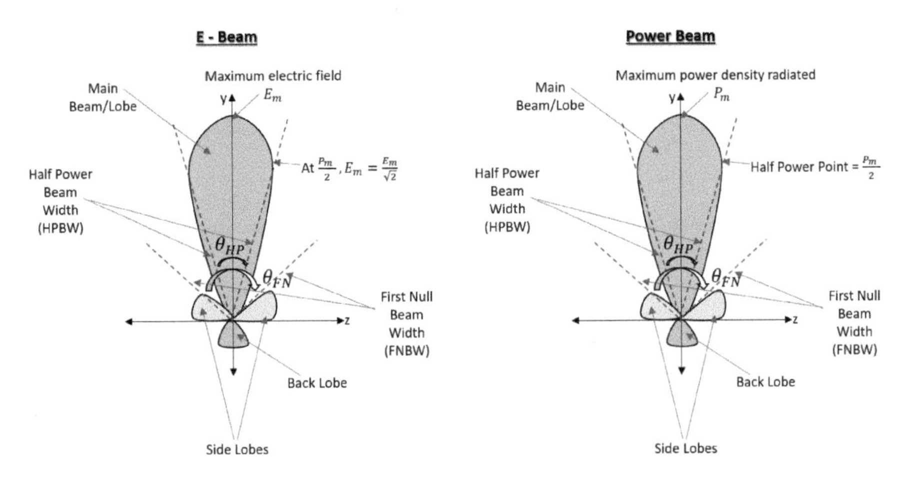

Figure 1.1 Radiation pattern of an antenna.

The radiation E-beam pattern indicates the magnitude of electric field strength observed at a constant distance r from the antenna in different directions. Most antennas, as indicated, are designed to transmit and receive the maximum signal in one particular direction at a specified center frequency and bandwidth. The power radiated in the small sidelobes and backlobes is wasted power and should be minimized. The sidelobes and backlobes in a transmitting antenna, in addition to being wasted radiated power, can lead to interference with other receiver antennas and systems. In receiver antennas, these can receive signals from unknown or unwanted source of signals (interference). The power beam has the same pattern as the E-beam, except that it is narrower since power is proportional to the square of the electric field. Hence, when power transmitted is $P_{\mathrm{m}}/2$, electric field is 0.707 E_{m}. The half-power beamwidth (HPBW) of the antenna is the region of the beam where the transmitted power is greater than or equal to half of maximum power transmitted, P_{m} W/m^2. That is greater than or equal to $P_{\mathrm{m}}/2$ W/m^2. In the E-field beam, this is the region in which the transmitted electric field is greater than or equal to 0.707 E_{m}, where E_{m} is the maximum transmitted electric field. In the region outside the HPBW, the transmitted signal of a transmitting antenna will be weak. It follows that any signal received by a receiver antenna will be weak when outside the HPBW.

How does an antenna work? The principle of antenna operation may be best explained by considering Faraday's law, which defines the induced voltage V in a coil as given by

$$V = \frac{\mathrm{d}(N\phi)}{\mathrm{d}t} = \frac{\mathrm{d}(NBA)}{\mathrm{d}t}$$
$$= N\left[B\frac{\mathrm{d}A}{\mathrm{d}t} + A\frac{\mathrm{d}B}{\mathrm{d}t}\right] \quad (1.1)$$

where N is the number of turns, B is the flux density, and A is the area of the coil. We note that there are two terms to the induced voltage. A receiving antenna may make use of one or both of these terms. When a time varying magnetic field cuts a conductor, it will induce a voltage, which will be proportional to $\mathrm{d}B/\mathrm{d}t$. This is how a wire antenna of a mobile telephone picks up signals. Alternatively, if a coil antenna is rotated in an electromagnetic field, there will be an induced voltage due to $\mathrm{d}A/\mathrm{d}t$, where the area of the coil cutting the signal is changing. Such antennas are used for direction finding; when the induced voltage is maximum, the antenna is directly facing the source that is emitting the electromagnetic signal. In military applications, such rotating antennas may be used to detect hidden transmitters and radar. In a transmitting antenna, a wire to which a time varying voltage (e.g. an frequency modulation (FM) radio signal) is applied will produce time varying magnetic flux density B in the space around it. At high frequencies, this B will produce an electric field intensity E. Both E and B together will now propagate through free space at the velocity of light.

1.2 Broadcast Frequency Bands

When we speak of frequency bands, we are describing the frequencies used in radio, TV, and other forms of electronic communications. These frequencies start at 1 Hz, then progress through the audio frequencies of 20–20,000 Hz, and then into the nine radio bands, all the way into the frequency range of visible light. From 1 to 300 GHz is generally considered to be microwave. The historic frequency band designations are as follows: extra low frequency, ELF (0–300 Hz); very low frequency, VLF (3–30 kHz); low frequency, LF (30–300 kHz); high frequency, HF (0.003–0.03 GHz); very high frequency, VHF (0.03–0.3 GHz); ultra-high frequency, UHF (0.3–3.0 GHz); super high frequency, SHF (3–30 GHz); extra high frequency, EHF (30–300 GHz); L-band (1.0–2.0 GHz), S-band (2.0–4.0 GHz), C-band (4.0–8.0 GHz), X-band (8.0–12.5 GHz), Ku-band (12.5–18.0 GHz),

K-band (18.0–26.5 GHz), and Ka-band (26.5–40.0 GHz). The millimeter (mm) waves range from 300 to 3000 GHz. There are new band designations too that are simpler than the historic designations: A (0–0.25 GHz), B (0.25–0.5 GHz), C (0.5–1.0 GHz), D (1.0–2.0 GHz), E (2.0–3.0 GHz), F (3.0–4.0 GHz), G (4.0–6.0 GHz), H (6.0–8.0 GHz), I (8.0–10.0 GHz), J (10.0–20.0 GHz), and K (20.0–40.0 GHz). We shall mostly use the historic band designations.

Communications, imaging, and radar antennas are normally designed for operation in the radio, TV, and microwave frequency bands. Consider a digital communication system. All image, video, and electronic mail data are represented by bits 1 and 0. For long distance communications, these rectangular pulses are modulated into sinusoidal signals. In FM, bit 1 may be represented by frequency f_1, and bit 0 by frequency f_2. In phase modulation, bit 1 may be represented by a sinusoidal signal at frequency f at phase 0°, whereas bit 0 may be represented by a sinusoidal signal at the same frequency but at phase 45°. Thus, in either case, the signal to be handled by the antenna is a sinusoidal electromagnetic signal. Now to pack more data (i.e. more 1s and 0s) into a short wave packet, higher frequencies have to be used since 1 bit typically takes $1/f$ s. This is the reason why communication systems carrying large amounts of information (e.g. two TV channels and 3000 telephone channels) must use the microwave frequency bands. However, low-data-rate amplitude modulation (AM) and FM radio stations use the lower frequency spectra. Antennas must be able to handle all these frequency bands. The type and size of antenna will depend on its frequency of operation. Infrared (wavelength range 10^4–10^6 m) used in night vision, visible light (wavelength range 720×10^9 m for deep red to 380×10^9 m for violet), ultraviolet light (wavelength range 10^7–10^{11} m used in sterilization), X-rays (wavelength range 10^{10}–10^{14} m) used in medical diagnostics like CT scans, and γ-rays (wavelength range 10^{11}–10^{16} m) need other kinds of radiators which are not dealt with here.

Figure 1.2 shows a simple communications receiver system and the electromagnetic spectrum. The particular VHF receiver shown in Figure 1.2 is designed to work at a frequency of about 200 MHz. The form in which data is received is the analog sinusoidal waveform. The 200 MHz waveform must be frequency downconverted (by the VHF receiver), processed (FM receiver) and converted into digital form by the digital-to-analog converter (DAC) before being stored in the converted form in a personal computer (PC). Such will be the sequence of steps adapted by an e-mail (electronic mail) received via a wireless system into your PC. The DAC is necessary only if

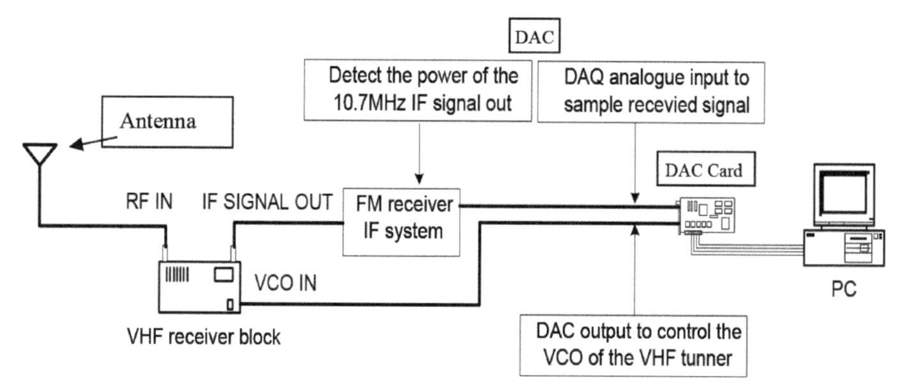

Figure 1.2 A communications receiver system.

the final form of the data should be in digital form, as in the case of e-mails. In the whole system, it is only in the final stage after the DAC that the waveform becomes a non-sinusoidal waveform of square pulse shape (e.g. 5 V for bit 1 and 0 V for bit 0).

Consider the issue of frequency bandwidth. A calling card is a credit card which enables you to make a telephone call from a public booth and to charge it on your private or office telephone. Note the enormous number of digits required. First, you dial the number you want to call (e.g. 298 1120) at area code, e.g. 818. Then your private telephone number (e.g. 7780 7698) to which you want the charge to be assigned.

Each number that you press is assigned two tones. For instance, if you press 2, then two electrical signals at frequencies 1336 and 697 Hz are transmitted along the telephone wire to the central exchange station. Then if it is wireless, both frequencies will be upconverted to frequencies around the carrier frequency of the wireless system (e.g. 900 or 1980 MHz). Thus, each number on dialing occupies two frequencies on the frequency spectrum. For the complete set of numbers, seven frequencies are required. Now this is only for the digits. When you start talking, more frequencies are required (if it is an analog FM system) to represent your voice tones. Hence, each voice link occupies a certain frequency bandwidth on the frequency spectrum. An antenna designed for a specific system (e.g. a mobile communication link at a carrier frequency of 800 MHz and a bandwidth of 100 kHz) will have to effectively launch and direct signals at that particular carrier frequency and over the specified bandwidth. The size of the antenna will depend on the frequency at which it is required to transmit or receive.

1.3 Basic Characteristics and Definitions of Terms

Wavelength. Wavelength is the distance traveled by one cycle of a radiated electric signal. The frequency of the signal is the number of cycles per second. It follows that frequency f is inversely proportional to the wavelength. Both wavelength and frequency are related to the speed of light c. The formula is given as follows:

$$c = f \times \lambda,$$

$$c = 3 \times 10^8 \text{ m/s (speed of light)},$$

$$f = \text{frequency (Hz)},$$

$$\lambda = \text{wavelength (m)}.$$

Sinusoidal waveform. Why do we mostly use sinusoidal signals? There are two simple reasons for this. First, consider the differentiation of a sine term

$$\frac{\mathrm{d}(\sin \omega t)}{\mathrm{d}t} = \omega \cos \omega t.$$

Although we get a cosine term, the shape of the resultant waveform is the same as that of the sine waveform. Communication electronic circuits are full of capacitors and inductors, which integrate or differentiate signals. This means that the waveform will not change if we use sinusoidal signals. Furthermore, we shall see that when high frequency signals are launched into free space, the signals propagate by means of the electric field intensity (E, V/m) and magnetic flux density (B, webers/m^2) maintaining each other through a process of spatial- and time-differentiation as defined by Maxwell's equations. Here again, using sinusoidal signals ensures that the basic waveforms of the signals carried by the electromagnetic wave do not change. If we had used a triangular waveform, for example, the resultant waveform after differentiation would be rectangular!

There is a second reason why a pure sinusoidal signal is attractive for communications: reduction of bandwidth. A single pure sinusoidal wave will only occupy a single frequency (spectrum) line in the frequency spectrum. Consider now a rectangular wave, say switching between 5 V (bit "1" in digital terms) and 0 V (bit "0"); if its Fourier transform is taken, the frequency spectrum will contain a collection of sinusoidal waves all at different frequencies. This means a rectangular pulse will occupy more space in the frequency spectrum, which is undesirable since the electromagnetic frequency spectrum is rapidly becoming filled up with many communications and remote sensing channels and systems. This is also one reason why the digital communication system is more attractive than the analog FM system. In the digital system, bits 1 and 0 may be modulated onto a single frequency carrier with a phase shift to differentiate bit 1 from bit 0. In an FM system, to carry a television channel for instance, 6 MHz bandwidth is required to transmit the picture and another 100 kHz or so for the voice.

Most of the antenna analysis presented in this book will assume a single-frequency electromagnetic signal. Of course all the results we obtain apply to a multiple-frequency signal as well (e.g. an FM signal), if we analyze the signal frequency by frequency (in its frequency domain), and then sum them all up to get back the resultant signal.

Radiation. Radiation is the emission of coherent modulated electromagnetic waves in free space from a single or a group of radiating antenna elements. In order to get directive radiation beams (patterns), a group of radiating antenna elements is used to maximize signal radiation in a given direction.

Array antenna. An array antenna consists of a group of radiating or receiving antenna elements.

Beamforming. In beamforming, we seek to determine the geometry of an antenna or the phase and current amplitudes (weights) for an array antenna to obtain a prescribed antenna beam (radiation pattern). This is also known as antenna synthesis.

Adaptive antenna. An adaptive antenna is an array of antennas that is able to change its radiation pattern to minimize the effects of noise, interference, and multipaths and to maximize signal reception in a given direction. It may also be used to transmit signals to a mobile receiver such that the incident signal strength at the mobile unit is always maximum. The adaptive nature of the antenna is due to the fact that it uses powerful signal processing techniques to do on-line, real-time adjustments of the radiation pattern to maximize its performance.

Switched-beam antenna. In a switched-beam antenna, a series of non-adaptive antenna beams are available at the transmitter/receiver site. By selecting a combination of beams, reception and transmission in a given direction may be maximized. This is not as dynamic as the adaptive antennas.

Smart antenna. In general, a smart antenna combines the adaptive antenna and switched-beam antenna technology. In communication systems, it is essentially an adaptive antenna since it intelligently adjusts the radiation pattern to maximize the performance of the antenna. In imaging systems, it may also include processing the signals to identify targets and to perform diagnostics on the object under observation.

Decibel notation. The formula for dB calculation when dealing with voltage levels is

$$\mathrm{dB} = 20\ \log(E_1/E_2). \tag{1.2}$$

Example 1.1. Given that

$$E_1 = 900\ \mathrm{mV/m},$$

$$E_2 = 100\ \mathrm{mV},$$

we find

$$\mathrm{dB} = 20\ \log(900/100),$$

$$\mathrm{dB} = 19.085.$$

The formula for dB calculation when dealing with power levels is given by

$$\mathrm{dB} = 10 \log(P_1/P_2) \tag{1.3}$$

since power is proportional to (voltage)^2 or (electric field)^2.

Radiation resistance. Radiation resistance is defined in terms of transmission, using Ohm's law, as the radiated power *P* from an antenna divided by the square of the driving currents *I* at the antenna terminals. From the perspective of the transmitter circuit connected to the antenna, the power radiated by the antenna is a power loss and is identified as the power dissipated in the radiation resistance.

Polarization. Polarization is the angle of the radiated field vector in the direction of maximum radiation. If the plane of the field is parallel to the ground, it is vertically polarized. When the receiving antenna is located in the same plane as the transmitting antenna, the received signal strength will be maximum. If the radiated signal is rotated at the operating frequency by electrical means in feeding the transmitting antenna, the radiated signal is circularly polarized. The circularly polarized signals produce equal received signal levels with either horizontal or vertical polarized receiving antennas.

1.4 Basic Antenna Parameters

All telecommunication systems use electromagnetic waves to carry information from a transmitter to the receiver. Similarly, even in digital computer systems, digital data may be transferred from the computer memory to the central processing unit by electromagnetic waves traveling along the copper conductors on the printed circuit board (PCB). In this chapter, we consider the launching of electromagnetic waves into free space and capturing the signal at the receiver using antennas. The basic concepts of some elementary antennas are described in the following sub-sections.

1.4.1 Antenna as a Spatial Filter: Radiation Pattern

The radiation pattern of an antenna gives an idea of the propagation of the radiated signal around the antenna. For an isotropic antenna (i.e. point radiating element), the radiation pattern will be spherical. The radiation pattern changes as the configuration of the radiating element changes. Figure 1.3 shows the radiating patterns of two directive antennas.

A major lobe (main beam) is the radiation lobe containing the maximum radiation power. Minor lobes are any lobes other than the main lobe. The minor (or side) lobes are unwanted since they radiate information (or electromagnetic energy) in unwanted and unintended directions. A backlobe is a minor lobe that occupies a direction opposite to that of the major lobe. Minor lobes should be minimized in order to get the antenna to radiate and receive only in the prescribed or intended direction.

Narrow or pencil beams are preferred for antennas that must be highly directive. In some applications like radio broadcasts where the radio program should be transmitted in all directions, the antenna radiation pattern should have a wide main beam that spans the entire 360° in the ground (i.e. H-) plane.

For a fictitious isotropic antenna (point radiating element), the radiation pattern will be perfectly spherical. In reality, what comes closest to an isotropic antenna is a short or infinitesimal dipole antenna. The radiation pattern depends on the antenna used and the combination of a number of antenna elements (as in array antennas). Let us say that the radiation pattern of a transmitting antenna at a distance r from the antenna is defined by $E(r,\theta,\phi) = E_{\mathrm{m}}\, f(\theta,\phi)$, where $E(r,\theta,\phi)$ is the maximum electric field at a distance r from the antenna, and the function $f(\theta,\phi)$ defines the radiating pattern. The transmitting antenna is assumed to be placed at the origin. Although we discuss the radiation pattern of an antenna for transmitting,

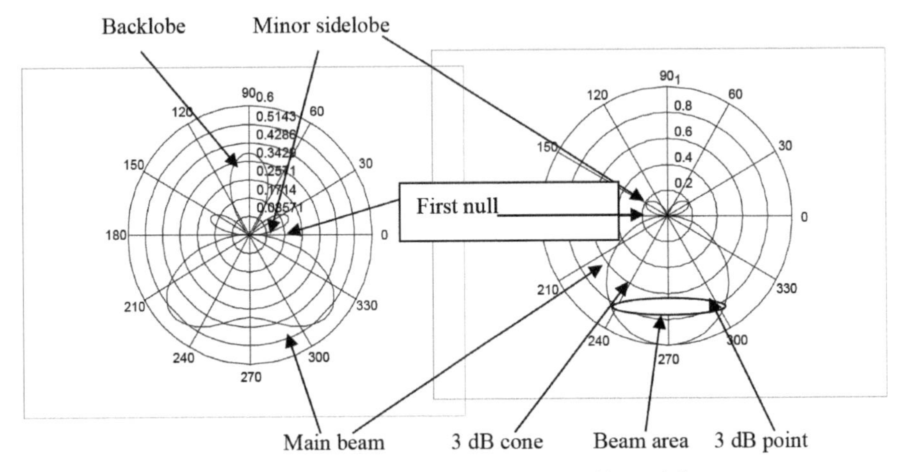

Figure 1.3 Antenna radiation patterns and beam lobes.

it is the same pattern when the antenna is receiving signals at the same frequency as when transmitting. Now E_{m} is the maximum of the maximum electric field measured in a circle of radius r around the antenna: E_{m} = Maximum$\{E(r,\theta,\phi)\}$. The time domain electric field at position (r,θ,ϕ) is given by $E(r,\theta,\phi)\cos(\omega t - kr)$. Therefore, the power density at (r,θ,ϕ) is given by $E(r,\theta,\phi) = P_{\mathrm{m}} f^{2}(\theta,\phi)$, where $P_{\mathrm{m}} = E_{\mathrm{m}}^{2}/2\eta\ \mathrm{W/m^{2}}$.

The fact that the radiation pattern has nulls in certain (θ,ϕ) directions and peaks in other directions implies that the antenna acts like a filter in the spatial (r,θ,ϕ) domain. Hence, an antenna may be considered as a spatial filter, a concept that will be used in antenna beamforming or antenna synthesis. We shall see that an antenna, in addition to being a spatial filter, is also a temporal filter with its response peaking at a certain resonant frequency and not allowing signals at other frequencies to be transmitted or received.

1.4.2 Antenna Gain and Beamwidth

Gain is the signal level produced (or radiated) by an antenna relative to that of a standard reference dipole antenna. It is used frequently as the figure of merit. Gain is closely related to directivity, which in turn is dependent upon the radiation pattern. High values of gain are usually obtained with a reduction in beamwidth. Gain can be calculated only for simple antenna configurations. Consequently, it is determined by measuring the performance relative to a reference dipole. The reference dipole is either the omni-directional isotropic antenna or the half-wave dipole antenna.

Hence the directivity, gain, and efficiency of antennas are scalar quantities that are used to express the antenna radiation characteristics. All these are defined for a particular antenna with respect to a reference isotropic antenna. If the total power radiated is P_r W, then the density of power radiated by an equivalent isotropic antenna at distance r is $P_r/4\pi r^2$ W/m^2. The directivity (D) of an antenna may alternatively be expressed as

$$D = \frac{P_m}{P_r/4\pi r^2} = \frac{r^2 P_m}{P_r/4\pi} = \frac{S_{max}}{P_r/4\pi} \tag{1.4}$$

where P_m = maximum power density radiated by the antenna in W/m^2; P_r = total power radiated by the antenna in W; $S_{max} = r^2 P_m$ maximum value of radiation intensity in W/unit solid angle; $P_r/4\pi$ = radiation intensity of an isotropic antenna. The directivity of the antenna will be entirely determined by the radiation pattern of the antenna. A narrower beam with fewer sidelobes will mean a highly directive antenna. The gain (G) of an antenna is given by

$$G = \frac{S_{max}}{P_T/4\pi} \tag{1.5}$$

$$< D \text{ for } P_r < P_T. \tag{1.6}$$

The gain of an antenna is usually less than the directivity of the antenna due to losses in the antenna. Sometimes the gain is specified in decibels: G_{dB} = 10 $\log_{10}$ (G) dBi. The efficiency of an antenna is given by

$$\eta_e = \frac{P_r}{P_T} = \frac{G}{D}. \tag{1.7}$$

When there is no loss in the antenna, i.e. $P_r = P_T$, the efficiency of the antenna is 100%. The gain G, sometimes called the isotropic gain, of an antenna does not mean that the antenna amplifies the overall power input. Indeed, if ohmic losses in the antenna are neglected, then the power delivered to the antenna is equal to the power radiated out, provided that there is no reflected energy at the antenna due to impedance mismatch. The gain of an antenna denotes the increase in maximum power density radiated by a given antenna A_1 compared to the power density radiated by an isotropic antenna A_i. An isotropic antenna A_i radiates equal power density over the entire surface of a sphere surrounding it, with the antenna at the center of the sphere. Thus, at a distance r from the antenna, the power density is $P_r/4\pi r^2$ W/m^2, where P_r is the total power radiated (in W) by the isotropic antenna. When the same amount of power P_r is radiated by another antenna A_1, the maximum power density it radiates will be $GP_r/4\pi r^2$ W/m^2, where G is the gain of antenna A_1. Some manufacturers define the gain of an antenna with reference to a half-wave dipole antenna instead of an isotropic antenna. In this case, the area $4\pi r^2$ over which an isotropic antenna radiates has to be replaced by the smaller area over which a dipole antenna radiates, which means that the dipole gain G_d of a given antenna will be smaller than its isotropic gain G_i, i.e. G.

The total efficiency of an antenna η_e is defined by taking into account the power losses at the antenna terminals and within the antenna structure itself. Power losses may be due to both reflections at the antenna terminals because of poor impedance matching between the antenna and the transmission line and the I^2R_o losses due to the ohmic resistance R_o of the antenna conductor and dielectric materials. Hence,

$$\eta_e = \eta_r\eta_c\eta_d, \tag{1.8}$$

where η_r = reflection efficiency, η_c = conduction efficiency, and η_d = dielectric efficiency.

HPBW θ_B in the plane of the antenna is the angular width of the radiation pattern, where the power level of the received signal is down by 50% (3 dB) from the maximum power density. Let Ω_B be the HPBW solid angle (in steradians) measured between the 3 dB points on the radiation pattern. We assume that all radiated power P_r is confined to the half-power beam; i.e. we ignore the radiation outside the HPBW angle (θ_B radians) as being small and negligible. For most antennas, the HPBW is approximately equal to FNBW/2, where FNBW is the first null beamwidth. The beam area is given by

$$A_b = r^2\Omega_B = r^2\frac{\pi}{4}\theta_B^2, \tag{1.9}$$

where

$$\Omega_B = \frac{\pi}{4}\theta_B^2. \tag{1.10}$$

If the total radiated power is P_r, then the maximum power density, assuming that the power density is uniform and constant over the entire beam area A_b,

$$P_m = P_r/A_b = P_r/r^2\,\Omega_B.$$

Therefore, the directivity of the antenna is given by

$$\begin{aligned} D &= \frac{P_{\max}}{P_{\mathrm{r}}/4\pi r^2} \\ &= 4\pi/\Omega_{\mathrm{B}}. \end{aligned} \tag{1.11}$$

For a lossless antenna $P_{\mathrm{T}} = P_{\mathrm{r}}$, and thus

$$G = D = 4\pi/\Omega_{\mathrm{B}}, \tag{1.12}$$

where Ω_{B} is the HPBW solid angle (in steradians) measured between the 3 dB points. This is an approximate formula and applies to all antennas. For an exact calculation of D and G, we must not assume that the power density over the beam area is constant and uniform. Since maximum power density $P_{\mathrm{m}} = (E_{\mathrm{m}}^2/2\eta)$, the maximum electric field radiated to a distance r from the antenna is given by $E_{\mathrm{m}} = (1/r)(DP_{\mathrm{r}}\,\eta/2\pi)^{1/2}$ V/m.

The FNBW is defined as the angular width between the two null points on either side of the main beam of the radiation pattern. The FNBW is always greater than the HPBW.

Example 1.2. An antenna has a gain $G = 61$ dB. Find θ_{B}.

$$G = 10^{61/10} = 125.89 \times 10^4,$$

$$\Omega_{\mathrm{B}} = \frac{4\pi}{G} = 0.0998 \times 10^{-4},$$

$$\theta_{\mathrm{B}} = \sqrt{\frac{4}{\pi}}\sqrt{\Omega_{\mathrm{B}}} = 0.0036 \text{ rad} = 0.204°.$$

Example 1.3. A satellite station in synchronous orbit has an antenna with a HPBW $\theta_{\mathrm{B}} = 0.2°$. At synchronous orbit, d = 36,000km is the distance between the satellite and the earth. Find the area of the earth covered by the satellite antenna beam

$$\theta_{\mathrm{B}} = 0.2° = 3.49 \times 10^{-3} \text{ rad},$$

$$\Omega_{\mathrm{B}} = \frac{\pi}{4}(3.49 \times 10^{-3})^2,$$

$$\text{area of spot} = r^2\Omega_B = 12.4 \times 10^9 \text{ m}^2.$$

For gain $G(\theta, \phi)$, or directivity $D(\theta, \phi)$, in a given direction (θ, ϕ) other than the direction of maximum radiation, the P_{m} of Equations (1.4) and (1.5) must be replaced by $P_{\mathrm{i}}(\theta, \phi)$, where $P_{\mathrm{i}}(\theta, \phi)$ is the power density radiated in the direction (θ, ϕ). $P_{\mathrm{i}}(\theta, \phi) = E(\theta, \phi)^2/(2\eta)$.

1.4.3 Effective Aperture

The concept of an effective aperture captures the idea of an antenna being open to capture power radiated by another (transmitting) antenna, which falls within its aperture. If P_{i} is the incident power density (W/m^2) at the antenna and P_{R} (W) is the total power received or captured by the antenna, the effective aperture is defined by

$$A_{\mathrm{em}} = \frac{P_{\mathrm{R}}}{P_{\mathrm{i}}} = \frac{\frac{1}{2}I^2 R_{\mathrm{T}}}{P_{\mathrm{i}}}. \tag{1.13}$$

For maximum power transfer

$$R_{\mathrm{A}} = R_{\mathrm{r}} + R_{\mathrm{o}} = R_{\mathrm{T}} \tag{1.14}$$

$$X_{\mathrm{A}} = -X_{\mathrm{T}} \tag{1.15}$$

where R_{A} = antenna resistance, R_{r} = antenna radiation resistance, R_{o} = antenna ohmic resistance, R_{T} = termination resistance, X_{A} = antenna reactance, and X_{T} = termination reactance. Hence, the current in the receiver circuit of the antenna is given by

$$\begin{aligned} I_{\mathrm{T}} &= V_{\mathrm{T}}/\left(Z_{\mathrm{A}} + Z_{\mathrm{T}}\right) = V_{\mathrm{T}}/2R_{\mathrm{T}} \\ &= V_{\mathrm{T}}/\left[2\left(R_{\mathrm{r}} + R_{\mathrm{o}}\right)\right] \end{aligned} \tag{1.16}$$

Hence, the effective aperture area of the antenna is given by

$$A_{\mathrm{em}} = \frac{V_{\mathrm{T}}^2}{8P_{\mathrm{i}}\left(R_{\mathrm{r}} + R_{\mathrm{L}}\right)} = \frac{V_{\mathrm{T}}^2}{8P_{\mathrm{i}}R_{\mathrm{r}}} \quad \text{for } R_{\mathrm{L}} << R_{\mathrm{r}} \tag{1.17}$$

For a short dipole of length L and an incident electric field E, using the following parameters $\left(R_{\mathrm{r}} = 80(\pi L/\lambda)^2\right.$, the radiation resistance ; $D = 1.5$, the directivity; $P_{\mathrm{i}} = E^2/2\eta_0 = E^2/2(377)\mathrm{W/m^2}$, the incident power density at the antenna; $V_{\mathrm{T}} = EL$, the voltage induced along the antenna wire), we get

$$\begin{aligned} A_{\mathrm{em}} &= 0.12\lambda^2\mathrm{m}^2 \\ &= \frac{\lambda^2}{4\pi}D \end{aligned} \tag{1.18}$$

for a short or infinitesimal dipole.

A useful equation for the gain of an antenna in terms of its maximum effective aperture area A_{em}, which we shall define in Chapter 2, is G

$=(4\pi A_{em})/\lambda^2$. Hence, $G \propto A_{em}$ and, further, $G \propto f^{\,2}$. Note that in general the gain of an antenna increases as the operating frequency f is increased; $f = c/\lambda$, where c is the velocity of light (approximately equal to 3×10^8 m/s) and λ is the wavelength. For an elemental dipole, $G = D = 1.5$, for a half-wave dipole $G = D = 1.64$.

1.4.4 Operation Zones

Three zones are defined for an antenna. These zones depend on the size of the antenna D (e.g. D = length of antenna for a wire type antenna) and the wavelength λ.

The near-field zone is the distance r from the antenna that satisfies the following relation:

$$r < r_1 = 0.62D\sqrt{\frac{D}{\lambda}}. \tag{1.19}$$

In the near-field region of an antenna is the region immediately surrounding the antenna where only the reactive field dominates.

The radiating near-field (Fresnel) zone is defined by distance r which lies in the range

$$r_1 < r < r_2 = 2D^2/\lambda. \tag{1.20}$$

In the radiating near-field or intermediate field region, the radiation field dominates and the angular field distribution is dependent on the distance from the surface of the antenna. For short or infinitesimal antennas where the size of the antenna is very small compared to the wavelength, this region does not exist.

In the far-field (Fraunhofer) zone, in which most antennas operate, the receiving antenna is far away from the transmitting antenna

$$r > r_2. \tag{1.21}$$

In the far-field region of an antenna, the angular field distribution may be assumed to be independent of the distance from the antenna. Although most communication antennas operate in the far-field zone, in biomedical imaging, the antennas (or coils) operate at frequencies of the order of 160 MHz and are only about 0.5 m away from the radiating source. In mobile communications, it is possible for the receiver antenna to pick up near fields when it is close to the base station, as in the case of indoor picocells.

1.4.5 Antenna as a Temporal Filter: Bandwidth

The bandwidth of an antenna may be defined as the frequency range over which its performance is satisfactory. Depending on the system for which an antenna is designed, it is important to match the beamwidth of the systems and the antenna bandwidth. For instance, in mobile communications, bandwidths of 25 kHz (the advanced mobile phone system (AMPS) system of USA) and 200 kHz (the global system for mobile (GSM) system of Europe) are common. The center frequency may be about 900 MHz. In this case, it is necessary to select an antenna which will operate at a center (resonant) frequency of say 900 MHz and provide a bandwidth of 25 kHz (AMPS) or 200 kHz (GSM). The frequency bandwidth is a general classification of the frequency band over which the antenna is effective. Wireless communications using code division multiple access (CDMA) at frequencies of 1900 and 2400 MHz demand the deployment of smart antennas with narrow bandwidths to enable range and capacity improvement through spatial filtering as well as temporal filtering. Data rates, which partly determine the bandwidth, will increase from about 1 Mega-bits/second (Mbps) to about 5 Mbps.

1) Broadband antennas perform over a center frequency with a bandwidth ratio of 10:1, which is the ratio of the maximum-to-minimum frequencies of acceptable operation. A 10:1 bandwidth indicates that the maximum frequency for acceptable operation is 10 times greater than the minimum frequency. Antennas operating at 5:1 bandwidth are generally suitable for most applications, including mobile communications, which require a bandwidth of about 70 MHz.
2) If the bandwidth is about 1:0.05, the antenna is called a narrow-band antenna. In a narrow-band antenna, the antenna conductor diameter to length ratio must be small. If the diameter of the antenna is increased, then it becomes a wide-band antenna. Narrow-band antennas tend to detune under ice and wind loading.

Figure 1.4 shows the antenna characteristics considered as a spatial filter with HPBW and FNBW. The antenna transmits effectively within the HPBW angle, and beyond the FNBW, we get the undesirable radiation through sidelobes. Figure 1.5 shows the characterization of an antenna as a temporal filter. Since an antenna can be considered as an RLC circuit, it has a certain frequency bandwidth. The bandwidth of an antenna may be conceptually understood by considering it as an RLC circuit. The capacitance forms between, say, the top wire and the bottom wire

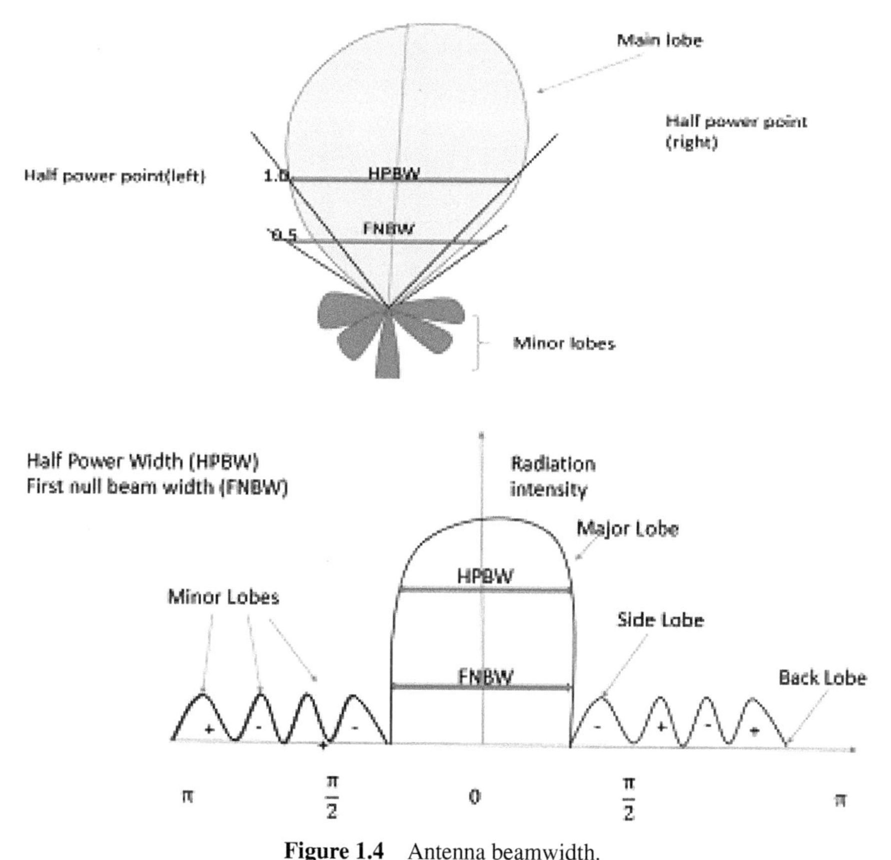

Figure 1.4 Antenna beamwidth.

of a half-wave dipole antenna. The wires will have both resistance and inductance.

Now any series RLC circuit will have a resonant frequency at which the current peaks. The resonant frequency in terms of the inductance and capacitance of the antenna is given by

$$f_0 = \frac{1}{2\pi\sqrt{L_A C_A}}. \tag{1.22}$$

This is the best frequency to operate the particular antenna. It is known that the response of the RLC circuit drops off from the resonant frequency. If we operate the antenna beyond the 3 dB points, then very little of the signal will get into the antenna and even less will be radiated out. Hence, we have the concept of bandwidth, which specifies the permissible frequency fluctuations

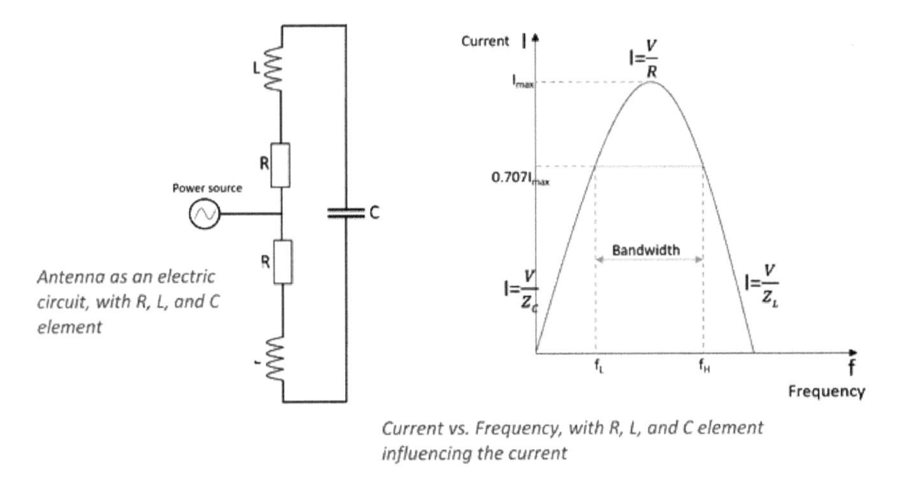

Figure 1.5 Antenna as an RLC circuit.

of a given antenna. Given the resonant frequency and the bandwidth, we will know the frequencies at which the antenna will give satisfactory performance. For a given communication or remote sensing system, the bandwidth of the antenna must be matched with the bandwidth of the signals.

1.4.6 Antenna Temperature

An antenna receives radiation from a variety of objects other than the transmitted signal, and its temperature is usually above absolute zero. $T_b >$ 0 K (−273 K).

These temperatures have adverse effects on communication systems. These additional signals associated with the antenna temperature appear as noise at the receiver electronics. However, in remote sensing systems, the radiation due to the finite temperature of objects other than the antenna can be captured and used to determine the material properties of the object. This is called passive remote sensing. Passive remote sensing is used in astronomy to study the properties of stars and far-off galaxies as well as in studying the geography of land through a passive sensing antenna attached to the fuselage of an aircraft or to a satellite.

In communication systems, the physical temperature of both the antenna and other hot objects produces electromagnetic noise which interferes with the signal received. Noise power is given by $p = kT_b$ W/Hz. Hence, for a bandwidth of Δf Hz, the total noise power is given by $P = p\Delta f$W.

Each one of the following objects is at a finite temperature and will radiate an electromagnetic signal over a broad spectrum.

Sky: 3 K; Mars: 164 K; Earth: 290 K; Man: 310 K.

Thus, the *noise temperature* at an antenna associated with noise results from a summation of temperatures at the antenna due to many hot objects, near and far away, and is given by the total noise temperature

$$T_{\mathrm{A}} = T_{\mathrm{B1}} + T_{\mathrm{B2}} + \cdots \tag{1.23}$$

Transmission line theory applied to conduction of noise along the line or waveguide, which connects the antenna to the receiver electronics, will give

$$T_{\mathrm{a}} = T_{\mathrm{A}} \mathrm{e}^{-2\alpha l} + T_0 \left(1 - \mathrm{e}^{-2\alpha l}\right) \tag{1.24}$$

where $T_0 = 273$ K, physical temperature of the system (room temperature), and α is the attenuation coefficient of the electrical circuit.

Hence, the total noise power at the receiver electronics is the sum of noise coming from the antenna (T_{a}) and the noise produced by the hot electronic circuitry (at a temperature T_{R}) itself. The total noise power at the receiver electronics operating with a bandwidth ΔfHz is given by

$$P_{\mathrm{N}} = k\left(T_{\mathrm{a}} + T_{\mathrm{R}}\right) \Delta f \tag{1.25}$$

For very high frequency antennas, with frequency $f > 10$ GHz, the temperature is given by $T = hf/k$, where Planck's constant $h = 1.055 \times 10^{-34}$ Js and Boltzmann's constant $k = 1.38 \times 10^{-23}$ J/K

1.4.7 Antenna Input Impedance

The antenna input impedance is the terminating resistance into which a receiving antenna will deliver maximum power. As with the gain of an antenna, the input impedance of an antenna can be calculated only for very simple formats and, instead, is determined by the actual measurement (see Figure 1.6).

The antenna impedance is given by

$$Z_{\mathrm{A}} = R_{\mathrm{A}} + \mathrm{j}X_{\mathrm{A}}, \tag{1.26}$$

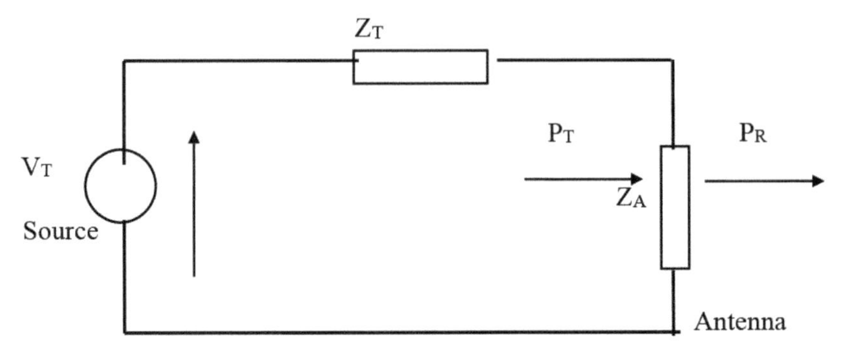

Figure 1.6 Power flow in an antenna.

where $R_A = R_r + R_o$. The radiation resistance R_r is related to the power radiated by the antenna and the ohmic resistance R_o is related to the heat loss in the antenna. The power dissipated in R_r is the radiated power P_r; hence, for an antenna, we seek to maximize R_r, whereas for other circuits like PCBs, we seek to minimize R_r since in PCBs, we want the information and energy to flow along the copper transmission line instead of being radiated out and interfering with other lines.

If the antenna is transmitting, and the source current that flows through the antenna is I_{rms}, then the radiated power $P_r = I_{rms}^2 R_r$ W. If the source voltage is $V_{s,rms}$ and the source impedance Z_s is properly matched to the antenna impedance Z_A (i.e. $Z_s = Z_A^*$), then $I_{rms} = V_{s,rms}/2R_A$. If the antenna is receiving, and its impedance is properly matched to the load (transmission line/receiver electronics) impedance Z_L (i.e. $Z_L = Z_A^*$), then the current in the receiver antenna is $I_{rms} = V_{s,rms}/2R_A$ where $V_{s,rms}$ is the voltage induced across the antenna by the incident electric field E. For a short dipole of length L, an approximate expression for the induced voltage is $V_{s,rms} = EL$. The received power or the power captured by the antenna $P_R = I_{rms}^2 R_A$ W. For an efficient antenna ($R_r >> R_o$), the antenna resistance is equal to the radiation resistance, i.e. $R_A = R_r$. Thus, in general, the power dissipated in R_r is equal to the power radiated by a transmitting antenna. In a receiving antenna, the power dissipated in R_r is the power captured by the receiving antenna; it is the same power (P_R) that is delivered to the receiver electronics.

For a small wire antenna of length L, we have

$$R_r = 800(L/\lambda)^2\Omega. \tag{1.27}$$

For a small circular loop antenna (magnetic dipole) of area A (= π (radius of loop)^2), the radiation resistance is

$$R_{\mathrm{r}} = 31,200(A/\lambda)^2\Omega. \tag{1.28}$$

The antenna reactance X_{A} is related to the electromagnetic energy stored in the vicinity of the antenna. Note that the input impedance of the antenna is a function of frequency, since X_{A} increases linearly with frequency if the antenna is inductive and decreases with frequency if it is capacitive. Similarly, the resistance is also frequency-dependent. Hence, when we match an antenna to the impedance Z_{T} of a transmission line or coaxial cable, it is matched at a particular center frequency and bandwidth. If the frequency of operation is changed, then unless matching is done again, power transfer to and from the antenna will be poor.

1.5 Reciprocity

1.5.1 The Friis Transmission Equation

The reciprocity theorem tells us that if a transmitter is connected to antenna A_x and the receiver to antenna A_y, the signal (i.e. electric field) strength observed by the receiver electronics will be the same if the transmitter is now connected to A_y and receiver to A_x. This also implies that the transmit and receive radiation patterns of an antenna are identical.

Thus, for example, if antenna A_x transmits maximum signal strength along the *z*-axis (θ = 90° and ϕ = 90°) and zero radiation along θ = 30° and ϕ = 90°, then if A_x is used to receive signals, it will best receive signals coming toward it along the *z*-axis and it will not pick up any signal coming from the θ = 30° and ϕ = 90° direction.

Let the gain of the transmitting antenna in the direction of the receiver antenna be G_x and that of the receiver antenna in the direction of the transmitter antenna be G_y; P_{T} is the total power input to the transmitting antenna. Ignoring the losses in the antennas (G = D), the total power radiated is P_{T} W. The power density at the receiver antenna is $G_xP_{\mathrm{T}}/4\pi r^2$ W/m^2. The effective aperture of the receiving antenna in the direction of the transmitting antenna is $A_{\mathrm{em}} = D_y \cdot \lambda^2/4\pi = G_y \cdot \lambda^2/4\pi$. Hence, the total power received by the second antenna at a distance r from the transmitting is given by

$$P_{\mathrm{R}} = \left(G_xP_{\mathrm{T}}/4\pi r^2\right)A_{\mathrm{em}} = \left(G_x\mathrm{P}_{\mathrm{T}}/4\pi r^2\right) \cdot \left(G_y\lambda^2/4\pi\right)\mathrm{W} \tag{1.29}$$

or

$$P_R/P_T = G_x G_y \lambda^2/(4\pi r)^2 \tag{1.30}$$

This is called the Friis transmission formula, and it relates the power P_R that appears at the radar receiver amplifier to the total power P_T transmitted by the radar. We have assumed that both antennas are lossless, that the impedances are all properly matched, and that both antennas are correctly polarized.

The gains G_x and G_y are, in general, given by $G_x = G_{x0}\, {f_x}^2(\theta, \phi)$ and $G_y = G_{y0}\, {f_y}^2(\theta, \phi)$, where G_{x0} and G_{y0} are the maximum gains of the two antennas (e.g. $G_{x0} = G_{y0} = 1.5$ if both antennas are elemental, short dipole or $G_{x0} = G_{y0} = 1.64$ if both antennas are half-wave dipoles). The functions $f_x(\theta, \phi)$ and $f_y(\theta, \phi)$ are the normalized electric (and magnetic) field radiation patterns of the two antennas in the spherical coordinate direction (θ, ϕ). The maximum value of the radiation pattern functions is 1.0.

In mobile or wireless communication systems, the antennas are considered as short dipoles that may be approximated to isotropic (omni-directional) antennas. Hence, $G = D = 1$. This means that the received power is

$$P_R = P_T(\lambda/4\pi r)^2 \text{ W} \tag{1.31}$$

or

$$P_{RdB}(r_0) = 10\log_{10}(P_T\ (\lambda/4\pi r)^2 \text{ dB}. \tag{1.32}$$

In this equation for received power, we have not considered the signal loss over the path traveled by the signal and shadowing effects. Path loss is due to losses in the finite conductivity ground over which the signal travels. Shadowing is due to objects like trees and buildings that tend to obstruct part of the signal. Thus, the received power will be less than the ideal received power P_R. An expression for received power with respect to the power received at the boundary of a cell in cellular communication systems is the following:

$$P_{RdB}(r) = P_{RdB}(r_0) - 10\ L_1\ log_{10}(r/r_0) + L_2 \text{ dB m}. \tag{1.33}$$

The path loss component L_1 typically varies from 3 to 4 in an urban macrocellular environment and from 2 to 8 for a microcellular environment. The cell size is 1 km for macrocellular systems, 100 m for outdoor microcellular systems, and 1 m for indoor picocellular systems. The loss component due to shadowing is a zero mean square Gaussian random variable. The transmitted power P_T for a typical mobile phone is less than or equal to 1 W. There is now an attempt to take this power level down to 0.1

or 10 mW by reducing the cell size. By using high frequency electromagnetic signals for wireless mobile communications, path loss is increased, and, thus, it is possible to reuse frequency (frequency reuse) in another adjacent cell. Thus, to reduce cell size, the present operating frequency of 800/900 MHz will be increased, say, toward 1885–2200 MHz.

1.5.2 The Radar Equation

Consider now a radar system. A transmitting antenna of gain G_x radiates a pulse of sinusoidal waves (continuous wave (CW) radar) or linear-frequency modulated waves (Chirp radar) that impinge on a target of radar cross-sectional area (RCS) of σ m^2. The same antenna or another antenna of gain G_y captures some portion of the signal reflected by the target. When the same antenna is used to capture reflected signal, the radar is called a monostatic radar. When a second antenna is used at a slightly different location to capture the reflected signal, it is called a bistatic radar.

If the target is at a distance r_1 from the transmitting antenna, then the power density at the target is, as before, $\sigma G_x P_{\mathrm{T}}/4\pi {r_1}^2$ W/m^2. If the second antenna is at a distance r_2 from the target, then the power density appearing at the receiving antenna is $\sigma G_x P_{\mathrm{T}}/(4\pi r_1 r_2)^2$ W/m^2. If the maximum effective aperture of the receiving antenna is expressed as

$$A_{\mathrm{em}} = G_y \cdot \lambda^2/4\pi, \tag{1.34}$$

then the received power is given by

$$P_{\mathrm{R}} = \left[G_y \cdot \lambda^2/4\pi\right]\left[\sigma G_x P_{\mathrm{T}}/\left(4\pi r_1 r_2\right)^2\right] \tag{1.35}$$

Therefore,

$$P_{\mathrm{R}}/P_{\mathrm{T}} = \left[\sigma G_x G_y \cdot \lambda^2/(4\pi)^3\left(r_1 r_2\right)^2\right] \tag{1.36}$$

When $r_1 = r_2 = r$, i.e. for the monostatic case,

$$P_{\mathrm{R}}/P_{\mathrm{T}} = \left[\sigma G_x G_y \cdot \lambda^2/(4\pi)^3(r)^4\right] \tag{1.37}$$

This equation is called the radar equation. Note that in this case, the power received is proportional to r^4, which indicates that the receiver gets very little power to process. Hence, radar transmitter powers are generally high, say an average power of 500 W and peak powers of the order of 5 kW, to ensure that the receiver gets enough reflected signal power to detect the presence of a target. In stealth technology, to evade detection, missiles and aircraft are designed to geometrically and material-wise reduce the RCS.

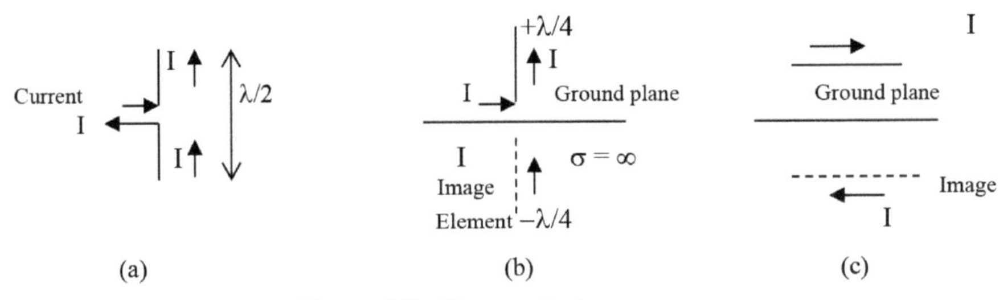

Figure 1.7 Elemental wire antennas.

1.6 Types of Antennas

1.6.1 Elemental Current Antennas

Elemental current antennas are generally conducting wires that are arranged such that the signal power radiated is maximized. An inexpensive, but effective, type of wire antenna is the half-wave dipole antenna, shown in Figure 1.7(a). The input impedance of a half-wave dipole antenna is 73 + j42.5 Ω and becomes 70 + j0 Ω if the length is slightly reduced. Hence, its input impedance almost exactly matches the 70 Ω characteristic impedance of a coaxial cable. The half-wave dipole antenna is formed by bending out length $\lambda/4$ from each end of a two-wire transmission line, where λ is the wavelength of the signal to be transmitted (or received). The total length of the antenna is $\lambda/2$, as shown in Figure 1.7(a).

An alternative way of forming a half-wave dipole antenna is to use a quarter-wavelength ($\lambda/4$) long wire placed over a perfectly conducting plane (Figure 1.7(b)). This quarter-wavelength monopole, resulting in a half-wave dipole, is based on the theory of images. The theory of images states that when a positive electric charge is placed above a perfectly conducting plane, the plane can be replaced by a negative electric charge placed at the mirror image point of the positive source charge. Hence, an electric current moving vertically upwards above a perfectly conducting plane will have associated with it a mirror image current that is also moving vertically upwards. The input impedance of such a quarter-wavelength monopole (36 Ω) is half that of the half-wavelength dipole (73 Ω), but its directivity is twice that of the half-wave dipole since its HPBW above ground (39°) is half that of the half-wave dipole (78°).

For over-the-horizon transmission of signals, as in the case of over-the-horizon radar, based on the reflection of electromagnetic signals from the ionosphere, horizontal dipoles such as those shown in Figure 1.7(c) are used.

The ionosphere is at a height of about 300 km above the earth. In the case of horizontally placed current carrying wire, the mirror image current flows in the opposite direction to that of the source current placed above the ground. These horizontally held antennas are also used by hobbyists to communicate to someone who is continents away on the earth. The electromagnetic communication signal travels all the way round the globe by repeatedly bouncing off the ionosphere and the earth.

The elemental type of antennas is designed to operate at frequencies ranging from 10 kHz to 1 GHz. The size of the antenna (length L) is related to the wavelength by $L/\lambda = 0.01\text{–}1$.

Similar to the half-wave dipole antenna is the loop antenna, which forms a magnetic dipole. A conducting wire is bent into a circular loop. When a high frequency source is connected across a small cut in the loop, the circular loop will radiate out the signal from the source. The radiation fields at (r,θ,φ) and radiation resistance of a loop antenna are given by

$$H_\theta = \left(\frac{ka}{2}\right)^2 I\frac{\mathrm{e}^{-jkr}}{r}\sin\theta, \tag{1.38}$$

$$E_\phi = -\eta H_\theta, \tag{1.39}$$

$$R_\mathrm{r} = 20\pi^2\left(\frac{2\pi a}{\lambda}\right)^2 \Omega, \tag{1.40}$$

where a is the radius of the loop and k $(=2\pi/\lambda = \omega/c)$ is the wave number.

1.6.2 Traveling Wave Antennas

Traveling wave antennas are generally long conductors or helical loops or conductors formed in a zigzag fashion over which the electric current–voltage wave travels close to the velocity of light. As the waves travel along the long conductors, they also radiate signals out into free space, thus forming an antenna. These antennas are used in the frequency range of 1–10 MHz and the size of the antennas (L) is typically of the order of $L/\lambda = 1\text{–}10$. In Figure 1.9(a) is shown a traveling wave antenna and in Figure 1.9(b) its radiation beam.

1.6.3 Array Antennas

Array antennas are formed by a series of wire or aperture antennas, generally arranged along a single line (linear array) or in a rectangular grid (planar

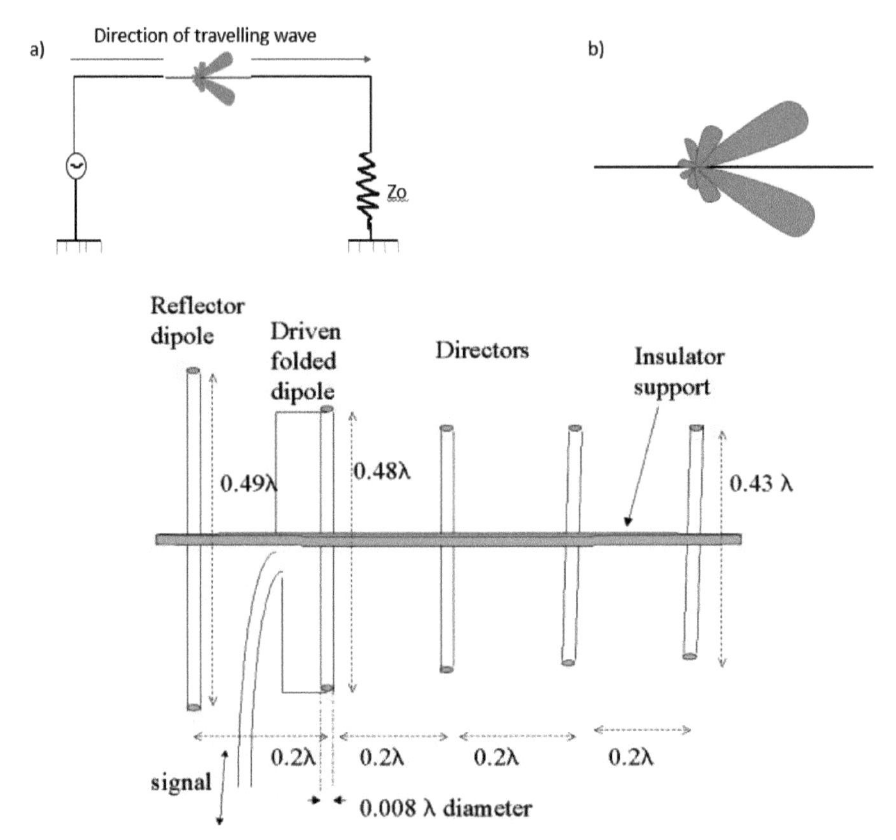

Figure 1.8 (a) A traveling wave antenna. (b) The radiation beam of the traveling wave antenna. (c) A Yagi–Uda array antenna.

array). In other words, an array antenna is a collection of single antenna elements. One popular type of array antenna is the Yagi antenna used for television signal reception, shown in Figure 1.8(c). The advantage of having a series of antenna elements radiating or receiving simultaneously is twofold: increased radiation power and better ability to direct the antenna radiation pattern (beam) in a specific direction. A folded dipole antenna, instead of a half-wave dipole antenna, is used to double the amount of power received; however, the input impedance of the folded dipole is about 388 Ω.

Such antennas are used in the frequency range of 5 MHz to 50 GHz, and the typical dimension of the antenna is $L/l = 1–100$.

1.6.4 Aperture Antennas

Aperture antennas are used at very high frequencies and are formed by having apertures or windows through which the electromagnetic signals may be radiated. A widely used type of aperture antenna is the rectangular horn antenna, where electromagnetic signals traveling through a waveguide are radiated out by opening up the waveguide into a horn shape (Figure 1.9(a)). These may be operated in the reverse direction to receive signals in a highly directive manner. The large parabolic reflectors used in satellite communication antennas function like apertures. Normally, a wire antenna or a horn antenna is placed in front of the parabolic dish to collect the signals focused by the parabolic aperture (receiving) or to project the signals from an oscillator on to the parabolic dish (transmitting). The parabolic reflector arrangement is shown in Figure 1.9(b), where a horn antenna is placed at the focal point of the parabolic reflector.

The low noise block converter (LNBC) provides low noise amplification and frequency downconversion (e.g. from 8 GHz downconverted to 800 MHz). The downconverted signal may be carried over coaxial cable without high ohmic losses. Apertures tend to be highly directive and enable better focusing. In radio astronomy, an array of parabolic reflector type antennas is used to observe signals radiated from a very small region in the galaxy.

1.7 Waves Along Conductors and in Free Space

The launching or radiation of an electromagnetic wave from a wire antenna is shown in Figure 1.10(a).

In Figure 1.10(b) is shown an antenna operating as a transmitter, with the signal packets being launched out into free space. In Figure 1.10(c) is shown an antenna operating in a receiver mode, where the electromagnetic waves or signals emitted from a transmitter antenna impinges on the receiver antenna and is captured and transmitted into the receiver electronics by the antenna. An antenna may simultaneously act as a transmitter and receiver, where it should be ensured that the frequency of the transmitted signals and frequency of the received signals are different to avoid interference between the two signals.

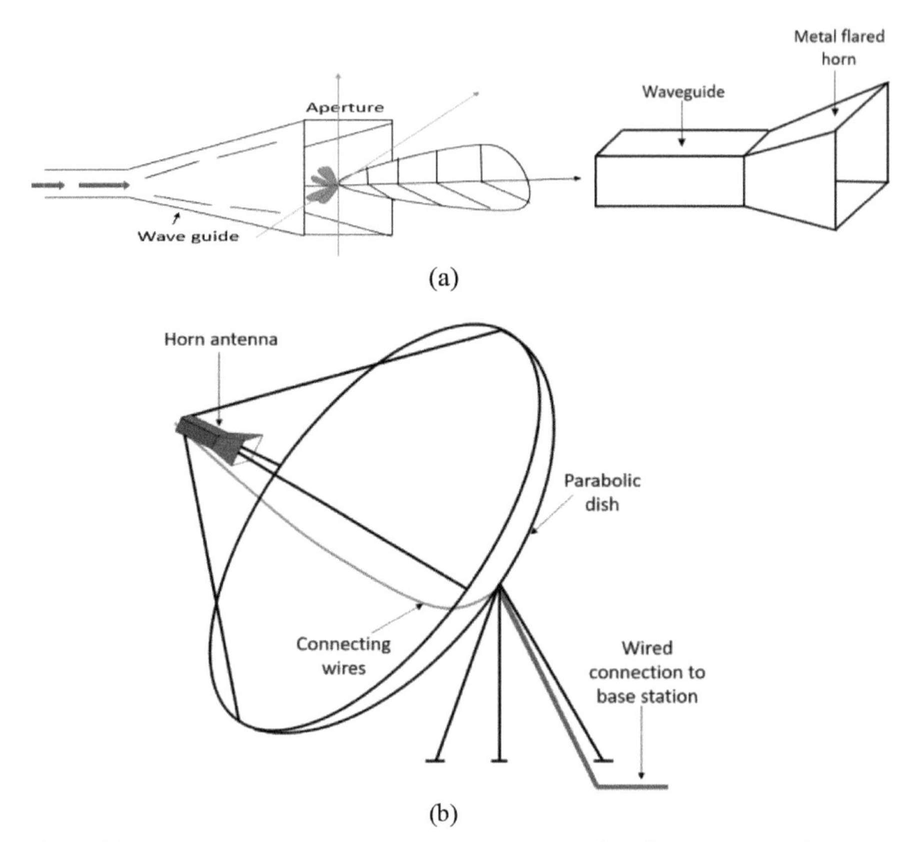

Figure 1.9 Aperture antennas. (a) Horn antenna. (b) Parabolic reflector antenna with a horn antenna and a parabolic reflector.

1.8 Maxwell's Equations and Electromagnetic Waves

1.8.1 Introduction

The basic theory that governs antenna performance and antenna signal processing is the electromagnetic theory defined by the four Maxwell equations. The set of four Maxwell's equations and the Lorentz force law are five elegant equations of natural law that underlie the entire discipline of electrical and communication engineering. Whether it is a low frequency device like an induction motor (60 Hz) or a high frequency (e.g. 10 GHz) device like the traveling wave tube (TWT) amplifier used at microwave satellites, whether it is a low frequency (50 or 60 Hz) system like the electric power system or a high frequency (e.g. 2 GHz) system like the wireless

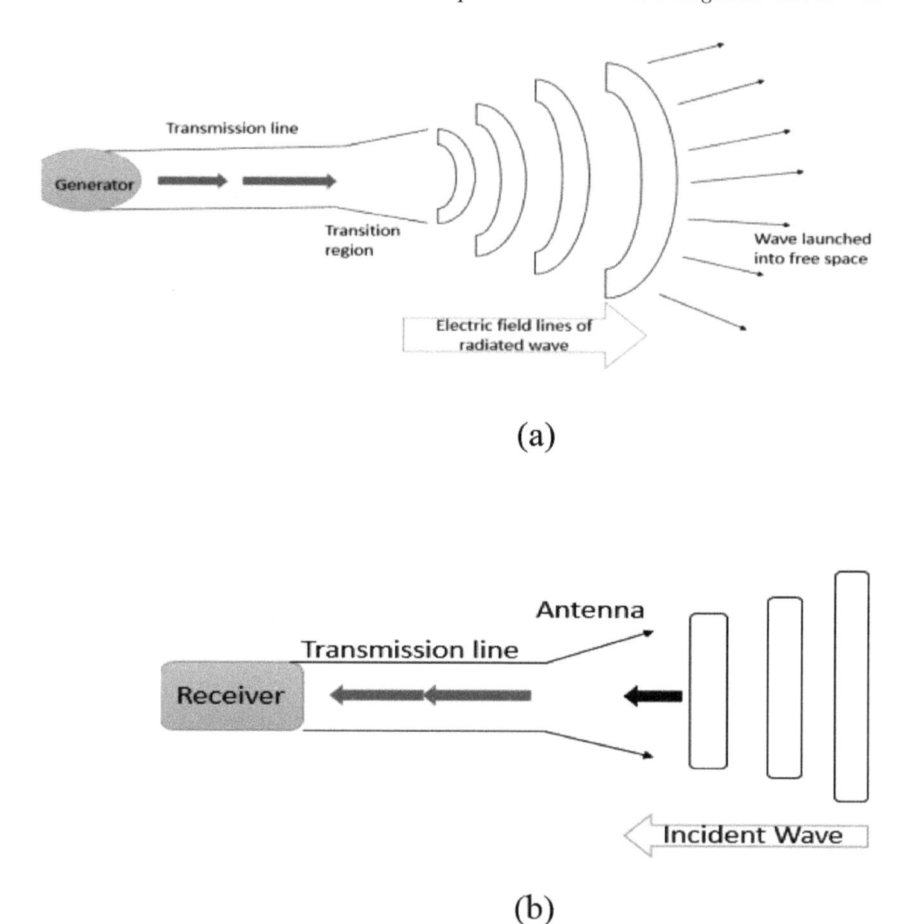

Figure 1.10 (a) Formation and launching of electromagnetic wave packets: Antenna in transmitter mode. (b) Antenna in receiver mode.

communication system, all these devices and systems are characterized and controlled by the Maxwell equations. In order to get exact equations that define the operating characteristics of a passive device like the antenna or an active device like the transistor, we must solve the Maxwell equations specifically applied to these devices. In this book, we shall be looking at the solution of Maxwell equations for a variety of antennas and the use of these solutions for antenna design and signal processing. Maxwell equations may be applied to all macroscopic electromagnetic devices and systems. The four

Maxwell equations are

$$\nabla \cdot \mathbf{D} = \rho, \tag{1.41}$$

$$\nabla \cdot \mathbf{B} = 0, \tag{1.42}$$

$$\nabla \times \mathbf{H} = \mathbf{J} + \frac{\partial \mathbf{D}}{\partial t}, \tag{1.43}$$

$$\nabla \times \mathbf{E} = -\frac{\partial \mathbf{B}}{\partial t}, \tag{1.44}$$

where $\mathbf{E}$ = vector electric field intensity in V/m, $\mathbf{B}$ = vector magnetic flux density in webers/m^2, $\mathbf{D}$ = vector electric flux density in coulombs, $\mathbf{H}$ = vector magnetic field intensity in A/m, $\mathbf{J}$ = vector electric conduction current density in A/m^2, and ρ = electric charge density in coulombs/m^3. The Maxwell equations are linear since they contain E and B and their derivatives to the first power only.

The Lorentz force equation is

$$\mathbf{F} = q(\mathbf{E} + \mathbf{v} \times \mathbf{B}), \tag{1.45}$$

where q = electric charge in coulombs and $\mathbf{v}$ = the velocity of the electric charge in m/s.

The Lorentz force defines the physical effects of the unseen electric and magnetic fields. We observe that it is the electric field intensity E and the magnetic flux density B that are the physical parameters evidently existing; both magnetic field intensity H and electric flux density are fictitious quantities that are mathematically defined to make the field definitions complete. We have $\mathbf{H} = \mathbf{B}/\mu$ and $\mathbf{D} = \varepsilon\mathbf{E}$ and $\mathbf{J} = \sigma\mathbf{E}$, where material properties μ, ε, and σ are the permeability, permittivity, and electric conductivity of the material in which the fields exist. The conduction current $\mathbf{J}$ (=$\sigma\mathbf{E}$) will not exist in regions of zero conductivity, i.e. in dielectric material like free space. However, $\partial\mathbf{D}/\partial t = \partial\varepsilon\mathbf{E}/\partial t = \mathbf{J}_D$, called the displacement current density, may flow in dielectric materials too. It is due to this displacement current that we can have wireless communications. Along the antenna conductor itself we have conduction currents, but once the signals are launched out from the antennas, the conduction current becomes zero, although the signal energy and information is now carried by displacement currents in free space. The displacement currents flow along the closed electric field lines of each wave packet.

1.8.2 Electromagnetic Waves

Antennas are devices over which time varying electric currents flow (in wire antennas) or time varying electric and magnetic fields appear at the aperture (e.g. the open end of a waveguide antenna). These time varying currents or electric and magnetic fields at the antenna terminal will produce time varying fields just outside the antenna. In this section, we shall show that these fields will constitute a propagating electromagnetic wave. With time varying fields, **E** and **B** are no longer independent as in the case of static or low frequency fields, but they are mutually coupled through the four Maxwell equations. We shall consider free space that is electric charge free (i.e. $\rho = 0$) and has zero conductivity (i.e. $J = \sigma E = 0$ in free space). Thus, the four Maxwell equations (1.41)–(1.44) reduce to

$$\nabla \cdot \mathbf{E} = 0, \tag{1.46}$$

$$\nabla \cdot \mathbf{B} = 0 \tag{1.47}$$

$$\nabla \times \mathbf{B} = \frac{1}{u^2}\frac{\partial \mathbf{E}}{\partial t} \tag{1.48}$$

$$\nabla \times \mathbf{E} = -\frac{\partial \mathbf{B}}{\partial t} \tag{1.49}$$

where we have set the velocity of electromagnetic waves as

$$u = (\mu_0\varepsilon_0\mu_r\varepsilon_r)^{-1/2} \tag{1.50}$$

The velocity of the wave in free space where $\mu_r = 1, \varepsilon_r = 1$ becomes the velocity of light in free space, $u = c = (\mu_0\varepsilon_0)^{-1/2} = 2.9998 \times 10^8 \mathrm{m/s}$. These four equations are not only linear but also homogeneous since there are no electric charges or electric currents present, and every term contains either an **E** or a **B**. One property of linear homogeneous equations is that the sum of two different solutions is also a solution. Thus, a general solution to these equations will consist of a sum of static as well as time varying fields. The physical origin of such static fields would be distributions of static charges and steady currents. We shall assume that such a distribution of steady electric charges and currents cannot exist in free space.

Consider now what the first two Maxwell's equations (1.48) and (1.49) have simplified for the case of uniform plane wave (UPW): $\mathbf{u}_y \mathrm{d}E_x/\mathrm{d}z = -\mu\partial\mathbf{H}/\partial\mathrm{t}$ and $-\mathbf{u}_x\mathrm{d}H_y/\mathrm{d}z = \varepsilon\partial E_x/\partial t$. Differentiating the second equation with respect to z and substituting from the first equation for $\mathrm{d}E_x/\mathrm{d}z$, we get

$$\frac{\partial^2 H_y}{\partial z^2} = \frac{1}{u^2}\frac{\partial^2 H_y}{\partial t^2} \tag{1.51}$$

which defines a magnetic wave traveling in the *z* direction. Similarly, we get

$$\frac{\partial^2 E_x}{\partial z^2} = \frac{1}{u^2}\frac{\partial^2 E_x}{\partial t^2}. \tag{1.52}$$

For an electromagnetic source oscillating at a harmonic frequency of ω, setting $\partial/\partial t$ = jω, we can show that Equations (1.51) and (1.52) have the harmonic traveling wave solutions

$$B_y = B_0 \cos(\omega t \pm kz), \tag{1.53}$$

$$E_x = E_0 \cos(\omega t \pm kz), \tag{1.54}$$

where k is the wave number and is related to the wavelength λ by $k = 2\pi/\lambda = \omega/u$. Note that a half-wavelength long antenna designed for operation at a radian frequency of f ($=\omega/2\pi$) will have a length $u/2f$ in a medium with relative permittivity ε_r (e.g. $\varepsilon_r = 80$ for seawater), and $c/2f$ for free space. Hence, the antenna designed for operation in seawater will have to be shorter by a factor of about 0.11 ($=\varepsilon_r{}^{1/2}$) to that used in free space for the same frequency of operation.

The electric and magnetic fields have waves (and hence electromagnetic energy and electronic information) traveling in the +*z* and $-z$ directions. This is normally the case with a simple wire antenna used in mobile phones; signals are radiated in both directions deduced from the idea of an omni-directional monopole antenna. If we want the waves to be directed in one direction only, say in the +*z* direction, we must find a way of redirecting the energy traveling in the $-z$ direction to the +*z* direction. This may be done by simply placing a reflecting plane just behind the antenna to turn back the $-z$ directed waves to the +*z* direction as well; by doing this, all the energy put into the antenna is radiated in the +*z* direction instead of being divided into half for transmission in both directions.

An alternative form of expressing the solution of the wave Equations (1.51) and (1.52) is

$$E_x = E^+ \exp(-kz) + E \exp(kz), \tag{1.55}$$

$$H_y = (E^+/Z) \exp(-kz) - (E/Z) \exp(kz), \tag{1.56}$$

where the wave or intrinsic impedance $Z = E^+/H^+ = E^-/H^- = (\mu/\varepsilon)^{1/2}\ \Omega$. Note that the electric and magnetic fields of the wave are in phase with each other. For free space $\mu = \mu_0$, $\varepsilon = \varepsilon_0$ and thus $E^+/H^+ = \mu_0 E_0/B_0 = (\mu_0/\varepsilon_0)^{1/2}$. From this substitution, recalling that $u = c = (\mu_0\varepsilon_0)^{1/2}$ for free space, we also find that

$$E_0 = cB_0. \tag{1.57}$$

In the foregoing discussion, we assumed that the medium through which the electromagnetic wave travels is lossless, i.e. conductivity $\sigma = 0$. If this is not the case, our analysis for the lossless case applies, except that we have to replace the permittivity ε by a complex permittivity ε^c. This comes about as follows. For lossy media, the conductivity is reflected in the following Maxwell's equations:

$$\nabla \times \mathbf{H} = \mathbf{J} + \frac{\partial \mathbf{D}}{\partial t}, \tag{1.58}$$

which may be rewritten as

$$\nabla \times \mathbf{H} = \boldsymbol{\sigma}\mathbf{E} + \frac{\varepsilon \partial \mathbf{E}}{\partial t}, \tag{1.59}$$

where the first term on the right-hand side (RHS) is zero if the medium is lossless. Thus, in the discussion so far, only the second term on the RHS, which may be expressed as $\mathrm{j}\omega\varepsilon E$ for a harmonic signal, existed. Now for the complete Equation (1.69), the RHS may be rewritten as $(\sigma + \mathrm{j}\omega\varepsilon)E = \mathrm{j}\omega(\varepsilon - \mathrm{j}\sigma/\omega)E = \mathrm{j}\omega\varepsilon^c E$, where $\varepsilon^c = (\varepsilon - \mathrm{j}\sigma/\omega)$. Once we make this switch from permittivity ε by a complex permittivity ε^c, the foregoing discussion for the lossless case may be used for the lossy case. The final result for waves in the lossy medium case is given by the following two equations:

$$E_x = E_0 \exp(\pm\alpha z) \cos(\omega t \pm \beta z), \tag{1.60}$$

$$B_y = B_0 \exp(\pm\alpha z) \cos(\omega t \pm \beta z), \tag{1.61}$$

where α is the loss attenuation constant and β is the phase constant. The amplitude of the wave is attenuated as it propagates through the medium; after traveling over a distance called the skin depth $z = \delta = 1/\alpha$, the wave amplitude would have been attenuated by a factor e^1. Beyond this distance δ, the signal will be too weak for a receiver system to pick up. The solution of the wave equation in lossy media may also be expressed as follows:

$$E_x = E^+ \exp(-\gamma z) + E \exp(\gamma z), \tag{1.62}$$

$$H_y = (E^+/Z) \exp(-\gamma z) - (E/Z) \exp(\gamma z), \tag{1.63}$$

where propagation constant $\gamma = \alpha + j\beta = [j\omega\mu(\sigma + j\omega\varepsilon)]^{1/2}$ and intrinsic impedance $Z = [j\omega\mu/(\sigma + j\omega\varepsilon)]^{1/2}$. E^+ and E are two constants associated with the amplitude of the forward (positive z directed) and backward (negative z directed) waves, respectively, at $z = 0$. When the medium is a very good conductor such that $\sigma >> \omega\varepsilon$, then $\alpha = \beta = (\omega\mu\sigma/2)^{1/2}$.

1.8.3 Energy in the Electromagnetic Field

Consider the following two Maxwell's equations:

$$\nabla \times \mathbf{H} = \mathbf{J} + \frac{\partial \mathbf{D}}{\partial t}, \tag{1.64}$$

$$\nabla \times \mathbf{E} = -\frac{\partial \mathbf{B}}{\partial t}. \tag{1.65}$$

Subtract the dot product of Equation (1.64) with **E** from the dot product of Equation (1.75) with **H**:

$$\mathbf{H} \cdot \nabla \times \mathbf{E} - \mathbf{E} \cdot \nabla \times \mathbf{H} = -\mathbf{H} \cdot \frac{\partial \mathbf{B}}{\partial t} - \mathbf{E} \cdot \frac{\partial \mathbf{D}}{\partial t} - \mathbf{E} \cdot \mathbf{J} \tag{1.66}$$

Equation (1.66) may be rewritten as

$$\nabla \cdot (\mathbf{E} \times \mathbf{H}) = -\mathbf{H} \cdot \frac{\partial \mathbf{B}}{\partial t} - \mathbf{E} \cdot \frac{\partial \mathbf{D}}{\partial t} - \mathbf{E} \cdot \mathbf{J} \tag{1.67}$$

Taking the volume integral of Equation (1.67), we get

$$\iint_s (\mathbf{E} \times \mathbf{H}) \mathrm{d}s = -\iiint_{\mathbf{v}} \cdot \frac{\partial \mu \mathbf{H}^2}{2\partial t} \mathrm{d}v - \iiint_{\mathbf{v}} \cdot \frac{\partial \varepsilon \mathbf{E}^2}{2\partial t} \mathrm{d}v - \iiint_v \mathbf{E} \cdot \mathbf{J} \mathrm{d}v \tag{1.68}$$

There are four terms in the energy Equation (1.68); these are physically interpreted as follows:

1) The $\mathbf{P} = \mathbf{E} \times \mathbf{H}$ term on the left-hand side (LHS) is called the Poynting vector. It denotes the flow or radiation of electromagnetic energy out of a surface s. The average power in an electromagnetic wave is found from $((1/2)\mathrm{Re}(\mathbf{E} \times \mathbf{H}^*))$, where the asterisk stands for conjugate.
2) The $-(\partial\mu\mathbf{H}^2/2\partial t)$ term indicates the rate of decay (note the negative sign) in the magnetic energy $\mu H^2/2$ J/m^3 stored in a volume v. It is this decay of energy around an antenna that produces the radiated power $\mathbf{E} \times \mathbf{H}$ W/m^2 flowing out from the antenna. The term $\mu H^2/2$ J/m^3 is associated with the energy stored in inductive circuits or devices.

1) The $-(\partial\varepsilon\mathbf{E}^2/2\partial t)$ term indicates the rate of decay of electric energy $\varepsilon E^2/2$ J/m^3 stored in a volume v. The term $\varepsilon E^2/2$ J/m^3 is what is important in the rare capacitive type of antennas. It is this term that exists in electronic transistors providing power amplification as well as in capacitors in the form of stored energy.
2) The fourth term $E.J = \sigma E^2$ W/m^3 (since $J = \sigma E$) is associated with the ohmic power loss due to the finite conductivity of the medium through which the wave travels. Typical (conductivity in S/m, relative permittivity) values encountered in wireless communication channels are (4, 80) for seawater, (0.001, 80) for fresh water, (0.02, 30) for swampy land, (0.004, 13) for forest terrain, (0.002, 14) for rocky terrain, and (0.001, 10) for sandy terrain. The values do not reflect the frequency dependence of the parameters.

For sinusoidal signals, we may replace $\partial/\partial t$ by jω, and thus the energy transported by the electromagnetic wave is $U = (1/2)\varepsilon E^2 + (1/2)\mu H^2$ J/m^3. Since $E = (\mu_0/\varepsilon_0)^{1/2}\, H = cB$, the energy carried by the electric field is equal to the energy carried by the magnetic field. We get

$$U_{\text{elect}} = \frac{1}{2}\varepsilon_0 E^2, \qquad U_{\text{mag}} = \frac{1}{2}\varepsilon_0 c^2 B^2, \tag{1.69}$$

and the total energy $U = \varepsilon_0 E^2 = \mu_0 H^2$. The amplitude of the Poynting vector

$$P = E^2/\eta = c\varepsilon_0 E^2 \text{ W/m}^2 \tag{1.70}$$

and the average power transported by the electromagnetic wave is (1/2)Re$\{E \times H^*\}$= $(1/2)E^2/\eta = (1/2)c\varepsilon_0 E^2$ W/m^2.

It can be shown that for an electromagnetic wave propagating in a good conductor, nearly all of the wave's energy is carried by the magnetic component. In the case of an isotropic homogeneous medium, the Poynting vector becomes

$$\mathbf{P} = \mathbf{E} \times \mathbf{H} \tag{1.71}$$

and the energy density becomes

$$U = \frac{1}{2}\left(\varepsilon E^2 + \mu H^2\right). \tag{1.72}$$

Note that these expressions reduce to those obtained previously when $\varepsilon = \varepsilon_0$, $\mu = \mu_0$, and $H = B/\mu_0$.

The ratio of electric to magnetic energy density is therefore

$$\frac{\frac{1}{2}\varepsilon E_0^2}{\frac{1}{2}\mu H_0^2}. \tag{1.73}$$

Now for a good conductor

$$\sigma >> \omega\varepsilon, \tag{1.74}$$

and the intrinsic impedance

$$Z = [j\omega\mu/(\sigma + j\omega\varepsilon)]^{1/2} \approx [j\omega\mu/\sigma]^{1/2}. \tag{1.75}$$

Since $E_0/H_0 = \eta$, from Equations (1.73) and (1.75), we get

$$\frac{\frac{1}{2}\varepsilon E_0^2}{\frac{1}{2}\mu H_0^2} = \frac{\varepsilon\omega}{\sigma}. \tag{1.76}$$

Now $\varepsilon\omega/\sigma$ is a very small quantity for a good conductor for all frequencies up to and including optical frequencies. Thus, to a very good approximation, the energy is located in the magnetic component of a wave propagating in a conductor. In contrast, we earlier found that when an electromagnetic wave propagates through a vacuum, its energy is shared equally between the electric and magnetic components. Hence, for media-like seawater, which acts like a good conductor at most undersea communications frequencies, antennas should be designed to capture the magnetic fields more than the electric fields. In free space, most antennas like the electric dipole are designed to capture the electric field, whereas some antennas like the loop antenna are designed to capture the magnetic field.

1.9 Points to Note When Purchasing or Designing Antennas

In this section, we shall describe aspects of antenna design and design parameters with reference to a particular mobile communication system. Figure 1.11 shows an aircraft in flight. It is required that we design an antenna that can be mounted on top of the aircraft to communicate with a satellite.

Figure 1.11(a) shows two positions of the required azimuth scanning beam for the aircraft antenna to preserve contact with the satellite as it flies past the satellite. Figure 1.11(b) shows the elevation beam of the same antenna, required for the aircraft antenna to make contact with the satellite positioned on the LHS of the aircraft. The following points should be observed when designing or selecting an antenna:

1) Have a thorough understanding of the system in which you will be using the antenna. Take, for instance, the challenging problem of designing an antenna for satellite to aircraft communication (Taira et al., 1991).

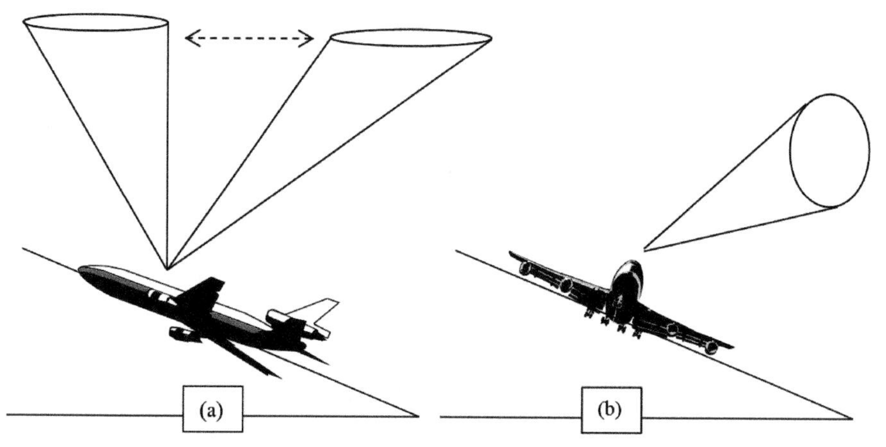

Figure 1.11 Antenna for aircraft to satellite communications.

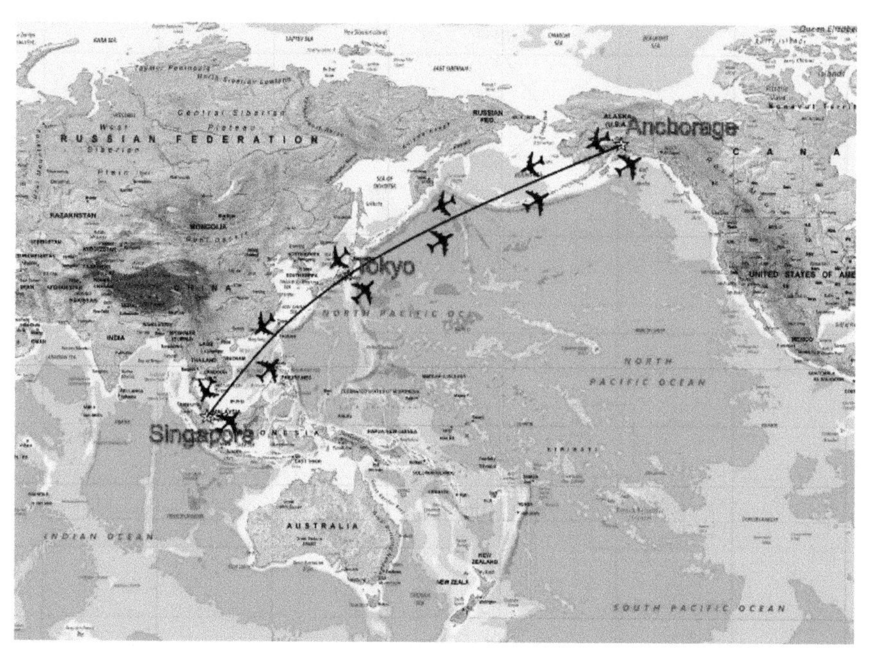

Figure 1.12 The flight route for which the aircraft–satellite communication antenna must be designed.

The flight route is, as shown in Figure 1.12, from Singapore to Tokyo to Anchorage, and then back again taking the same route. A geostationary satellite like the ETS-V satellite orbits at an altitude of 36,000 km above the earth. An aircraft might cruise at an altitude of about 20 km. Therefore, the antenna radiation pattern must be sufficiently tilted upwards so that the aircraft antenna can *look up* at the satellite; this means that the antenna beam must have an elevation beam. Further, the ETS-V satellite is stationary at 150°E, longitudinal. Consider an aircraft flying from Singapore (Changi airport) to Tokyo (Narita airport), and then from Tokyo to Anchorage. When the aircraft is flying Singapore–Tokyo–Anchorage, the satellite is to the RHS of the aircraft. Hence, the antenna radiation pattern should be focused to the RHS of the aircraft. Over the Singapore to Tokyo route, which is the first portion of the flight route, the satellite station is to the right front end direction of the aircraft. As the aircraft flies past the satellite, the beam should now look back at the satellite. Hence, the antenna beam should have an azimuth beamwidth scanning azimuth angle 110° to about azimuth angle 51°. The elevation of the beam should be about 30° for the Singapore to Tokyo route. For the Tokyo to Anchorage route, the elevation of the beam should be about 5°, and the azimuth beam should scan from 150° to 130°. Figure 1.13(a) shows the antenna azimuth beams for the forward flight from Singapore to Anchorage.

Similarly, on the way back from Anchorage to Singapore (Figure 1.13(b)), the aircraft antenna beam should have the following LHS looking beams: from Anchorage to Tokyo, the azimuth ranges from 300° to 330°, and the elevation is 5°; from Tokyo to Singapore, the azimuth of the beam should span 290°–230°, and the elevation of the beam above the plane of the aircraft is about 36°. Obviously to design such an antenna requires not only sound antenna hardware but also signal processing software to keep changing the direction of the radiation pattern in which the antenna has to communicate. In practice, a two-element array antenna is used to get the elevation of the beam and an eight-element array antenna is used to get the azimuth control of the beam.

However, LHS and RHS communication with such tight control of beams requires two identical antennas placed on either side of the aircraft; the array antenna on the RHS of the aircraft will be used during the Singapore to Anchorage flight. The second array will be used during the Anchorage to Singapore flight. Phase dividers are used to change the phase of the signal supplied to each element of the array antenna to get a scan beam that changes with the movement of the aircraft antenna. Such challenging antenna design problems demand optimum design.

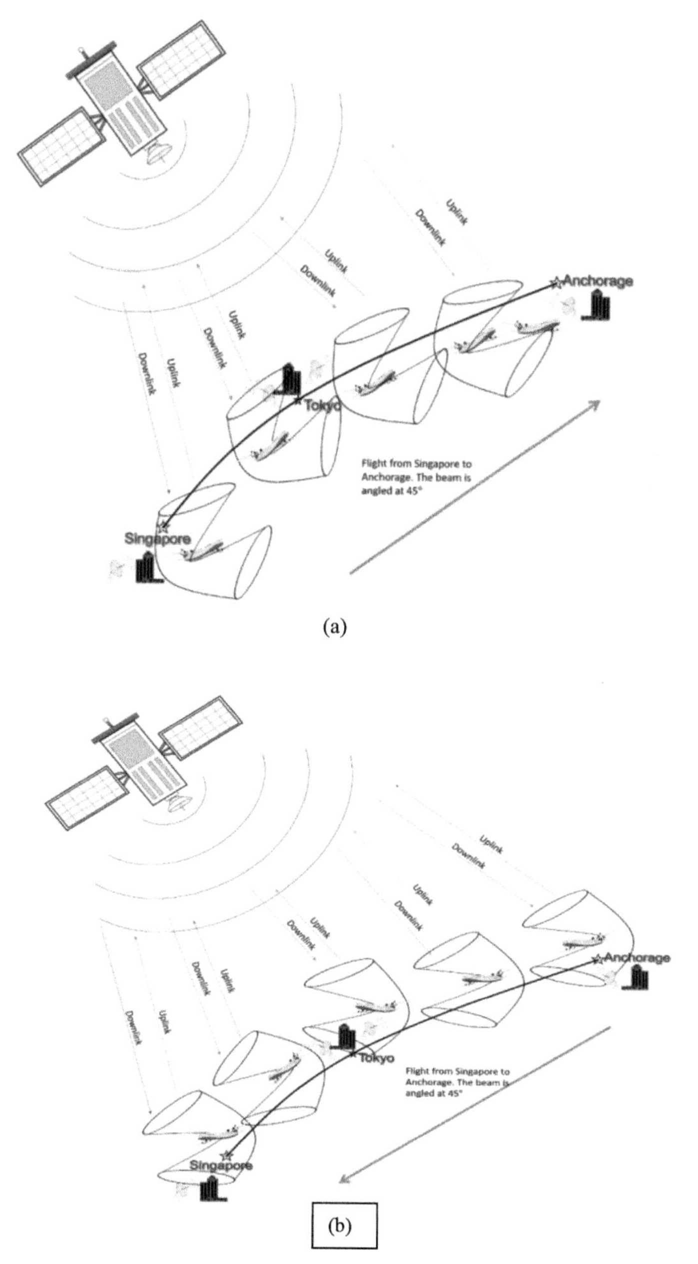

Figure 1.13 Aircraft-to-satellite communication antenna radiation patterns.

2) Design and select the antenna to match the transmitter and receiver frequency, as well as the bandwidth of the signals. In the aircraft to satellite communication system, for instance, the operating frequency may be 1.6/1.5 GHz. Hence, the antenna must be able to perform well at carrier frequencies of 1.6 and 1.5 GHz, recalling that at each carrier frequency, the signal frequency will vary over the bandwidth of about 7% of the carrier frequencies. Normally, mechanically light and small patch antennas tend to have bandwidths of about 3%.
3) Decide on the mechanical restrictions to be imposed on the antenna. It will be impractical to mount an antenna weighing above 20 kg or so on an aircraft. The mechanical structure of the antenna should be able to withstand vibrations in the range of 5–2 Hz. To reduce wind resistance, in mobile applications, it may be necessary to use bandwidth-inefficient microstrip patch antennas.
4) Determine the antenna performance in changing atmospheric conditions. In a typical flight from Singapore to Anchorage, the antenna will be expected to perform in temperatures varying from $-60°$ to 75°F. The voltage standing wave ratio (VSWR) of the antenna can vary by as much as 40% for such drastic temperature variations, giving rise to the antenna being poorly matched to the transmitter/receiver electronic circuitry.
5) Where the antenna beam has to be steered, determine the effect of beamsteering on antenna gain.
6) Study the gain of the antenna. Antenna gain tends to deteriorate when the beam (radiation pattern) is turned. The gain may vary from say 20 at the ideal angle (e.g. 90°) to 12 at another angle (e.g. 60°). Beamsteering tends to increase the number of sidelobes as well. Furthermore, the gain of the antenna may differ by about 10% for the two carrier frequencies; normally, the gain is maximized for the receiver carrier frequency. Hence, the gain at the transmitting frequency tends to be smaller than the designed value.
7) Study the VSWR of the antenna. For an array antenna, the VSWR of the array should be closer to 1. A single element of the array may have a VSWR of about 2.5.
8) Evaluate the gain-to-temperature (G/T) ratio for the antenna. At different elevation angles, the noise will be different. Evaluation of the carrier-to-noise (C/N) ratio is also important to determine the characteristics of the electronics required; in general, for mobile communication systems, the C/N value will vary dramatically from position to position.

9) It is now customary to determine the amount and type of signal processing necessary to make the antenna meet all the required performance criteria. Instead of talking just about antennas, we speak of *adaptive antennas* or *smart antennas*. Hence, the signal processing aspects of the antenna should be considered. Signal processing may be performed using smart antennas with signal processing hardware or software.

1.10 Smart Antennas and Electromagnetic Signal Processing

A smart antenna is made up of a collection of antenna elements arranged in an array (e.g. along a straight line) or in a series of arrays (e.g. along a rectangular grid) and the electromagnetic radiation (and reception) of the array antenna is electronically controlled to minimize noise and interference and to maximize its radiation toward a desired point or direction. Thus, in a smart antenna signal processing devices are attached to each element of the array so as to make the antenna radiation pattern or beam dynamic in space and time. A smart antenna is a unit consisting of both the physical antenna structure (elements) and computer codes to perform signal processing on the electronic signals received by the physical antenna elements (receiver antenna) or the signals being transferred to the physical antenna elements (transmitting antenna). An antenna can simultaneously transmit and receive signals on the same array of antenna elements as long as the frequencies are different and will not interfere with each other. The adaptive antenna arrays form the smart antennas. Sometimes les flexible switched-beam antennas and dynamically phased arrays are also considered as smart antennas.

One of the unique features of this book is its detailed handling of the science of the electromagnetic signals that form the integral components of the wireless systems, and to model and process them in a form that permits accurate and fast processing of the signals to develop an intelligent system that handles the electromagnetic signals as the electric signals of the brain and its tie to both the electric nerve and heart signals, all essential for the healthy and efficient function of the biological, and in this present case the electronic system.

An array antenna is made up of N antenna elements that are spatially arranged at specified locations about a common fixed point. Using the principle of superposition, the radiation pattern of an array antenna is the summation of the radiation pattern of its antenna elements. The beams may be formed and directed by changing the electronic current phase, current amplitude, or both current phase and amplitude of the currents at each antenna element. Control of current phase and amplitude is performed to maximize gain in a specific direction, to minimize sidelobe, and to increase the range of the beam. In mobile communications, the ability to electronically control the antenna beam may be exploited to improve the antenna and communication system performance.

The spatial arrangement of the antenna elements in an array antenna has received less attention in recent times. Obviously the geometry of the array antenna will have an impact on the beam generated. The linear, circular, and planar arrays are the three most commonly used geometries in array antennas, but it has been found that the geometry of the array does not have a significant effect on the efficiency of the antenna beam. It is the beamforming algorithm, which may be implemented on software or hardware, that largely determines the efficiency of the array antenna.

Beamforming of array antennas allows for achieving diversity in wireless communications without having to resort to more than one location for antennas. Antennas located in different locations in a neighborhood have been used to increase diversity and thus to combat fading in microwave mobile radio systems.

The following are two common forms of array antennas:

1) *Switched-beam system.* An electronic switch that switches out or, in some of the antenna elements in an array antenna, is used in switch beam systems (SBS) to select the best beam for a given user scenario. Although it is not easy to null any interfering signal that is close to the desired user signal with SBS, it is a much cheaper and simpler system than the smart antenna system.
2) *Adaptive antenna system.* An adaptive antenna system is able to achieve better performance. The weight in an adaptive array system (AAS) may be adjusted to point the tip of the beam toward the desired signal, and interferers are nulled by adjusting electronic weights that control the amplitude or phase of the currents at the antenna elements. It may not be as fast as the SBS, but it provides a more optimized beam for a given situation of users and interferers.

For convenience, the array antennas are sub-divided into two classes. The first is the phase array antennas where only the phase of the currents is changed by the weights. The second class of array antennas is the smart antennas where both the amplitude and phase of the currents are changed to produce a desired beam. One major disadvantage with the phase array systems is that the optimum weights to maximize the signal-to-noise ratio (SNR) are not computed due to the limit on the degree of freedom. However, it is a much faster system than the smart antennas.

Acknowledgment

The help of Eleasha Yalehen and Martha Elap with drawings.

2

Elementary Antenna Theory

P.R.P. Hoole

Abstract

In this chapter, the derivation of antenna radiated electric and magnetic fields from Maxwell's equations are presented for the short electric dipole (current carrying wire), the magnetic dipole (current carrying loop), and the half-wave length dipole. The radiated power, gain, antenna input impedance, radiation beam patterns, and beam characteristics are also presented in detail. Impedance matching, near and far fields, and radiated signal reflection are also presented in this chapter with the mathematical analysis for each.

2.1 Introduction

Consider the two Maxwell's equations that couple the electric field intensity **E** and magnetic flux density **B**. The magnetic field intensity **H** = **B**/μ. The magnetic flux density itself will be expressed as the curl of the magnetic vector potential **A**. A vector is completely defined only after both its curl and divergence are defined. To complete the definition of vector **A**, its divergence will be set to $\nabla\cdot\mathbf{A} = -\mu\varepsilon\ \partial\phi/\partial t$, where ϕ is the scalar potential.

2.1.1 Maxwell's Equations

The study of electromagnetic (EM) fields requires the use of the four Maxwell's equations given below:

$$\nabla \times \mathbf{H} = \mathbf{J} + \mathrm{j}\omega\varepsilon\mathbf{E}, \tag{2.1}$$

$$\nabla \times \mathbf{E} = -\mathrm{j}\omega\mu H, \tag{2.2}$$

$$\nabla \cdot \mathbf{B} = 0, \tag{2.3}$$

$$\nabla \cdot \mathbf{E} = \rho/\varepsilon. \tag{2.4}$$

Maxwell's equations relate the two basic physical parameters E (electric field intensity) and B (magnetic flux density) to the sources J (current density) and ρ (electric charge density). In telecommunication, radar, and biomedical imaging, the presence and effects of ρ are considered negligible. Thus, J is the primary source of information about the voice and video signals being transmitted or the object being imaged.

2.1.2 The Magnetic Vector Potential A for an Electric Current Source J

The vector potential **A** is useful in solving for the EM fields generated by an electric current source J. Since the magnetic flux **B** is solenoidal, we have

$$\nabla \cdot \mathbf{B} = 0. \tag{2.5}$$

Hence, **B** can be represented as the curl of another vector because it satisfies the vector identity

$$\nabla \cdot \nabla \times \mathbf{A} = 0, \tag{2.6}$$

where **A** is an arbitrary vector. Thus, we define

$$\mathbf{B}_A = \mu \mathbf{H}_A = \nabla \times \mathbf{A}, \tag{2.7}$$

where the subscript A denotes that the field is due to the vector potential **A**.

Substituting Equation (2.7) into Equation (2.2), we get

$$\nabla \times \mathbf{E}_A = -\mathrm{j}\omega\mu \mathrm{H}_A = -\mathrm{j}\omega \nabla \times \mathbf{A}. \tag{2.8}$$

Rearranging Equation (2.8), we have

$$\nabla \times [\mathbf{E}_A + \mathrm{j}\omega \mathbf{A}] = 0. \tag{2.9}$$

Using the vector identity

$$\nabla \times (-\nabla \phi_{\mathrm{e}}) = 0, \tag{2.10}$$

and comparing Equations (2.9) and (2.10), we get

$$\mathbf{E}_A + \mathrm{j}\omega \mathbf{A} = -\nabla \phi_{\mathrm{e}}, \tag{2.11}$$

where ϕ_e represents an arbitrary scalars potential function which is a function of position.

Taking the curl on both sides of Equation (2.7) and using the vector identity

$$\nabla \times \nabla \times \mathbf{A} = \nabla (\nabla \cdot \mathbf{A}) - \nabla^2 \mathbf{A}, \tag{2.12}$$

we get

$$\nabla \times (\mu \mathbf{H}_A) = \nabla (\nabla \cdot \mathbf{A}) - \nabla^2 \mathbf{A}. \tag{2.13}$$

For a homogenous medium, Equation (2.13) reduces to

$$\mu \nabla \times (\mathbf{H}_A) = \nabla (\nabla \cdot \mathbf{A}) - \nabla^2 \mathbf{A}, \tag{2.14}$$

which on equating to Equation (2.1) leads to

$$\mu \mathbf{J} + \mathrm{j}\omega\mu\varepsilon \mathbf{E}_A = \nabla (\nabla \cdot \mathbf{A}) - \nabla^2 \mathbf{A}. \tag{2.15}$$

Substituting Equation (2.11) into Equation (2.15), we get

$$\begin{aligned} \nabla^2 \mathbf{A} + k^2 \mathbf{A} &= -\mu \mathbf{J} + \nabla (\nabla \cdot \mathbf{A}) + \nabla (\mathrm{j}\omega\mu\varepsilon\phi_{\mathrm{e}}) \\ &= -\mu \mathbf{J} + \nabla (\nabla \cdot \mathbf{A} + \mathrm{j}\omega\mu\varepsilon\phi_{\mathrm{e}}), \end{aligned} \tag{2.16}$$

where $k^2 = \omega^2\mu\varepsilon$.

To define a vector entirely, its curl and divergence should be defined. In Equation (2.7), the curl has been defined. Now in order to simplify Equation (2.16), we let

$$\nabla \cdot \mathbf{A} = -\mathrm{j}\omega\mu\varepsilon\phi_{\mathrm{e}}. \tag{2.17}$$

Hence, Equation (2.16) reduces to

$$\nabla^2 \mathbf{A} + k^2 \mathbf{A} = -\mu \mathbf{J}, \tag{2.18}$$

which is known as the Lorentz condition.

Hence, we have defined the vector potential **A** completely. Also the electric field in Equation (2.11) can now be written as

$$\mathbf{E}_A = -\nabla\phi_{\mathrm{e}} - \mathrm{j}\omega\mathbf{A} = -\mathrm{j}\omega\mathbf{A} - \mathrm{j}\frac{1}{\omega\mu\varepsilon}\nabla (\nabla \cdot \mathbf{A}). \tag{2.19}$$

The electric field **E** can now be calculated if the vector potential function **A** can be found.

In the presence of the source ($J \neq 0$) and $k = 0$ (static case), the wave equation reduces to

$$\nabla^2 A_z = -\mu J_z. \tag{2.20}$$

This form is called Poisson's equation and it has a parallel equation for a scalar potential ϕ, relating it to the charge density ρ:

$$\nabla^2\phi = -\frac{\rho}{\varepsilon}. \tag{2.21}$$

The solution for Equation (2.21), from Coulomb's law for electrostatics, is given by

$$\phi = \frac{1}{4\pi\varepsilon}\iiint_v \frac{\rho}{r}\,\mathrm{d}v', \tag{2.22}$$

where r is the distance from any point on the charge density to the observation point. Since Equation (2.20) is similar in form to Equation (2.21), by comparing it to the solution of the second-order differential equation, or the Poisson equation, for the scalar potential ϕ, the solution for **A** in general is

$$\mathbf{A} = \frac{\mu}{4\pi}\iiint_v \mathbf{J}\frac{\mathrm{e}^{-jkr}}{r}\,\mathrm{d}v'. \tag{2.23}$$

Hence, knowing the current density distribution and the limits of integration, the vector potential function could be found. The procedure to find the electric field E and magnetic field H is given in Section 2.4.

In all our analyses, unless otherwise stated, we shall assume that the current or signal we are dealing with is sinusoidal in the time domain. Hence, in the phasor domain analysis, we represent the time domain term by $\exp(\mathrm{j}\omega t)$. This term gets canceled out on both sides of Maxwell's equations and, hence, need not be carried over in antenna analysis. However, when the final complete solution is required, the $\exp(\mathrm{j}\omega t)$ term must be attached at the end of the solution. Thus, if the final solution looks like $E_0 \exp(-\mathrm{j}kR)$, then the complete phasor domain solution is $E_0\exp(-\mathrm{j}kR)\exp(\mathrm{j}\omega t)$. Taking the real of this, the time domain solution is $E_0 \cos(\omega t - kR)$. The wave number $k = \omega/c = \omega(\mu_0\varepsilon_0)^{-1/2}$, where ω, c, μ_0, and ε_0 are the radian signal frequency, velocity of light, free space permeability, and free space permittivity, respectively. The values of some important physical constants are given below:

permittivity of free space = $\varepsilon_0 = 8.854 \times 10^{-12}$ F/m;
permeability of free space = $\mu_0 = 4\pi \times 10^{-7}$ H/m;
speed of light in free space = $c = 2.998 \times 10^{8}$ m/s;
wave impedance in free space $\eta = 120\pi = 376.7\ \Omega$;
gravitational constant = 6.67×10^{-11} Nm^2/kg^2;

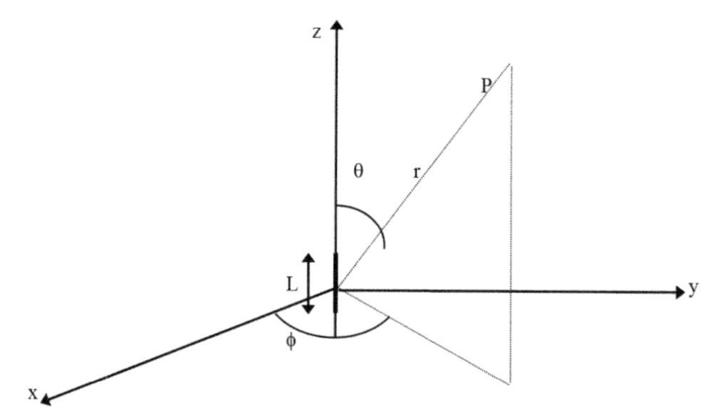

Figure 2.1 Infinitesimal Hertzian dipole antenna.

Boltzmann's constant = 1.38×10^{-23} J/K;
Planck's constant = 6.63×10^{-24} Js;
electron mass = 9.11×10^{-31} kg;
proton mass = 1.67×10^{-27} kg.

Note that even for non-sinusoidal time domain waveforms, this analysis is valid since any time domain signal could be translated into the frequency domain and represented by a summation of sinusoidal waves.

2.2 Infinitesimal Wire Antennas (Hertzian Dipole $L < \lambda/50$): The Elemental Dipole

2.2.1 Electromagnetic Fields Radiated by a Hertzian Dipole

Consider an infinitesimal length of conductor, of length dz = L, carrying a current I at frequency ω. The antenna geometry is shown in Figure 2.1.

The conductor is placed at the origin along the z-axis. We would like to determine the EM field radiated by such a short antenna. Such an antenna is an omni-directional antenna since it radiates in all directions; it may be made directive when used with a dish reflector or in an array. Referring to Figure 2.1, the problem is to find electric field intensity E and magnetic field intensity H (or magnetic flux density B) at observation point P. The medium in which the element is placed is free space. Hence, the wave number $k = \omega/c$, where c is the velocity of light

$$\mathbf{A}_z = \frac{\mu}{4\pi}\int \frac{\mathbf{J}}{r}\,\mathrm{e}^{-\mathrm{j}kr}\,\mathrm{d}R$$

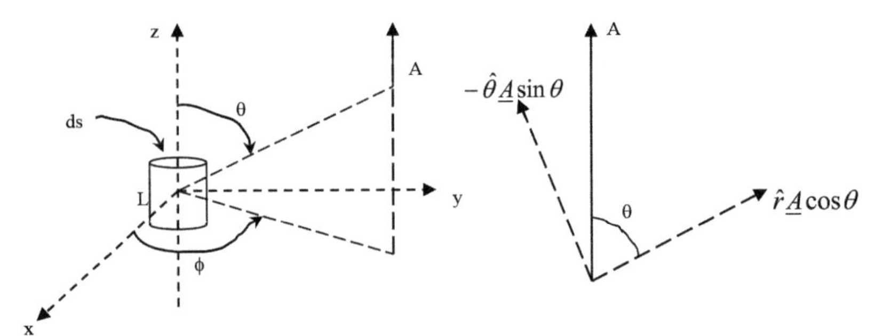

Figure 2.2 Dipole geometry and current distribution.

$$= \mathbf{u}_z \frac{\mu}{4\pi} \int_{-L/2}^{L/2} \frac{I\mathrm{e}^{-\mathrm{j}kr}}{r}\, \mathrm{d}z = \mathbf{u}_z \frac{\mu I L}{4\pi r} \mathrm{e}^{-\mathrm{j}kr}, \tag{2.24}$$

where we have changed the current density J (A/m^2) to current I (A) by observing that $J = I/S$, where S is the cross-sectional area of the conductor and the volume element dR = S·dz; hence, J·dR = I·dz. Furthermore, the current along an infinitesimal dipole is assumed to be uniform, i.e. the current along the antenna is not a function of z. That is, why we were able to pull current I out of the integral. In most finite length antennas, like the half-wave dipole antenna, current is distributed along the wire in, say, a cosine or sine function shape. Thus, this cosine or sinc shaped function needs to be integrated with respect to z for a finite length antenna. Referring to Figure 2.2, we have

$$\begin{aligned} A_r &= A_z \cos\theta \\ &= \frac{\mu I L}{4\pi r} \mathrm{e}^{-\mathrm{j}kr} \cos\,\theta, \end{aligned} \tag{2.25}$$

$$A_\theta = -A_z \sin\theta = -\frac{\mu I L}{4\pi r} \mathrm{e}^{-\mathrm{j}kr} \sin\,\theta, \tag{2.26}$$

$$A_\phi = 0. \tag{2.27}$$

When we take the curl of vector potential $\mathbf{A}$ in spherical coordinates, only the $\mathbf{u}_\phi$ component is not zero. Therefore, using

$$\mathbf{B}_A = \mu \mathbf{H}_A = \nabla \times \mathbf{A}, \tag{2.28}$$

the magnetic field intensity is given by

$$\mathbf{H} = \mathbf{H}_\phi = \frac{1}{\mu} \nabla \times \mathbf{A} = \frac{1}{\mu r} \left(\frac{\partial (r A_\theta)}{\partial r} - \frac{\partial A_r}{\partial \theta} \right) \mathbf{u}_\phi, \tag{2.29}$$

from which we get

$$\mathbf{H}_\phi = \frac{\mathrm{j}kIL}{4\pi}\left[\frac{1}{r} + \frac{1}{\mathrm{j}kr^2}\right] \sin\,\theta\, \mathrm{e}^{-\mathrm{j}kr}\mathbf{u}_\phi. \tag{2.30}$$

The electric field intensity may now be obtained from the magnetic field intensity, using Maxwell's equation relation

$$\mathbf{E} = -\frac{\mathrm{j}}{\omega\varepsilon}\nabla \times \mathbf{H}. \tag{2.31}$$

Hence,

$$\mathbf{E} = \frac{1}{\mathrm{j}\omega\varepsilon}\left[\frac{1}{r\,\sin\,\theta}\frac{\partial}{\partial\theta}(H_\phi\,\sin\,\theta)\,\mathbf{u}_r - \frac{1}{r}\frac{\partial}{\partial r}(rH_\phi)\,\mathbf{u}_\theta\right] = \mathbf{E}_r + \mathbf{E}_\theta, \tag{2.32}$$

$$\begin{aligned}\mathbf{E}_r &= \frac{IL\mathrm{e}^{-\mathrm{j}kr}}{4\pi\mathrm{j}\omega\varepsilon}\left(\frac{1}{r\,\sin\,\theta}\right)\frac{\partial}{\partial\theta}\left[\sin^2\theta\left(\mathrm{j}\frac{k}{r} + \frac{1}{r^2}\right)\right]\,\mathrm{u}_r \\ &= \frac{IL\mathrm{e}^{-\mathrm{j}kr}\,\cos\,\theta}{2\pi\varepsilon}\left[\frac{1}{cr^2} + \frac{1}{\mathrm{j}\omega r^3}\right]\,\mathbf{u}_r\ \mathrm{V/m},\end{aligned} \tag{2.33}$$

$$\mathbf{E}_\theta = \frac{IL}{4\pi\mathrm{j}\omega\varepsilon}\left(-\frac{1}{r}\right)\frac{\partial}{\partial r}\left[r\,\sin\,\theta\,\mathrm{e}^{-\mathrm{j}kr}\left(\frac{\mathrm{j}k}{r} + \frac{1}{r^2}\right)\right]\,\mathbf{u}_\theta \tag{2.34}$$

$$\begin{aligned}&= IL\frac{\mathrm{e}^{-\mathrm{j}kr}\,\sin\,\theta}{4\pi\mathrm{j}\omega\varepsilon}\left[-\frac{k^2}{r} + \mathrm{j}\frac{k}{r^2} + \frac{1}{r^3}\right]\,\mathbf{u}_\theta \\ &= \frac{IL\mathrm{e}^{-\mathrm{j}kr}\,\sin\,\theta}{4\pi}\sqrt{\frac{\mu}{\varepsilon}}\left[\mathrm{j}\frac{k}{r} + \frac{1}{r^2} + \frac{1}{\mathrm{j}kr^3}\right]\,\mathbf{u}_\theta\mathrm{V/m},\end{aligned} \tag{2.35}$$

where $\mathbf{u}_r$, **u**, and **u** are the spherical coordinate unit vectors. Considering only the terms that are $1/r$ dependent far-field terms, we get

$$\frac{E_\theta\,(1/r)}{H_\phi(1/r)} = \sqrt{\frac{\mu}{\varepsilon}} = \eta,$$

the intrinsic impedance of the medium containing the dipole, $\eta = 120\pi\Omega$ for free space. It should be remembered that to complete the above set of equations, the term exp(jωt) should be added to each equation. Hence, the complete phase term will be exp(j($\omega t - kr$)). The total radiation power cannot vary with distance. Hence, power density must vary as $1/r^2$; this rapid decrease in signal strength as the signal propagates in space is due to the spreading out of the antenna beam and not due to power dissipation. In communication design, this power loss is treated as path loss

$$\mathbf{E}_r = \eta \frac{IL \cos\theta}{2\pi} \left[\frac{1}{r^2} + \frac{1}{jkr^3} \right] e^{-jkr} \mathbf{u}_r, \tag{2.36}$$

$$\mathbf{E}_\theta = j\eta \frac{kIL \sin\theta}{4\pi} \left[\frac{1}{r} + \frac{1}{jkr^2} - \frac{1}{k^2 r^3} \right] e^{-jkr} \mathbf{u}_\theta, \tag{2.37}$$

$$E = 0. \tag{2.38}$$

The exp($-jkr$) term is a rotational term in the spatial domain r, similar to the exp($j\omega t$) term in the time domain. Hence, the radiated signal rotates in the time and spatial domains. The rotation in the spatial domain, with a period of λ m (where wavelength $\lambda = 2\pi/k$), causes sudden dips in the received signals as observed in mobile communications.

It is seen from Equations (2.36)–(2.38) that the infinitesimal current element will generate a magnetic field component that is perpendicular to both the radial and tangential electric field components, respectively.

The E_r field has the radiating near-field and evanescent-field components. These are components that vary inversely as the square and cube of the distance r, respectively. The E_θ field has, in addition, a component which varies as the inverse of the distance, and this is termed the far-field component. In the far-field region, E_θ and H_ϕ are connected by a simple coefficient, namely the free space impedance η. This is not true in the near-field region where both E_r and E_θ are related to H_ϕ.

The near-field region is defined as the distance d from the source such that

$$d < \frac{2D^2}{\lambda}, \tag{2.39}$$

where D is the dimension of the antenna along the direction vector. Beyond this distance, the radiating near fields and evanescent fields disperse very rapidly.

In radar imaging, we use the electric dipole to model the discrete scattering elements. Hence, the complete field equations of Equations (2.36) and (2.37) become important in order to determine the dependence of distance on image synthesis in the near-field region. The Doppler frequency shift tends to significantly modify the received power.

2.2.2 Electric Field Radiation Pattern of an Electric Dipole

2.2.2.1 The E-Plane Radiation Pattern

The field pattern of an electric dipole in a plane vertical to the *xy*-plane (E-plane) at a distance of 1 m (length of dipole $\lambda/50$, frequency = 100 MHz) is considered.

In Figure 2.3(a), the variation of the magnitude of the E_r component on an equi-phase surface is shown. In other words, this is the amplitude variation of E_r on a spherical wave front of an electric dipole at a distance of 1 m. It is apparent that the amplitude is dependent on the angle of reception and it is zero at an angle of 90° away from the *z*-axis. The same observation is true in Figure 2.3(b), where the nulls are shifted along the *z*-axis.

Figure 2.3(c) shows the variation of the field amplitude with receiving angle, when the strength is dependent on both E_r and E_θ simultaneously. It is seen that the radiation pattern of the resultant field is almost spherical, like that of a fictitious isotropic or point radiator.

The variation of amplitude due to the resultant signal, as shown in Figure 2.3(c), is very small for the entire range of reception angle.

These observations are important, in that the angle of reception plays a role in the received signal amplitude, and, thus, prior knowledge on this perturbation is necessary to perform corrections on the received signal amplitude. If the modulating factor against angle is known *a priori*, the correction could be incorporated into the signal processing algorithm to automatically correct the received signal strength.

The correction will depend on whether the signal received is due to E_r, E_θ, or the resultant and can be implemented using the normalized factors plotted in Figure 2.3(a)–(c), respectively. For the case where the circularly polarized signal is received, the correction is not necessary since the response is omni-directional.

2.2.2.2 The H-Plane Radiation Pattern

The field pattern of an electric dipole at a distance of 1 m in the horizontal plane (H-plane) (length of dipole $\lambda/50$, frequency = 100 MHz) is considered. The field on the horizontal plane is entirely due to the E component (at θ = 90°, E_r will not be present) and it is as shown in Figure 2.3(d). Amplitude correction is not required as the response is omni-directional.

Far-field or radiation-field components of **E** and **H** are given by

$$H_\phi = \mathrm{j}\frac{kIL\mathrm{e}^{-\mathrm{j}kr}\ \sin\ \theta}{4\pi r}, \tag{2.40}$$

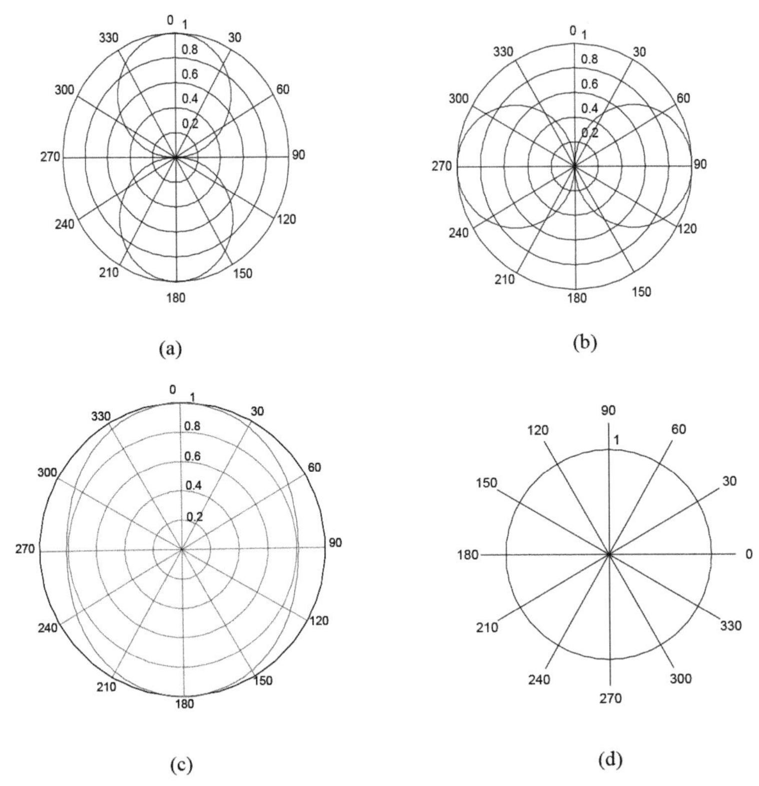

Figure 2.3 Normalized field pattern (a) of E_r in the vertical plane (E-plane), (b) of E in the vertical plane (E-plane), (c) of resultant electric field, and (d) on the horizontal plane (H-plane).

$$E_\theta = \mathrm{j}\eta \frac{kIL\mathrm{e}^{-\mathrm{j}kr} \sin\,\theta}{4\pi r}. \tag{2.41}$$

When we divide Equation (2.40) by Equation (2.41), we get the wave impedance,

$$Z = \frac{E_\theta}{H\phi} = \eta. \tag{2.42}$$

Note that the radiation (or far-field) components are the parts of the fields that are dependent on $1/r$ and will dominate the fields at far distances. Thus, the $\mathbf{E}_r$ components are not significant at far distances. The far-field magnetic and electric field amplitudes decrease as $1/r$ and the phase changes as $-kr$ (i.e. $-2\pi r/\lambda$). The power transmitted by the electric and magnetic fields radiated by an antenna are given by the Poynting vector $\mathbf{P} = \mathbf{E} \times \mathbf{H}$. Therefore, the direction of power flow is perpendicular to the plane containing **E** and **H**. It is of interest to note that the $\mathbf{E}_r \times \mathbf{H}$ will result in a power term with a j component; this means that the $\mathbf{E}_r \times \mathbf{H}$ power is an imaginary (or reactive) power term. It is the $\mathbf{E}_\theta \times \mathbf{H}$ power that transfers real (communication or remote sensing) signal power

$$\text{Power density radiated } P_\text{i} = (1/2)\ |\mathbf{E}_\text{rad} \times \mathbf{H}^*_\text{rad}|$$
$$= (1/2)\ \left|E_\theta\left(\frac{1}{r}\right) H_\phi \left(\frac{1}{r}\right)\right|\ \text{W/m}^2 \quad (2.43)$$

ignoring the near-field ($1/r^3$) and intermediate-field ($1/r^2$) components.

Expressing this in a more concise form, the radiated power density is given by

$$P_\text{i} = \tfrac{1}{2}(\mathbf{E} \times \mathbf{H}^*) \ = \ Re\ \tfrac{1}{2}\ |H_\phi E_\theta|$$
$$= \frac{1}{2\eta}\ |E_\theta^2|\ \ \text{W/m}^2. \quad (2.44)$$

Note that radiated power intensity P_i will become half the maximum power intensity radiated at the points on the radiation pattern where electric field intensity is $0.707E_0$, where E_0 is the maximum electric field intensity radiated.

Radiation intensity is given by

$$S = \text{power/unit solid angle}$$
$$= \frac{4\pi r^2 P_\text{i}}{4\pi} = r^2 P_\text{i} = \frac{\eta}{2}\left(\frac{kIL}{4\pi}\right)^2 \sin^2\theta$$
$$\approx r^2 E_\theta^2. \quad (2.45)$$

Although we shall elaborate on the near-field phenomena in this chapter, it should be noted that the near-field power flow associated with $\mathbf{E}_r \times \mathbf{H}_\phi$, which is an imaginary (i.e. made of a j-) term, is a reactive power and it circulates around the surface of a fictitious sphere surrounding the antenna. This term is quite significant in the near-field regions of biomedical imaging, synthetic-aperture remote sensing, and wireless communications.

Since $H = E/\eta$ in the far field, $P_i = E^2/2\eta$ W/m^2 in the radiation- or far-field region. The power density P_i, note, is a function of $\sin^2\theta$. This means that if we define the directivity of any antenna with respect to the dipole antenna instead of the isotropic antenna, then the $\sin^2\theta$ term will tend to increase the P_i at any given point over the P_I value of an isotropic antenna with the same input power. Hence, the dipole directivity (or gain) of an antenna will be smaller than the isotropic directivity (and gain) of that antenna.

The average power radiated, replacing the wave number k by $k = 2\pi/\lambda$, is obtained by integrating P_i over the spherical surface surrounding the antenna, with the surface element given by $\mathrm{d}S = r^2 \sin\theta \, \mathrm{d}\theta \, \mathrm{d}\phi$

$$P_\mathrm{r} = \frac{1}{2}\frac{I^2L^2\sqrt{\mu/\varepsilon}}{4r^2\lambda^2}\int_{\phi=0}^{2\pi}\int_{\theta=0}^{\pi} \sin^3\theta \, r^2 \, \mathrm{d}\theta \, \mathrm{d}\phi$$

$$= \frac{\pi I^2 L^2\sqrt{\mu/\varepsilon}}{3\lambda^2} \text{ W}. \tag{2.46}$$

In the above integration, note that by integrating from 0 to π radians in the θ domain and from 0 to 2π radians in the ϕ domain, we integrate over a whole sphere placed around the antenna with its center coinciding with the center of the antenna. In working out the integration, the following rearrangement was made: $\sin^3\theta = \sin\theta \sin^2\theta = (1/2)\sin\theta(1-\cos 2\theta) = (1/2)\sin\theta - (1/4)(\sin(-\theta) + \sin 3\theta) = (3/4)\sin\theta - (1/4)\sin 3\theta$. The current I that appears in the power term is the average or uniform current that flows along the antenna. It is equal to the maximum time domain current for short dipoles with a uniform current distribution. If the current distribution was, instead, triangular with maximum current at the center of the dipole with the current tapering off to zero at either ends of the antenna, I should be replaced by $I/2$, and the radiated power is reduced.

Defining the radiation resistance using the circuit relation

$$P_\mathrm{r} = \tfrac{1}{2} I^2 R_\mathrm{r} \, W, \tag{2.47}$$

where the current I is the maximum time domain current. Therefore, equating the two expressions we have for average power radiated, we get

$$R_\mathrm{r} = \frac{2I^2L^2\sqrt{(\mu/\varepsilon)}\,(\pi/3)}{I^2\lambda^2}$$

$$= \frac{2}{3\lambda^2}\pi L^2\sqrt{\frac{\mu}{\varepsilon}} = 80((\pi L)/\lambda)^2 \ \Omega. \tag{2.48}$$

If the current distribution along the short dipole antenna is triangular in shape, as it usually is, then the current I in Equation (2.47) should be replaced by $(I/2)$, and the radiation resistance will be $20((\pi L)/\lambda)^2 \ \Omega$.

The power received by a receiving antenna

$$P_{\mathrm{R}} = A_{\mathrm{em}} P_{\mathrm{i}}, \tag{2.49}$$

where A_{em} is the effective aperture of the antenna and P_{i} is radiation power density at the antenna. The voltage induced along an infinitesimal short dipole antenna by the radiation electric field E is approximately given by

$$V = EL\ \mathrm{V}, \tag{2.50}$$

where E is the electric field associated with the power density that appears at the antenna of length L.

For maximum power transfer, the following condition must be achieved by using impedance matching techniques. Ignoring the ohmic resistance of the antenna, the termination impedance must satisfy the following condition:

$$Z_{\mathrm{T}} = R_{\mathrm{r}} - \mathrm{j}X_{\mathrm{A}} = Z_{\mathrm{A}}^{*}, \tag{2.51}$$

which states that the termination impedance of the cable at the antenna, Z_T, must be equal to the conjugate of the antenna impedance Z_{A}^{*}. The antenna impedance is made of the radiation resistance R_{r} and reactance X_{A}. Thus, the power received by the receiving antenna is given by

$$P_{\mathrm{R}} = \frac{V_{\mathrm{rms}}^2}{4\,R_{\mathrm{r}}} = \frac{E_{\mathrm{rms}}^2 L^2}{4R_{\mathrm{r}}} = AP_{\mathrm{i}} = A\frac{E_{\mathrm{rms}}^2}{\sqrt{\mu/\varepsilon}}, \tag{2.52}$$

where we have used the matched impedance condition for which the terminating load resistance $R_{\mathrm{L}}(= R_{\mathrm{T}}) = R_{\mathrm{r}}$, and the total terminal impedance $Z_{\mathrm{T}} + Z_{\mathrm{A}} = 2R_{\mathrm{r}}$ so that the power at the receiver is $I_{\mathrm{rms}}^2 R_{\mathrm{L}} = (V_{\mathrm{rms}}/2R_{\mathrm{r}})^2 R_{\mathrm{L}}$.

Hence, the effective area of the antenna

$$A_{\mathrm{em}} = \frac{3\lambda^2}{8\pi}\ \mathrm{m}^2$$

or, in general,

$$A_{\mathrm{em}} = \frac{D\lambda^2}{4\pi}\ \mathrm{m}^2. \tag{2.53}$$

For a lossless (i.e. ohmic resistance is zero) elemental dipole,

$$\begin{aligned} D &= \frac{r^2 \frac{1}{2}\left(E_\theta H_\phi\right)_{\max}}{P_{\mathrm{r}}/4\pi} \\ &= \frac{(r^2/2)\left(I^2 L^2/\left(4\pi\right)^2\right)\sqrt{\mu/\varepsilon}\ (2\pi/\lambda r)^2}{(\pi/3\ I^2 L^2 \sqrt{\mu/\varepsilon})/\lambda^2 4\pi} = \frac{3}{2}. \end{aligned}$$

Thus, we have

$$\frac{D}{A_{\text{em}}} = \frac{3/2}{(3/8\pi)\,\lambda^2} = \frac{4\pi}{\lambda^2} = \frac{G}{A_{\text{em}}}. \tag{2.54}$$

We have assumed that the receiving antenna is placed at the peak point of the transmitting antenna beam and that the axes of the transmitting and receiving antennas are parallel. In general, if the receiving antenna is placed at an angle θ in the polar coordinate, then $D(\theta) = D \sin^2 \theta$. Furthermore, if the receiver antenna axis is tilted at an angle δ from the transmitting antenna axis, then the effective aperture A_e is no longer equal to the maximum effective aperture A_{em} but reduced to a value given by $A_e = A_{em} \sin^2\theta$. Ignoring the losses in the transmitting antenna, we have

$$\text{gain } G = D = \frac{4\pi A}{\lambda^2}; \quad \frac{G_{F1}}{G_{F2}} = \frac{f_1^2}{f_2^2}. \tag{2.55}$$

Note that in much of our discussion, we have ignored the ohmic resistance R_o of the antenna, assuming it to be much smaller than the radiation resistance R_r. If both resistances are accounted for, then we have gain $G = (R_r/(R_r + R_o))$ D; thus, the gain G is less than the antenna directivity D.

These electrically short antennas are used for special reasons such as when windage resistance should be reduced in high-speed aircraft and warships, when obstacle clearance is a limiting factor, when concealment is required, or when an antenna is to be mounted on animals or birds for radio tracking. The radiation resistance of the small antenna is very small, and hence the ohmic resistance has to be accounted for when determining the antenna impedance and radiation efficiency. Impedance matching is critical when using electrically short antennas since the power captured by such antennas is also quite small and of the order of $(E^2\lambda^2)/(320\pi^2)$ W, where E is the incident electric field. When short antennas are to be mounted on vehicles or on the collars of animals, a monopole of length $\lambda/4$ may be placed vertical to an artificial conducting *ground* of length $\lambda/2$ on the collar. The closer an antenna is kept to a human body, the smaller its gain.

2.3 Antenna in Motion

When the transmitting or receiving antenna is moving, as in the case of mobile communications (see Figure 2.4), the received signal frequency is shifted due to the motion of the antenna. Consider a signal $E_0 \cos(\omega t - kR)$

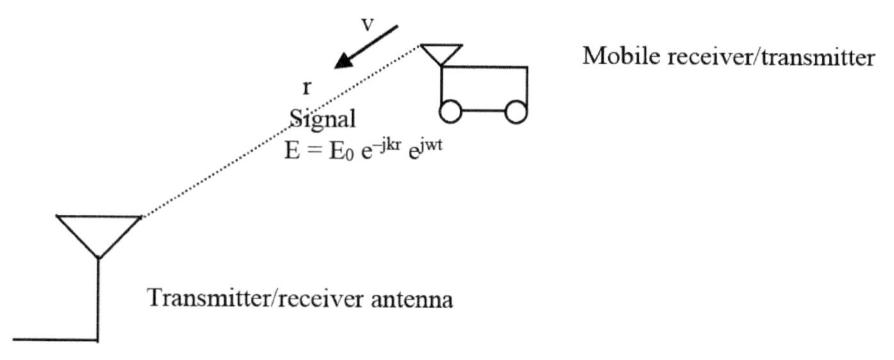

Figure 2.4 Mobile antenna and Doppler frequency shift.

being radiated by an antenna moving directly toward the receiver antenna at a velocity v.

The distance between the antennas is now a function of time and is given by= $r_0 - vt$.nce, the signal picked up by the receiver antenna is

$$E = E_0 e^{j[\omega_0 t - kr_0 + kvt]}. \tag{2.56}$$

Thus, we note that the effect of antenna motion is to shift the frequency of the signal. If the transmitter is moving toward the receiver, as in the case we have considered, the signal frequency increases by the Doppler frequency f_D:

$$\omega = \omega_0 + \omega_0 \frac{v}{c} \tag{2.57}$$

$$f_D = f_0 \frac{v}{c}. \tag{2.58}$$

Doppler frequency f_D is positive if velocity v is positive and it is negative if the velocity is negative. If the transmitter is moving away from the receiver, the frequency is shifted down. This in turn will modify the power of the received signal. Often some form of compensation is required in communication links to get rid of the effects of the Doppler shift. However, in remote sensing systems like radar, the Doppler frequency is treated as a useful parameter from which, for instance, the velocity of the transmitter or target may be estimated. It could be shown that if the transmitter is moving at an angle δ with respect to the straight line connecting the two antennas, the Doppler frequency shift will be

$$f_D = f_0 \frac{v}{c} \cos \delta, \tag{2.59}$$

indicating that the maximum frequency shift is when $\delta = 0$. The velocity v is the relative velocity between the transmitter and the receiver; it is positive if the distance r between the transmitter and receiver is getting smaller, and it is negative if r is increasing with time. The Doppler frequency shift tends to significantly modify the received power. From Equation (2.59), we observe that the effective aperture, and hence the received power, will vary with frequency. From Equation (2.54), we observe that the radiation resistance, and hence the radiated power, is also frequency dependent. In the fundamental Equations (2.46) and (2.47), the wave number $k = \omega/c = 2\pi f/c$, where c is the velocity of light (=2.998 $\times$ 10^8 m/s in free space). Thus, the magnitudes and phase of the radiated fields are frequency dependent.

2.4 Finite Length Wire Antenna (Dipole): The Half-Wave (λ/2) Dipole

2.4.1 Radiation from an Electric Dipole Antenna of Any Length L

Consider an antenna of length L, carrying a sinusoidal current I (see Figure 2.5). Along the transmission line connecting the antenna to the source, the current $I_c = I$. Along the dipole antenna, the current $I_d = I$. Assume that the dipole antenna is placed along the z-axis. With the center of the antenna at the origin, at the top end L_1, $z = L/2$, and at the bottom end of the antenna L_2, $c = -L/2$. There will be both electric fields E and magnetic fields H, each coupled to the other, launched out from the antenna in packets. We want to find the radiation electric field at point P. We shall consider the observation point to be sufficiently far away (i.e. $r >> L, \lambda$) to ignore the near- and intermediate-field terms. We shall first work out the electric field intensity for a finite length wire antenna of any length L, and then focus on radiation from the more popular half-wavelength antenna ($L = \lambda/2$). One reason for the popularity of the half-wavelength antenna is that its input impedance is about 73 Ω, which matches well with coaxial cables of line impedance 75 or 50 Ω

$$I(z) = \begin{cases} I_{\mathrm{m}} \sin k(L/2 - z), & 0 \le z \le L/2, \\ I_{\mathrm{m}} \sin k(L/2 + z), & -L/2 \le z \le 0. \end{cases} \tag{2.60}$$

For a short (Hertzian) dipole of length dz carrying current I, the radiation field using Equation (2.41) is given by

$$\mathrm{d}E_\theta = \eta\, \mathrm{d}H_\phi = \mathrm{j}\eta \frac{kI\, \mathrm{d}z\, \mathrm{e}^{-\mathrm{j}kR}}{4\pi R} \sin\theta\, \mathrm{d}z', \tag{2.61}$$

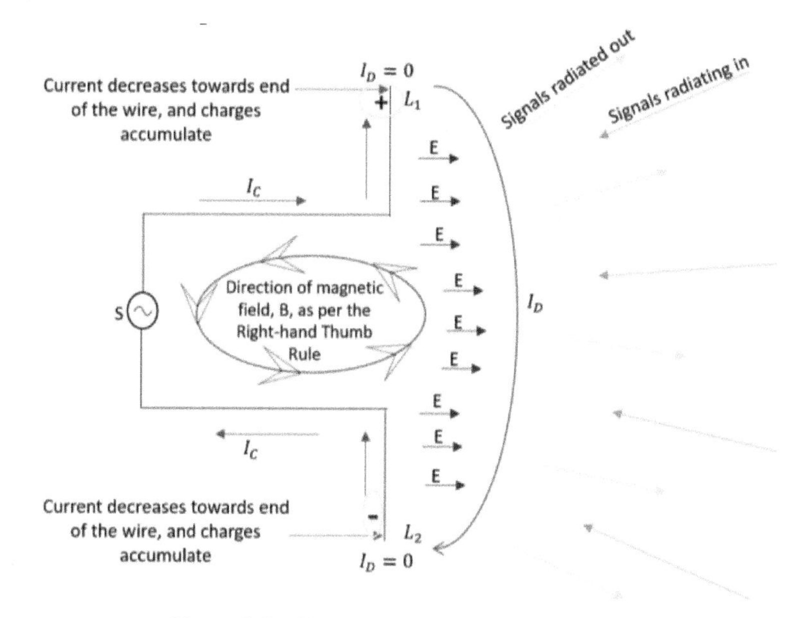

Figure 2.5 General L-length dipole wire antenna.

where

$$R = \sqrt{x^2 + y^2 + (z - z')^2}$$
$$= \sqrt{r^2 + (z'^2 - 2rz' \cos\,\theta)}. \tag{2.62}$$

The following relations have been used:

$$z \approx r\,\cos\,\theta, \qquad R \approx r\; - \;z' \cos\,\theta, \qquad z' << r.$$

Therefore, the incremental electric field at the observation point due to the small element dz on the dipole antenna may be rewritten as

$$\mathrm{d}E_\theta = \mathrm{j}\eta \frac{kI\mathrm{e}^{-\mathrm{j}kr}}{4\pi r}\;\sin\,\theta\;\mathrm{e}^{\mathrm{j}kz'\;\cos\,\theta}\mathrm{d}z'. \tag{2.63}$$

Thus, the resultant electric field is given by

$$E_\theta = \eta_0 H_\phi = \int_{-L/2}^{+L/2} \mathrm{d}E_\theta, \tag{2.64}$$

yielding

$$E_\theta = \eta_0 H_\phi = \mathrm{j}\eta \frac{I_\mathrm{m}\mathrm{e}^{-\mathrm{j}kr}}{2\pi r}\left[\frac{\cos(k(L/2)\,\cos\,\theta) \;-\; \cos(kL/2)}{\sin\,\theta}\right]. \quad (2.65)$$

A sketch of E for different lengths of dipole will show that the radiation pattern gradually changes as L is increased. When $L = 2\lambda$, the radiation pattern will no longer be maximum in the directions of $\theta = \pi/2, -\pi/2$ but split into four lobes with maximum directions close to $\theta = \pi/4, 3\pi/4, -\pi/4$, and $-3\pi/4$.

2.4.2 Radiation from a Half-Wave Electric Dipole Antenna: $L = \lambda/2$

A half-wave dipole antenna may be constructed by using two hollow aluminum conductors, each of $\lambda/4$ length. Both elements are vertically aligned with each other, with the transmitter source or the receiver line connected to the two ends adjacent to each other at the center. This is the center fed dipole antenna. Alternatively, a $\lambda/4$ length element may be placed over a ground plane. The image acts as the second half of the dipole.

The radiation (i.e. $1/r$) components of the magnetic field and electric field of the half-wave dipole are given by

$$H_\phi = \mathrm{j}I\,\mathrm{e}^{-\mathrm{j}kr}\frac{\cos\left((\pi/2)\cos\,\theta\right)}{2\pi r\,\sin\,\theta}\ \mathrm{A/m}, \quad (2.66)$$

$$E_\theta = \mathrm{j}kI\,\frac{\mathrm{e}^{-\mathrm{j}kr}\ \cos\left((\pi/2)\cos\,\theta\right)}{2\pi\varepsilon\omega r\sin\,\theta}\ \mathrm{V/m}. \quad (2.67)$$

The radiation resistance of a half-wave dipole is 73 Ω.

Therefore, the standard 75 Ω coaxial cable may be very closely matched to a half-wave dipole antenna without any additional matching circuits, which accounts for the popularity of the half-wave dipole antenna. The reactive impedance $X = \mathrm{j}42.5$ Ω if $L < \lambda/2$. The gain of the half-wave dipole is 1.64. Thus, the maximum gain (or directivity) of a half-wave dipole is 1.64, which is greater than the figure of 1.5 we had for the infinitesimal (or Hertzian) dipole.

In general, for a lossless half-wave dipole antenna, the direction (θ) dependent directivity and gain are given by

$$D(\theta) = G(\theta) = 1.64\left(\left|\frac{\cos\left((\pi/2)\cos\,\theta\right)}{\sin\,\theta}\right|\right)^2. \quad (2.68)$$

The effective aperture of the half-wave dipole antenna, with $D = 1.64$, is given by

$$A_{\text{em}} = \frac{1.64}{4\pi}\lambda^2 = 0.13\lambda^2. \tag{2.69}$$

The radiation patterns for an elemental dipole and a half-wave dipole, when compared, will show that the half-wave dipole has a narrower beam. Since $z = 0$ is a plane of symmetry for the half-wave dipole current (as well as some other wire antennas), the half-wavelength long ($\lambda/2$) wire may be replaced by a quarter-wavelength ($\lambda/4$) wire placed above a ground plane. The radiation pattern of the $\lambda/4$-length antenna will be the same as that of a $\lambda/2$-length wire antenna. The directivity will be doubled, but the radiated power, radiation resistance, and the input impedance will all be halved for a $\lambda/4$-length antenna placed above a ground plane. The ground plane can be made of six or more horizontally placed grounded wires of length $\lambda/2$ arranged in a spider-web-like pattern just under the quarter-wave antenna wire.

Example 1. The antenna wire radius = 0.4 mm. A 0.5 m long dipole is operating at 300 MHz. The conductivity of the antenna is given by $\sigma = 6 \times 10^7\ \Omega^1\ \text{m}^1$. Discuss its radiation efficiency

$$\text{Skin depth } d = \sqrt{2/\omega\mu\sigma}. \tag{2.70}$$

The ohmic resistance, for radius r and length L,

$$R_o = L/\sigma A = L/\sigma(2\pi r d). \tag{2.71}$$

Radiation resistance, since $L = 0.5$ m = $\lambda/2$ ($\lambda = 3 \times 10^8/300 \times 10^6 = 1$),

$$R_r = 73.2\ \Omega.$$

Radiation efficiency

$$\eta_{\text{R1}} = G/D \approx R_{\text{r}}/(R_{\text{o}} + R_{\text{r}}). \tag{2.72}$$

If frequency = 3.0 MHz, $\lambda = 100$ m $>> L$, then the antenna is an infinitesimal (Hertzian) dipole

$$R_r = 80(\pi L/\lambda)^2,$$

and the new efficiency

$$\eta_{\text{R2}} << \eta_{\text{R1}}.$$

At microwave frequencies, dielectric rods (called dielguides) of relative permittivity of the order of 4 (and conductivity of the order of

$10^{-7}\ \Omega^{-1}\ \mathrm{m}^{-1}$!) are used to direct the microwave signals to the parabolic reflector. The ohmic loss becomes too large for metallic conductors to be used at microwave and millimeter frequencies. However, as in the case of optical signals moving along a glass optical fiber, microwave and millimeter waves tend to cling on to dielectric rods, just as low frequency signals tend to cling onto metallic conductors. Thus, the metallic dipole antennas are mostly used for frequencies of the order of 1 GHz and lower; at higher frequencies, we need the hollow metallic or dielectric waveguides to construct effective antennas. Sometimes metallic dipole antennas are coated with low conductivity, high permittivity dielectric coats to increase antenna capacitance and hence antenna bandwidth, although this results in a reduction of radiation efficiency.

Example 2. Two antennas of 0.5 m length are operating at 300 MHz. One antenna radiates 500 W. The second antenna is 1 km away, at $\theta = 90°$, $\phi = 60°$. The axes of both antennas are parallel to each other. Determine the power received by the second antenna and the current induced in it.

The incident power density is given by (see Figure 2.8)

$$P_{\mathrm{i}} = \frac{1}{2\eta}\left|E_{\theta}\right|^{2}, \quad \text{where } \eta = 120\pi.$$

The radiated power from the transmitter $P_{\mathrm{r}} = I_{\mathrm{rms}}^{2}R_{\mathrm{r}} = \frac{1}{2}I^{2}R_{\mathrm{r}}$, where I_{rms} is rms current and I is maximum current.

$$I = \sqrt{\frac{2P_{\mathrm{r}}}{R_{\mathrm{r}}}} = \sqrt{\frac{2 \times 500}{73.2}} = 3.7 \text{ A}.$$

The radiation electric field

$$|E_{\theta}| = \frac{\eta I}{2\pi r}\left|\frac{\cos(k(L/2)\cos\theta) - \cos k(L/2)}{\sin\theta}\right|,$$

$\theta = \pi/2$, $kL/2 = \pi L/\lambda = \pi/2$; the length of the antenna $L = 0.5$ m, $\lambda = 1$ m. We have $\cos(kL/2) = 0$. The receiver is at $r = 1000$ m.

$$|E_{\theta}| = \frac{120\pi I}{2\pi r}\left|\cos\left(\frac{\pi}{2}0\right)\right| = (60/r)I = 0.221 \text{ V/m}.$$

The incident power on the receiver antenna is $P_{\mathrm{i}} = (1/2)E^{2}/120\pi$ W/m^{2} = 0.065 mW/m^{2}.

Hence, the total power received

$$P_{\mathrm{R}} = P_{\mathrm{i}}A_{\mathrm{em}} = \frac{1}{2 \times 120\pi}(0.221)^{2} \times 0.13 \times 1 \text{ W} = 8.4 \times 10^{-6} \text{ W}.$$

Recalculate P_R for (i) $\theta = \pi/4$, (ii) $\theta = \pi/6$, and (iii) f = 1280 MHz, $L = \lambda/2$, and $\theta = \pi/2$. The problem may be solved using the Friis equation as well. The receiver power is generally very small and of the order of μW, nW, or pW. Hence, efficient low noise amplifiers (LNA) are required to keep signal-to-noise ratio (SNR) in receiver electronics to acceptable values.

In the MF band (960, 1200, and 1400 kHz), 10 kW transmitters radiating signals from a dipole antenna at a height of 0.25λ will provide a radio coverage within a 140 km radius of ground wave service. Ground waves, as opposed to sky waves, are EM waves radiated by the antenna that skims along the surface of the earth, the most common mode of transmission at LF and MF bands. Due to losses in the ground, the radiation efficiency of the antenna could go down from 95% for a ground conductivity of 0.01 S/m to about 80% when the ground conductivity is 0.001 S/m. For higher frequency transmission, the ground wave becomes an ineffective transmission route and the sky waves should be used as in the case of cellular communications. Furthermore, in order to prevent distortion of the antenna radiation pattern by obstructions, the antenna should be installed with a clearance of 10 m if the area of the obstruction is 300 m^2, and a clearance of 80 m should be allowed if the obstruction area is about 2500 m^2. HF dipole antennas transmitting at 100 kW may provide coverage of an entire country if frequencies of 7.3, 5.4 and 3.4 MHz are used to combat the variations in the ionosphere height (approximately 300 km above the earth due to diurnal, seasonal, and the 11-year sunspot cycles). In this case, the antennas will have to be very long, and the radiated EM signal bounces off the ionosphere and is reflected to various regions of the country.

2.5 Radiation Resistance

A few points regarding the radiation resistance may help at this stage. The radiation resistance is not the ohmic resistance of the antenna conductor; it is associated with the power radiated out from the antenna. The ohmic resistance is related to the power dissipated or lost in the antenna conductor. The ohmic resistance is the (power loss in the antenna)/I^2, whereas radiation resistance is equal to (power radiated)/I^2, where I is the current along the antenna. The resistances of half-wave ($\lambda/2$) and quarter-wave ($\lambda/4$) antennas are 73 and 36 Ω; the radiation resistance is a function of the length of the antenna. As the dipole antenna length is increased, the radiation resistance tends to oscillate in between 70 and 130 Ω for different lengths of the dipole; however, at antenna lengths of 80λ and above the resistance tends to settle down at about 130 Ω. Antennas like the Marconi antenna are half-wave antennas, where only a quarter-wavelength long conductor is placed above the ground; the reflection (image) in the ground then makes it a half-wave antenna, but the input resistance of the Marconi antenna is only 36 Ω.

The height at which the antenna is mounted also has an effect on the radiation resistance. If a half-wave ($\lambda/2$) antenna is placed vertical above the ground, the radiation resistance will be close to 73 Ω at all heights above 0.5λ. Below this height, the radiation resistance tends to increase exponentially. This could cause problems with impedance matching with the coaxial cable connected to the antenna. If the half-wave antenna is placed horizontal above the ground, the radiation resistance tends to decrease rapidly toward zero at heights below 0.5λ. At heights above this, the radiation resistance of a horizontally placed half-wave dipole also tends to be close to 73 Ω. It is also useful to note that the electrical length of the antenna is generally about 5% greater than the physical length of the antenna, especially when the antenna conductor is not very thin. Hence, to construct a half-wave antenna at 200 MHz ($\lambda = 1.5$ m), the physical length of the antenna should be $0.95 \times 0.5\lambda = 0.7125$ m. It is very difficult to get an exact match between the physical and electrical lengths. If the physical length of the antenna is slightly longer than the electrical length, the antenna will be inductive (use a series capacitor at the antenna input to cancel out the inductive effect); if the physical length is shorter than the electrical length, then it will be capacitive (use a series inductor to cancel out the capacitive effect).

2.6 Impedance Matching

To improve the transmission line–antenna frequency characteristics, it is important to ensure that the impedances of the line and antenna are matched. Else, where there is impedance mismatch, part of the power delivered to a transmitting antenna will be reflected back to the source. In the case of a receiver antenna, all the power captured by the antenna P_T will not be delivered to the receiver electronics if the impedance of the antenna and that of the line/LNA combination are not properly matched. When a signal is going from a cable of impedance Z_1 to an antenna of impedance Z_2, the amount of reflected power is $((Z_2 - Z_1)/(Z_1 + Z_2))^2$. A variety of matching techniques are available, the quarter-wave transformer being one of the most popular methods to match an open wire line to a half-wave dipole antenna. As shown in Figure 2.8, in order for the transmitted power P_R to be maximized, there must be impedance matching at the transmitting antenna.

To receive as much as possible of the received power P_r at a receiving antenna, impedance matching must be done at the receiver antenna. A short circuited line (short circuit stub) is connected at the point where the line (or cable) is connected to the antenna such that the impedance at the antenna terminal is transformed from Z_0 to $Z_1 = (Z_0 R_{in})^{1/2}$, where R_{in} is the total impedance of the antenna. The antenna impedance itself varies with the frequency of operation. The antenna impedance is minimum if the operating frequency is the resonant frequency of the antenna. If the operating frequency deviates from the resonant frequency, the antenna impedance may become either capacitive (when operating frequency is smaller than the resonant frequency) or inductive (when the operating frequency is greater than the resonant frequency). These frequency-dependent changes must be kept in mind when impedance matching is done.

The SNR will be greatly increased where impedances are properly matched. Consider the antenna as receiving a signal $E + E_n$, where E is the desired signal at frequency f_0 and E_n is noise. The antenna and the line/LNA impedances are matched at frequency f_0 such that signal E is captured efficiently. If E_n is additive white Gaussian (random) noise (AWGN), it will have a frequency spread over the entire bandwidth of the antenna. Now by matching the circuit at f_0, we inevitably reduce the strength of E_n delivered to the LNA since only the portion of E_n at f_0 will be entirely transferred to the LNA. The other power components of E_n will be reflected back to the antenna. Hence, the SNR at the LNA is improved. However, note, if the noise is Rayleigh, then E_n is also at f_0; in the case of Rayleigh noise, common in wireless communications, E_n is a reflected part of E. Hence, for Rayleigh noise, an improvement of SNR should not be expected with impedance matching, although impedance matching is required to capture the desired signal E.

2.7 Radiation Safety

One important aspect to address when installing an antenna is the radiation level. Consider a mobile phone that transmits a signal at $P_r = 1$ W. Then the power density at a distance of 2 cm from the antenna, optimistically assuming isotropic radiation, is given by $P_r/4\pi r^2 = 1/4\ \pi(0.02)^2 = 20$ mW/cm^2, a figure which should be multiplied by the directivity (e.g. $D_{\lambda/2} = 1.64$) of the antenna in order to get a more accurate value. Is this power density within the safety limits for a human head that is close to the mobile phone antenna? It is the responsibility of the telecommunications authority to ensure

Table 2.1 Dielectric properties of biomaterials at different frequencies.

Material	27.12 MHz		63 MHz		350 MHz	
	σ	ε_r	σ	ε_r	σ	ε_r
Muscle	0.75	106	0.93	88	1.33	53.0
Blood	0.28	102	–	–	1.20	65.0
Skin	0.74	106	–	–	0.44	17.6
Brain	0.45	155	0.55	109	0.65	60.0
Fat	0.04	29	0.06	11.6	0.07	5.7

Table 2.2 Electrical impedance of biomaterials.

Frequency (MHz)	Muscle	Blood	Skin	Brain	Fat
27.12	16.7	26.0	16.8	20.6	60.2
	39°	31°	39°	31°	21°
63.00	22.5	–	–	27.3	84.6
	36°			28°	27°
350.00	40.5	39.8	70.4	45.5	144.6
	26°	22°	26°	15°	17°

that all transmitters confine to the limits set on the maximum radiation electric field intensity permitted or the maximum radiation power intensity permitted. Biological tissues have electrical properties that are frequency dependent. From Table 2.1, note that the conductivity of most materials increases with signal frequency. This means that as higher communication or remote sensing signals impinge on a human body, relatively more energy will be absorbed. More absorption indicates that more power is converted into heat loss, which could lead to the burning of tissues. Dead tissues are associated with cancer. Table 2.2 shows the impedances of various biomaterials. The complex impedance values of the biomaterials are calculated using the conductivity and relative permittivity values given in Table 2.1. These impedances determine the amount of radiation signals which penetrate into particular materials since the transmission coefficient of material interface is given by $2Z_2/(Z_1 + Z_2)$, where the signal is assumed to be going from a medium of intrinsic impedance Z_1 into a medium of impedance Z_2.

In Table 2.3, the radiation limits applied in general are shown. From country to country, these values may differ. Some authorities, like the former Soviet Union telecommunications authority, impose stricter limits. The maximum permissible electric field at extremely low frequencies used to be 25 V/m (rms) in Russia instead of the 87 V/m shown in Table 2.3; at microwave communication frequencies, the limit set was 0.01 mW/cm^2 instead of 10 mW/cm^2.

Table 2.3 Radiation limits (general public).

Frequency (MHz)	E (rms)	H (rms)	Power (W/m^2)
0.01	87	$0.23/\sqrt{f}$	
1–10	$87/\sqrt{f}$	$0.23/\sqrt{f}$	
10–400	27.5	0.073	2
400–2000	$1.375\sqrt{f}$	$0.0037\sqrt{f}$	$f/200$
2000–300,000	61	0.16	10 (mW/cm^2)

International Radiation Protection Association.

Depending on the regulations imposed by the telecommunications authorities, it is the duty of the antenna systems design engineer to ensure that the maximum radiation field from the antenna does not exceed the specified maximum limit. Therefore, electric field strength measurement very close to the antenna is necessary to ensure that these limits are observed. This is one important reason why in wireless communications, many base station (BS) antennas are required to provide service: using one BS antenna to serve an area larger than, say, 5 km^2 area will mean that the electric field radiated will need to be stronger to transmit clear signal to far distances. This may prove unsafe for people close to the BS of a wireless communication system.

The safety and health issues related to radio frequency radiation (RFR) remains a point of contention. The aforementioned safety limits of EM radio frequency signals is very much related to the heating effects of the RFR. The heating effects when the RFR power is so high that it heats up the body by 1°C or more after 30 minutes exposure at a specific absorption rate of 4 W/kg. Moreover, the limits are set by looking at short-term heating effects rather than the long-term effects of RFR. The following health impacts of RFR impress one with the urgent need for public education of RFR and the need for more research by experts in engineering, medicine, and health coming together as a team for investigation. The additional health effect, going beyond immediate burning effects of short-term heating, include: "increased cancer risk, cellular stress, increases in harmful free radicals and genetic damage, structural and functional changes of the reproductive system, learning and memory deficits, neurological disorders and negative impact on general wellbeing" include sleep impairment due to blue light from screens and RFR in the environment (A. Fitzgerald in Oxford Today). Thus, the 2–10 W/m^2 RFR limits depending on frequency urgently need further study. With 5G wireless systems and future 6G wireless systems, the RFR frequencies are set to be pushed toward the TeraHz (10^{12} Hz) range, between microwaves and infrared light, with wavelengths from 3 mm down to 30 μm.

2.8 The Effect of Antenna Height and Ground Reflection

Consider the case where wire antennas are being used in wireless communications. The stationary BS antenna will be placed on top of a building or tower of height, say, h_1 relative to the ground. The mobile phone or mobile station (MS) antenna will be closer to the ground at a height, say, h_2 relative to the ground. When a signal is being radiated from the MS antenna to the BS antenna, ignoring reflections due to all other objects except for the ground, two signals will arrive at the BS antenna for each transmission. Signal 1 will be the direct signal traveling from the MS antenna to the BS antenna, and signal 2 will be the ground-reflected signal which travels from the MS to the ground and is then reflected back to the BS antenna.

Let the reflection coefficient of the ground for vertically polarized signals be denoted by Γ_v. Then the resultant signal appearing at the BS antenna is given by

$$E = (k/r)E_0 e^{j\omega t} + \Gamma_v (k/r) E_0 e^{j(\omega t + \Delta\delta)} \tag{2.73}$$

or

$$E = (k/r)E_0 e^{j\omega t}\left(1 + \Gamma_v e^{j\Delta\delta}\right) \tag{2.74}$$

where $\Delta\delta = k\Delta r$ is the phase difference due to the extra distance Δr that the reflected signal has traveled. The (k/r) factor appears from the forms that we got for radiation fields in Sections 2.3 and 2.4. If the direct distance between the MS and BS antennas is r, then

$$\Delta r = \left((h_1 + h_2)^2 + r^2\right)^{1/2} - \left((h_1 - h_2)_n^2 + r^2\right)^{1/2} \tag{2.75}$$

which for $r >> h_1 + h_2$ reduces to $\Delta r = (2h_1 h_2)/r$.

Hence, the resultant electric field signal picked up at the BS antenna is given by

$$E = (k/r)E_0 e^{j\omega t}\left(1 + \Gamma_v(\cos\Delta\phi + j\sin\Delta\phi)\right) \tag{2.76}$$

or

$$E \cong (k/r)E_0 e^{j\omega t}\left(1 + \Gamma_v + j\Gamma_v \Delta\phi\right) \tag{2.77}$$

For small values of $\Delta\phi$, $\Delta\phi = k\Delta r = k\,(2h_1 h_2)/r$.

The (conductivity σ, relative permittivity ε_r) values of seawater $(80, 4)$, rural earth $(14, 0.002)$, urban ground $(3, 0.0001)$, turf $(5, 0.01)$, and dry sandy loam (3,0.03) are such that in wireless communications, the reflectivity coefficient of ground is close to unity at UHF frequencies. For ground with reflection coefficient $\Gamma_\mathrm{v} = -1, E \cong -\mathrm{j}(k/r)E_0\mathrm{e}^{\mathrm{j}\omega t}\Delta\phi = -\mathrm{j}\left(2h_1h_2\right)\left(k^2/r^2\right)E_0\mathrm{e}^{\mathrm{j}\omega t}\mathrm{V/m}$. Therefore, the power density $\left(\mathrm{in W/m^2}\right)$ received is given by

$$P_\mathrm{i} = E^2/2\eta = 2(h_1h_2)^2(k^4/r^4)(E_0^2/\eta). \tag{2.78}$$

Therefore, we note that in wireless, mobile communication antennas, the radiated electric field varies as $1/r^2$ instead of $1/r$ with distance, and the power varies as $1/r^4$ instead of $1/r^2$ due to the adverse effects of the ground-reflected path signal interfering with the direct signal. In order to reduce the fading due to the reflected path, more directive beam antennas could be used in order to avoid aiming signals toward the ground. To get such directive radiation beams, we need to use more than one antenna element to achieve a smart antenna with a beam adaptively being changed as the MS antenna keeps moving in the spatial domain.

2.9 Inverse Doppler Effect in the Near-Field Region

When there is relative motion between the transmitter and the target scatterer, the frequency observed at the receiver is different from that which is transmitted. This is called the Doppler phenomenon. The Doppler frequency shift in the near-field region is different from that experienced by an observer in the far-field region since the Doppler frequency shift depends on the distance between the source and the observation point. Hence, the radial and tangential electric field (magnetic field) components of the EM radiation exhibit different Doppler frequency shifts. The Doppler frequency shift of the radial electric field component is given by

$$\omega_\mathrm{dr} = \frac{k^2v(r_0 - vt)}{[1 + (kr_0 - kvt)^2]}. \tag{2.79}$$

The Doppler frequency shift of the tangential electric field component is given by

$$\omega_\mathrm{dt} = \frac{k^3v[k^2(r_0 - vt)^2 - 2](r_0 - vt)^2}{[k^4(r_0 - vt)^4 - k^2(r_0 - vt)^2 + 1]}, \tag{2.80}$$

where k is the wave number, r_0 is the initial distance of the scatterer from the transmitter, and v is the relative velocity between the transmitter and the scatterer. Figure 2.8 shows the variation of the Doppler frequency of the radial electric field component when a scatterer, which is initially at a distance of 100 m, moves past the transmitter at a velocity of 100 m/s.

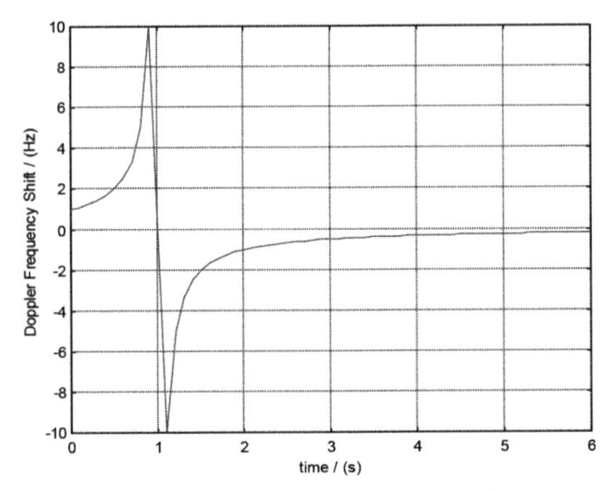

Figure 2.6 Doppler frequency shift.

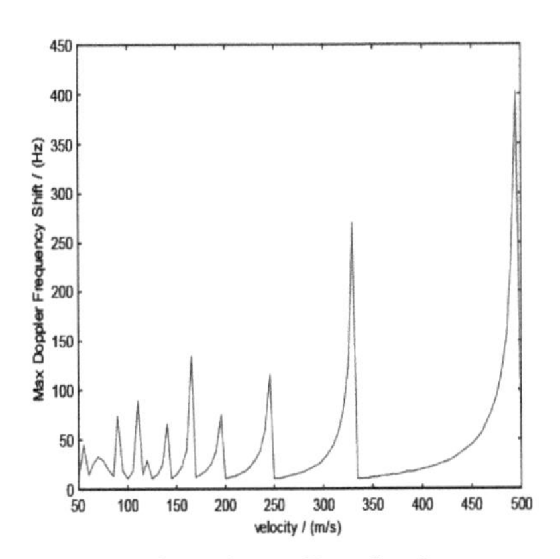

Figure 2.7 Variation of maximum Doppler frequency with velocity.

From Figure 2.6, for $t < 1$ s, it is seen that when the scatterer is approaching the transmitter, the Doppler frequency should increase. But from Figure 2.6, it is also seen that, very close to the transceiver ($0.9 \leq t \leq 1$), the Doppler frequency shift is actually decreasing. This is the so-called inverse Doppler effect. A similar effect is seen in the time interval $1 \leq t \leq 1.1$. The variation of the maximum Doppler frequency shift with velocity of the radial electric field component is shown in Figure 2.6. From Figure 2.7, it is seen that the maximum Doppler frequency shift of the radial electric field component varies erratically with velocity.

A similar inverse Doppler effect is exhibited by the tangential electric field component as well. In imaging of moving objects in the near-field

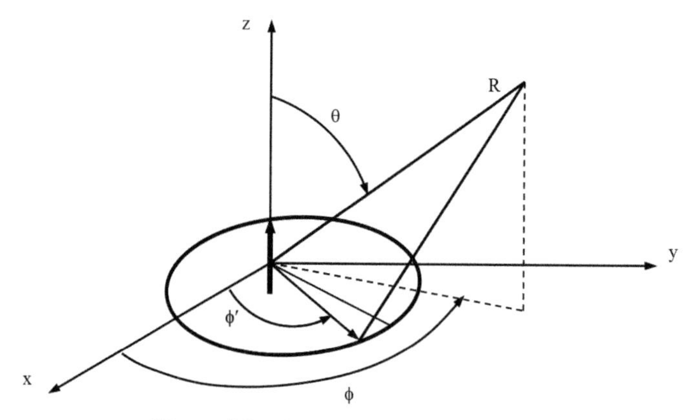

Figure 2.8 Small circular current loop.

region, this inverse Doppler effect has to be taken into consideration in the imaging routines. In this thesis, the relative velocity between the transceiver and scatterers was assumed to be zero.

2.10 The Magnetic Dipole: Loop Antenna

Consider the current loop shown in Figure 2.8.

The components of the magnetic field **H** produced by the loop are

$$H_r = \mathrm{j}\frac{ka^2 I_0 \cos\,\theta}{2r^2}\left[1+\frac{1}{\mathrm{j}kr}\right]\mathrm{e}^{-\mathrm{j}kr}, \tag{2.81}$$

$$H_\theta = -\frac{(ka)^2\, I_0\ \sin\,\theta}{4r}\left[1+\frac{1}{\mathrm{j}kr}-\frac{1}{(kr)^2}\right]\mathrm{e}^{-\mathrm{j}kr}, \tag{2.82}$$

$$H_\phi = 0. \tag{2.83}$$

Using Equation (2.1) with **J** = 0, the electric field components could be found as

$$E_r = E_\theta = 0, \tag{2.84}$$

$$E_\phi = \eta\frac{(ka)^2\, I_0 \sin\,\theta}{4r}\left[1+\frac{1}{\mathrm{j}kr}\right]\mathrm{e}^{-\mathrm{j}kr}, \tag{2.85}$$

where η is the intrinsic impedance of free space. Equations (2.81)–(2.85) are also the EM fields radiated by a magnetic dipole. The fields radiated by a magnetic dipole characterize the radiation fields of a hydrogen nucleus stimulated by a radio frequency pulse in magnetic resonance imaging (MRI).

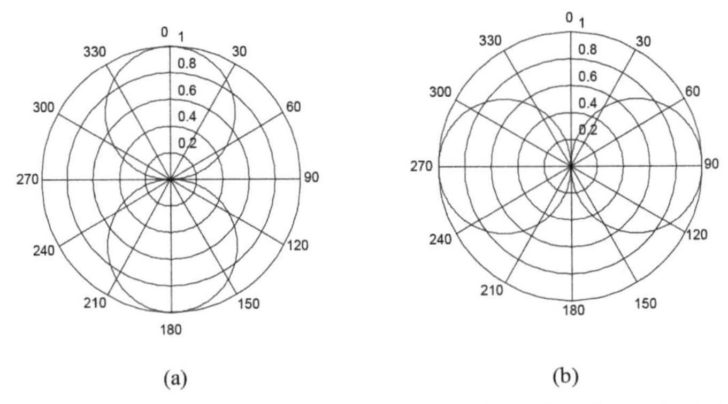

Figure 2.9 Normalized field pattern of (a) H_r and (b) H on the vertical plane.

2.10.1 Magnetic Field Pattern of a Magnetic Dipole

Field patterns of a magnetic dipole in a plane vertical to the xy-plane at a distance of 1 m (length of dipole $\lambda/50$, frequency = 100 MHz) are considered.

The magnetic field equations and the associated field patterns for a magnetic dipole are equivalent to those of the electric field of an electric dipole. The polar field patterns of the magnetic field components H_r and H are shown in Figure 2.9(a) and (b), respectively. These radiation patterns will be used in the analysis of the resonating nuclear magnets.

2.10.2 The Helical Broadband Antenna

A helical antenna, shown in Figure 2.10, is a conductor that is wound into a helical shape and is fed properly. It may be thought of as a combination, or an array, of small loop antennas.

The symbols used to describe the helix are defined as follows:

D = diameter of helix (between centers of coil material);
C = circumference of helix = πD;
S = spacing between turns;
α = pitch angle;
L = length of one turn;
N = number of turns;
A = axial length;
λ = wavelength.

An infinitely long helix is a transmission line, which can support an infinite number of modes. Corresponding to these modes, a finite length

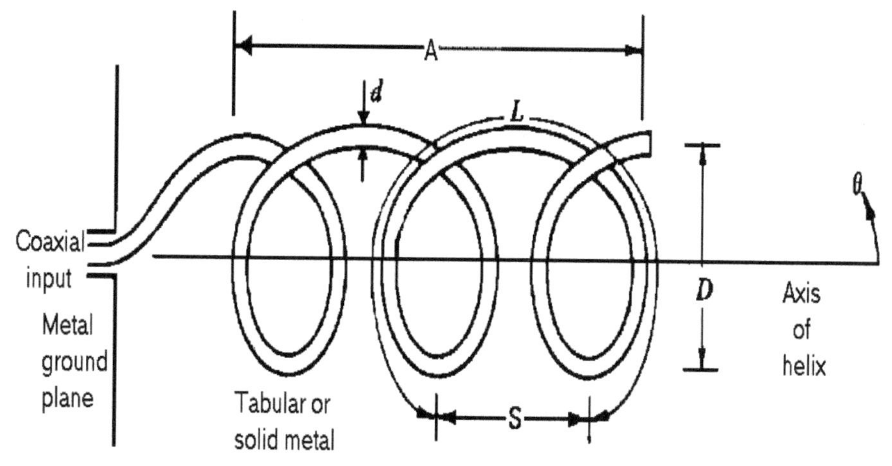

Figure 2.10 Geometry and dimensions of a helical antenna.

helix can radiate in a number of modes. Two of these modes are the normal mode and the axial mode. The normal mode yields radiation that is at right angles to the axis of the helix. This occurs when the helix diameter is much smaller than one wavelength. The axial mode provides maximum radiation along the axis of the helix. This will occur when the helix circumference is one wavelength. Although the helical antenna radiates in different modes depending on conditions, the axial mode is often of most importance. The axial mode helical antenna is used in such diverse applications as printed circuit board probing and in satellite reception. The radiation patterns of the two different modes are as shown in Figure 2.11.

In the axial mode of radiation, the maximum radiation is along the axis of the helix, that is, the helix radiates as an end-fire antenna. The axial mode occurs when the helix circumference is of the order of one wavelength. The axial mode carries a nearly pure traveling wave; therefore, the effect of the ground plane may be neglected. Also, the size of the ground plane is not very critical, but it should be made wider than half a wavelength. The conductor diameter d has very little effect on the axial mode helix antenna properties, so a nominal value d may be chosen. The helix may be fed using a coaxial cable with the center conductor attached to the helix and the outer conductor attached to the ground plane that is a solid metal (copper) square plane. However, it is not practical to implement a copper plate larger than half a wavelength, which is approximately 30 cm for a 500 MHz helical antenna.

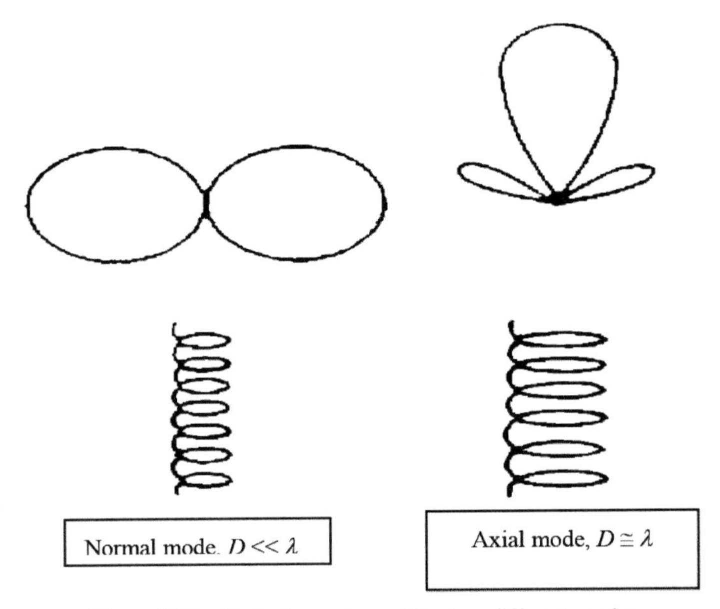

Figure 2.11 Radiation pattern of the two different modes.

The empirical formula for the half-power beam width (HPBW) of a helical antenna is as follows:

$$\mathrm{HPBW} = \frac{52}{(C/\lambda)\sqrt{N(S/\lambda)}}. \tag{2.86}$$

If we choose a typical configuration of $C = \lambda$, $\alpha = 12°$, and $N = 12$, then

$$S = C \tan \alpha,$$

a formula which holds for $12 \leq \alpha \leq 15°$, $N > 3$, and $3/4\lambda < C < 4/3\lambda$.

We let $N = 12$, $\alpha = 12° \Rightarrow S = 1.2 \tan 12° = 0.255$ m. Hence,

$$\mathrm{HPBW} = \frac{52}{(C/\lambda)\sqrt{N(S/\lambda)}} = 32.6°.$$

An empirical formula for the input resistance of a helical antenna is

$$R_{\mathrm{in}} = 140(C/\lambda)\ \Omega, \tag{2.87}$$

which yields an input impedance of 140 Ω for the 250 MHz helical antenna.

2.11 Effect of Ground on Antenna Radiated Electric Fields

So far in this chapter, all the analysis and discussions have assumed that the antenna is placed in free space and that it is far away from the ground. Although the fields near elevated microwave antennas may closely approximate this idealized situation, the fields of most antennas are affected by the presence of the ground. The reason for this is that at microwave frequencies, the antenna radiation beam is quite narrow, and the radiated wave energy will therefore not impinge on the ground if the antenna is well above the ground and pointed away from the ground. This is not the case for antennas with wide or isotropic beams. Any EM energy from the radiating element directed toward the ground undergoes a reflection). The amount of reflected energy and its direction in which the reflected wave travels are determined by the geometry and constitutive parameters (i.e. conductivity and permittivity) of the ground. Since the reflected wave will now interfere with the main or direct wave from the antenna, there the overall radiation pattern of the antenna will be different from the free space radiation pattern of the antenna. Furthermore, the reflected energy will be frequency dependent since the effective conductivity of the ground (i.e. $\sigma + j\omega\varepsilon$) increases with frequency.

To analyze the situation where the reflection at the ground must be taken into account, we may replace the ground by the image of the antenna, as shown in Figure 2.12. It is seen from Figure 2.22 that the reflected wave (or ray) and the image ray have the same path lengths (i.e. AD = BD). At any observation point *P*, the resultant EM field strength will be the vector sum of the field strengths of the direct ray and the reflected ray. The reflected ray travels an extra distance of BC further than the direct ray, and thus at point *P*, it would have a phase different from that of the direct ray. Although the amplitudes of the two rays will differ as well, we ignore this difference by assuming that AD is much greater than BC. It is the phase difference between the two rays that gives rise to signal cancelation or fading as in mobile communication systems. The application of the principle of images to find the resultant field strength at any point *P* applies not only to a perfectly conducting ground but also to ground of any conductivity. We will use this image principle to find the total field strength $\mathbf{E}_\theta$ and $\mathbf{E}_\mathrm{R}$ components of a vertical and horizontal dipole transmitter at a distance D m from the receiver; the heights of the transmitting and receiving antennas from the ground are H_1 m and H_2 m, respectively.

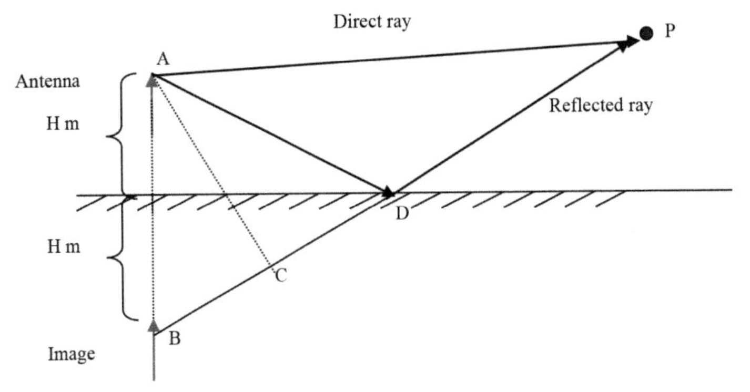

Figure 2.12 Illustration of the use of image concept to find total field strength at point *P*.

2.11.1 The Vertical Dipole

Figure 2.13 shows a vertical dipole antenna of length *L* and carrying a current *I* placed above a ground plane. The dipole antenna is assumed to be an infinitesimal dipole antenna. Figure 2.14 shows the electric field components of the direct and reflected rays.

The direct ray component of the electric field **E** at the receiver is given as

$$\mathbf{E}_{\theta \mathrm{d}} = \mathrm{j}\eta_1 \frac{kIL \sin \theta_{\mathrm{d}}}{4\pi} \left[\frac{1}{r_1} + \frac{1}{\mathrm{j}kr_1{}^2} - \frac{1}{k^2 r_1{}^3} \right] \mathrm{e}^{-\mathrm{j}kr_1} \, \mathbf{u}_\theta, \tag{2.88}$$

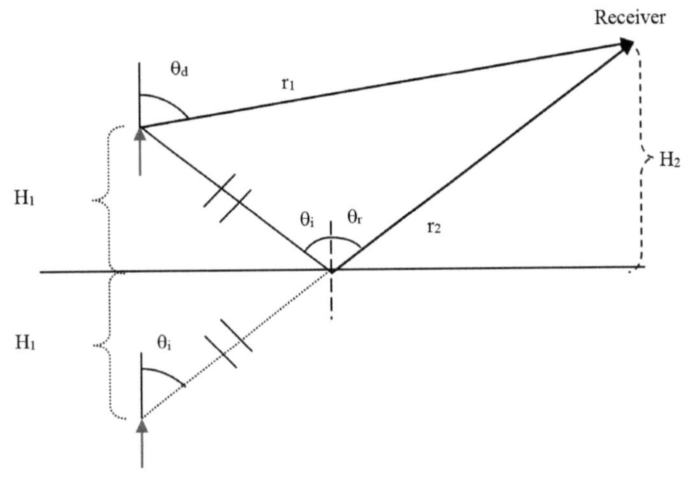

Figure 2.13 Multipath effects for a vertical dipole antenna.

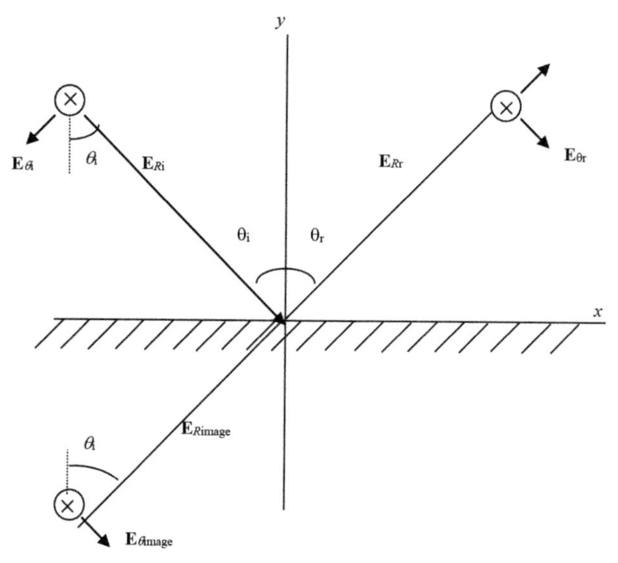

Figure 2.14 Reflected E field components $\mathbf{E}_\theta$ and $\mathbf{E}_\mathrm{R}$ for a vertical dipole above a ground plane with finite conductivity.

$$\mathbf{E}_{R\mathrm{d}} = \eta_1 \frac{IL \cos \theta_\mathrm{d}}{2\pi} \left[\frac{1}{r_1{}^2} + \frac{1}{\mathrm{j}kr_1{}^3} \right] \mathrm{e}^{-\mathrm{j}kr_1} \, \mathbf{u}_r. \tag{2.89}$$

The $\mathbf{E}_\theta$ and $\mathbf{E}_R$ components of the image shall be resolved into their x - and y components in order to be able to get the resultant components by adding them to the fields of the direct ray. Thus, we seek to obtain the $\mathbf{E}_x$ and $\mathbf{E}_y$ fields from the image. Subsequently, we shall get the resultant $\mathbf{E}_x$ and $\mathbf{E}_y$ components at the observation point (i.e. the receiver). Once we know the x - and y -components, we can get the reflected $\mathbf{E}_\theta$ and $\mathbf{E}_R$ through these $\mathbf{E}_x$ and $\mathbf{E}_y$ components. From Figure 2.14, for the electric fields at the receiver, we get

$$\mathbf{E}_{x\mathrm{r}} = \mathbf{u}_x \left(E_{R\mathrm{image}} \sin \theta_\mathrm{i} + E_{\theta\mathrm{mage}} \cos \theta_\mathrm{i} \right) \tag{2.90}$$

$$\mathbf{E}_{y\mathrm{r}} = \mathbf{u}_y \left(E_{R\mathrm{image}} \cos \theta_\mathrm{i} - E_{\theta\mathrm{image}} \sin \theta_\mathrm{i} \right) \tag{2.91}$$

where $E_{R\mathrm{image}}$ and $E_{\theta\mathrm{image}}$ are given by Equations (2.96) and (2.97), respectively, by replacing r_1 by r_2 and θ_d by θ_i. We can also express the fields at the receiver in terms of x - and y -components as follows:

$$\mathbf{E}_{x\mathrm{total}} = \mathbf{E}_{x\mathrm{r}} + \mathbf{E}_{x\mathrm{d}} \tag{2.92}$$

$$\mathbf{E}_{y\mathrm{total}} = \mathbf{E}_{y\mathrm{r}} + \mathbf{E}_{y\mathrm{d}} \tag{2.93}$$

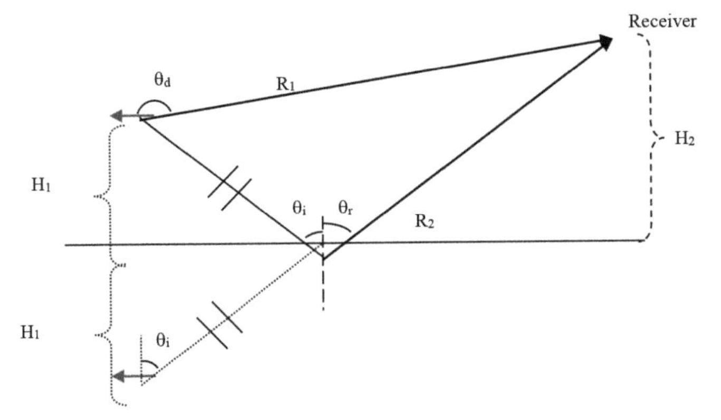

Figure 2.15 Horizontal dipole above a ground plane.

where

$$\begin{aligned}\mathbf{E}_{x\mathrm{d}} &= \text{direct ray } x \text{ -component at the receiver}\\ &= \mathbf{E}_R \sin\theta_\mathrm{d} + \mathbf{E}_\theta \cos\theta_\mathrm{d} \qquad (2.94)\\ \mathbf{E}_{y\mathrm{d}} &= \text{direct ray } y - \text{component at the receiver}\\ &= \mathbf{E}_R \cos\theta_\mathrm{d} - \mathbf{E}_\theta \sin\theta_\mathrm{d} \qquad (2.95)\end{aligned}$$

2.11.2 The Horizontal Dipole

The formulae derived for the vertical dipole can be used for the horizontal dipole (see Figure 2.15) to get the resultant signal at the receiver except that the θ_i used in the formula for $\mathbf{E}_{imageR2}$ and $\mathbf{E}_{rimageR2}$ need to be replaced by $\pi/2 + \theta_I$ as shown below:

$$\mathbf{E}_{\theta\mathrm{image}} = \mathrm{j}\eta_1 \frac{kIL\ \sin(\pi/2+\theta_\mathrm{i})}{4\pi}\left[\frac{1}{r_2} + \frac{1}{\mathrm{j}kr_2^2} - \frac{1}{k^2 r_2^3}\right] \mathrm{e}^{-\mathrm{j}kr_2}\ \mathbf{u}_\theta,$$

$$\mathbf{E}_{R\mathrm{image}} = \eta_1 \frac{IL\ \cos(\pi/2+\theta_\mathrm{i})}{2\pi}\left[\frac{1}{r_2^2} + \frac{1}{\mathrm{j}kr_2^3}\right] \mathrm{e}^{-\mathrm{j}kr_2}\ \mathbf{u}_r.$$

The direct ray field components are given by

$$\begin{aligned}\mathbf{E}_{x\mathrm{d}} &= -(\mathbf{E}_R \cos\theta_\mathrm{d} + \mathbf{E}_\theta \sin\theta_\mathrm{d}),\\ \mathbf{E}_{y\mathrm{d}} &= \mathbf{E}_R \sin\theta_\mathrm{d} + \mathbf{E}_\theta \cos\theta_\mathrm{d}.\end{aligned}$$

Simulation results for vertical and horizontal dipole as the distance between transmitter and receiver varies are given in Figures 2.16 and 2.17.

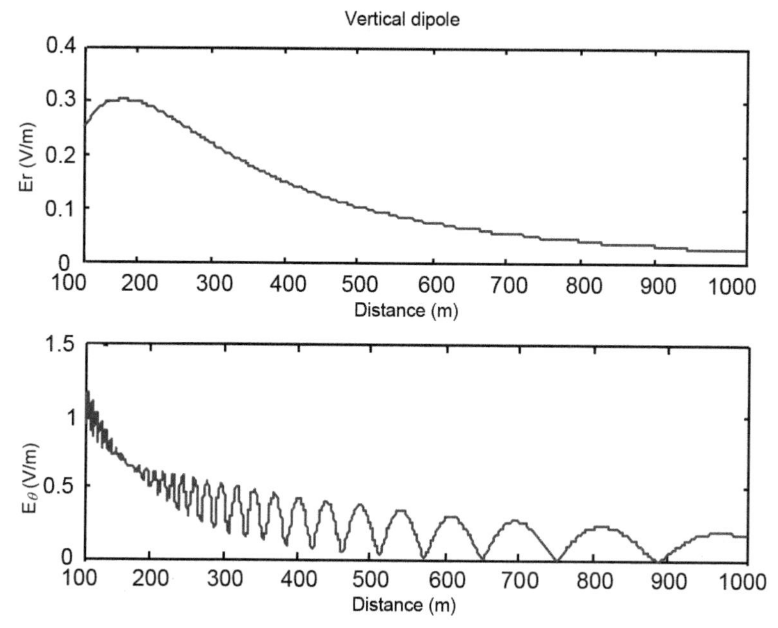

Figure 2.16 Fading with vertically polarized antenna.

The transmitter and receiver are both kept at a height of 78 m, as in the case of a typical airport air-traffic control tower.

The plots show $\mathbf{E}_\theta$, $\mathbf{E}_R$, $\mathbf{E}_{total}$, $\mathbf{E}_x$, $\mathbf{E}_y$ components at the receiver for a signal frequency of 120 MHz. The distance between the transmitter and the receiver varies from 100 to 300 m. The frequency of 120 MHz was chosen since it is one of the 760 channel frequencies used for air-traffic control. The conductivity of ground was assumed to be σ = 99999999 $\approx$ α, the conductivity of a perfect conductor ground. The transmitter current is 1 A, and the relative permittivity of the ground ε_{R} = 25. Figure 2.16 shows the results for a vertically polarized antenna, and Figure 2.17 shows the results for a horizontally polarized antenna. The vertically polarized antenna demonstrates rapid fading of signal with distance, known as Rayleigh fading due to the reflected signal destructively interfering with the direct signal.

2.12 Frequency Independent Antennas

In certain special applications, antennas that may be used over very wide frequency bandwidth may be required. This means that the input impedance

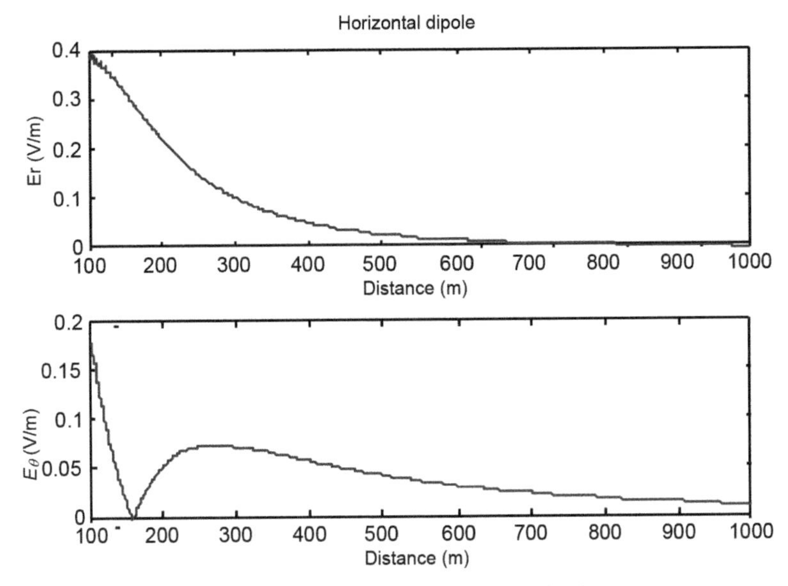

Figure 2.17 Fading with horizontally polarized antenna.

and radiation pattern of the antenna must be constant over a large frequency range. An example of such an antenna is the Vivaldi antenna.

The curved guiding structure is defined by $y = A \exp(kx)$. Waves travel along the curved structure and are radiated out. When the spacing between the two curved conductors is small, very little radiating takes place, and the signals cling to the conductors as they travel. However, as they near the end of the curved structure, the distance between conductors is large, and the EM energy is released to be radiated away. The gain of the Vivaldi antenna is due to the signal phase velocity inside the curved structure being equal or greater than that in the medium outside the antenna. When the outside medium is free space, the phase velocity of the signal is 3×10^8 m/s, and the velocity inside the antenna is greater than or equal to this. Since high frequency signals have a higher phase velocity than a low frequency signal, an impulse applied to the frequency independent antenna will appear as a linear frequency modulated (FM) signal (i.e. a chirp signal) at the output of the antenna. The higher frequency signals appear first at the output with the lowest frequency component appearing last at the output, and hence the chirp waveform at the output of the antenna.

Other forms of frequency independent antennas are the biconical dipole, equiangular spiral, conical spiral, and log-periodic structures. In general, to obtain such a frequency independent antenna, its geometrical features are entirely defined by angles. In such an antenna, its physical surface must satisfy the equation $r = f(\theta) \exp(k\phi)$ in the spherical coordinate system (r,θ,ϕ). Absorbing material may be placed on one side to get unidirectional radiation with a planar equiangular spiral antenna, which radiates circularly polarized waves.

Acknowledgment

Ng Joo Seng and Seow Chee Kiat developed the MATLAB^{TM} programs contained on disk (Program 2.1). Sections 2.9 and 2.10 are based on Naveendra (1999) and Section 2.11 on Ng and Seow (1998).

3

Focused Beam Antennas

P.R.P. Hoole

Abstract

This chapter presents the most powerful antennas which allow for narrow, focused, and steerable antenna beams. This is primarily achieved by using two or more individual antenna elements and arranging them in different geometrical positions and controlling the amplitude and electronic phase angle of the signals received by the array antenna or being delivered to the array for transmission. Two-element array is analyzed to get the fundamentals of array antennas resolved, with the dipole wire antennas forming the elements of the array. Further, the N linear array antenna, where the antenna elements are placed in a straight line, is described as analyzed. Following this, the aperture antennas are discussed which produce narrow, directed single main beams. This includes the patch antenna and the horn antenna. These individual patch and horn antennas may be used in an array, instead of the dipole wire antenna, to obtain narrower steerable beams, as in wireless mobile systems and large dish antennas used for space research.

3.1 Introduction

The wire antennas we looked at in the last chapter are largely antennas with wide beams; the half-power beam width (HPBW) is of the order of a few tens of degrees, but in many applications, like line of sight microwave communications, satellite communications, and radar, the antenna should have an HPBW of the order of 0.2–1°. In such cases, we need narrow beam antennas. Two widely used narrow beam antennas are the array antennas, which also permit electronic control of the beam direction (beamsteering), and the aperture antennas.

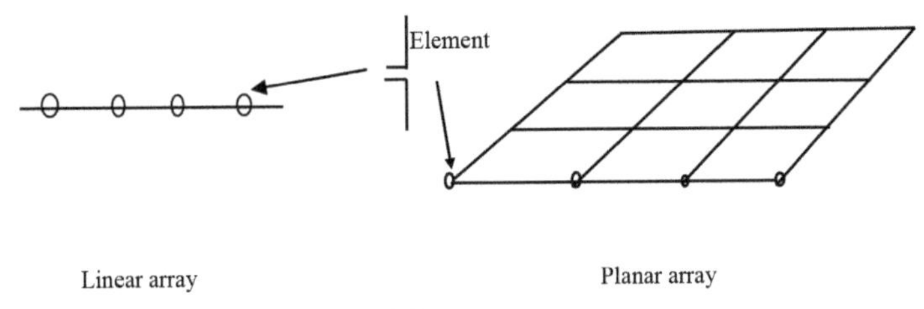

Figure 3.1 Array antennas.

In an array antenna, a number of radiating elements are arranged together to improve power gain and directivity. The individual antenna elements may be arranged along a straight line (linear array) or in a square grid (planar array) (see Figure 3.1). Other forms of arrangements are also possible; in radio astronomy, for instance, large parabolic reflector type aperture antennas are arranged in a star formation.

Consider an array antenna made up of nine wire antennas. Each element is connected to a source voltage of $V_S = \mathrm{Re}(Ve^{j\omega t})$; the current flowing in each wire antenna element is $I = V/Z$. Then the resultant radiated electric field $E_T = 9E$, where E is the electric field radiated by each wire antenna. The resultant power density of the radiated signal is $\frac{1}{2}(9E)^2/120\pi = 40.5E^2/120\pi$ W/m^2. Here $\eta = 120\pi\Omega$ is the intrinsic or wave impedance of free space. If there were only one wire antenna, then the radiated power density would have been only $0.5E^2/120\pi$W/m^2. By using an array of nine antennas (making up an array antenna) instead of a single antenna, we have increased the radiation power by (40.5/0.5), i.e. 81 times! Similarly, if the same array antenna were used to receive signals, its reception efficiency will be 81 times better than that of a single antenna element. Two further advantages of the phased array antenna are 1) the antenna beam can be electronically steered without any need for a mechanical or inertial system to turn the antenna and 2) the array antenna can be placed on airframes without altering the aerodynamics of the vehicle.

Two requirements which should be met when designing array antennas are 1) fields from separate elements must be in phase at the receiving antenna, and 2) currents in all elements must be identical in magnitude. Unless these requirements are met, for instance, if the phases of the received signals are not in phase at each element, then there may be destructive interference at the receiver. When array antennas are used for transmission, they may have two types of elements: 1) driven elements (these are antenna elements to which the signal source V_S is connected); 2) parasitic elements (these are antenna elements to which the source is not directly connected). They receive power from the driven elements through coupling. Some parasitic elements are placed in front of driven elements to narrow the radiation beam and are called directors. Other parasitic elements are placed behind the driven element to prevent signals going in the backward direction and to focus all radiation energy in one single forward direction. These parasitic elements are called reflectors. The large parabolic dishes placed behind the horn or wire antennas in microwave antennas are reflectors. The driven element–parasitic element combination array antennas are also used as receiver antennas.

3.2 Array Antennas: Two-Element Linear Array

An array antenna is a system comprising a number of radiating elements, generally similar. A linear array antenna is a number of antenna elements placed in a straight line. It yields a narrow, steerable fan beam. To get a narrow, steerable pencil beam, we need to have a planar array in which, for instance, the antenna elements should be arranged on a square grid. Mechanically steerable 100 $\times$ 50 cm fixed-pencil beam array antennas are used in most fighter aircraft, usually mounted on the nose of the aircraft and costing about US$ 2 million. In this section, we shall consider a simple case, one in which the radiators are all identical, all oriented in the z-direction, all aligned on the y-axis, and all equally spaced. The z-directed dipoles are placed on the x-axis, with an interelement separation of d. We shall see that when similar antennas are arranged in various configurations with proper amplitude and phase relations, we obtain desired radiation characteristics, e.g. direction and width of main beam, sidelobe levels, directivity.

3.2.1 Two-Element Hertzian Dipole Array Antenna

We shall consider briefly two infinitesimal (Hertzian) dipoles used to form an array antenna. The currents flowing into the two elements are out of phase by an electrical angle δ. The phase shift may be obtained by using both elements from a single source so that the current magnitude is I for both antenna elements, but by installing a phase shifter in the cable run between the first and the second element, the current supplied to the second element may be phase shifted by any desired angle δ (see Figure 3.2).

We shall show that a microprocessor, for instance, controlling the steps of phase shift δ can steer the radiation pattern or beam of the antenna in any required direction. A simpler method will be to place a capacitor in series along the line, or a microstrip line of length $\lambda/4$, between the two elements

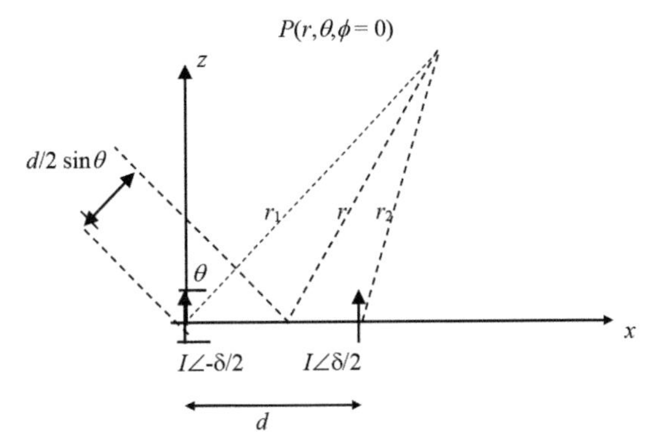

Figure 3.2 Two-element array with the observation point on the *zx*-plane.

to give a phase shift of 90°. The distance between the two elements is *d*. For convenience, we assume that the observation point is on the *zx*-plane (ϕ = 0) and that it subtends an angle θ with the *z*-axis. By allowing the observation point $P(r,\theta,\phi = 0)$ to be on the *zx*-plane, we look for a resultant radiation field which will not be a function of ϕ, thus making the initial analysis simple; later, when we consider the radiation field of a two-element array made of half-wave dipoles, we shall allow the observation point to be at $P(r, \theta, \phi)$.

Furthermore, the observation point is assumed to be in the far-field zone of the antenna so that only the $E(1/r)$ and $H(1/r)$ components exist. Hence, the electric field we shall be interested in is the $\mathbf{u}_\theta$ component; the vector sign will be dropped in the following analysis, with the understanding that the electric field we obtain is the E_θ component.

Using radiation field for an infinitesimal wire antenna, Equation (2.47) in Chapter 2 in particular,

$$E = \mathbf{u}_\theta \, \mathrm{j}\eta \frac{kIL}{4\pi \, r_1} \mathrm{e}^{-\mathrm{j}kr} \sin\theta, \tag{3.1}$$

and we may write the electric field intensities radiated by the two elements as follows:

$$E_1 = \mathbf{u}_\theta \, \mathrm{j}\eta \, \frac{kIL}{4\pi \, r_1} \mathrm{e}^{-\mathrm{j}(kr_1+\delta/2)} \sin\theta_1, \tag{3.2}$$

$$E_2 = \mathbf{u}_\theta \, \mathrm{j}\eta \, \frac{kIL}{4\pi \, r_2} \, \mathrm{e}^{-\mathrm{j}(kr_2-\delta/2)} \sin\theta_2. \tag{3.3}$$

Hence, the resultant field at the observation point P is given by

$$E_p = E_1 + E_2$$
$$= \mathbf{u}_\theta \mathrm{j}\eta \frac{kIL}{4\pi}\left[\frac{\mathrm{e}^{-\mathrm{j}(kr_1+\delta/2)} \sin\theta_1}{r_1} + \frac{\mathrm{e}^{-\mathrm{j}(kr_2-\delta/2)} \sin\theta_2}{r_2}\right], \quad (3.4)$$

where we have assumed that $r_1, r_2 >> d$ and $\theta_1 \approx \theta_2 \approx \theta$.

Amplitude: As far as the amplitude of the resultant field is concerned, we may assume that the three distances are equal since the differences which appear will make very little difference when r is very large (far-field zone). Hence, in the denominator, we let $r_1 = r_2 = r$.

Phase: As far as the phase is concerned, for a high frequency signal, even a small difference in distance in the exponential term will make a large difference in the phase of the resultant signal. The phases of the electric fields radiated by the two elements may interact destructively or constructively to significantly decrease or increase the resultant field magnitude. Indeed we could get total signal fading if the two signals are out of phase by π radians at the observation point. Hence, for the phase term appearing in the exponential, we set

$$r_1 = r + d/2 \sin\theta \text{ and } r_2 = r - d/2 \sin\theta.$$

Thus, the resultant electric field is given by

$$E_p = u_\theta \mathrm{j}\frac{\eta k I L \mathrm{e}^{-\mathrm{j}kr}}{4\pi r} \sin\theta \left[\mathrm{e}^{-\mathrm{j}(kd\sin\theta+\delta)/2} + \mathrm{e}^{\mathrm{j}(kd\sin\theta+\delta)/2}\right]. \quad (3.5)$$

Using $\mathrm{e}^{\mathrm{j}x} = \cos x + \mathrm{j}\sin x$, we may express the resultant field for an observation point on the zx-plane ($\phi = 0$) as

$$E_p = u_\theta \mathrm{j}\frac{\eta k I L \mathrm{e}^{-\mathrm{j}kr}}{4\pi r} \sin\theta \left[2\cos\left(\frac{kd\sin\theta + \delta}{2}\right)\right] \quad (3.6)$$

$$= \text{single element pattern} \times [\text{AF}], \quad (3.7)$$

where the array factor AF is given by

$$\text{AF} = 2\cos\left(\frac{kd\sin\theta + \delta}{2}\right). \quad (3.8)$$

The array factor AF is largely dependent on d (the distance between the elements) and δ (the phase angle difference between the electric currents in the two elements). For linear array antennas, the array factor is the same in the E-plane and in the H-plane. (For the geometry we are discussing, the E-plane is the *zx*-plane and the H-plane is the *xy*-plane.) The array factor is symmetrical about the axis on which the array of elements is placed. When the radiation pattern is to be sketched by hand, it is common to sketch the H-plane pattern since in the H-plane the radiation pattern is identical to the array factor. In the E-plane, the radiation pattern is the array factor multiplied by the sin θ factor that appears in the E-plane elemental pattern; this is more difficult to sketch.

In order to illustrate the generation of the AF pattern, we could try sketching the radiation patterns and compare with the computer generated patterns shown in Figure 3.3, for a two-element antenna with the following antenna parameters:

1) $d = \lambda/4$, $kd = \pi/2$, $\delta = \pi$, $\mathrm{AF} = 2 \sin(\pi/4 \sin \theta)$;
2) $d = \lambda/2$, $kd = \pi$, $\delta = 0$, $\mathrm{AF} = 2 \cos(\pi/2 \sin \theta)$;
3) $d = \lambda/4$, $kd = \pi/2$, $\delta = \pi/2$, $\mathrm{AF} = 2 \cos(\pi/4(\sin \theta + 1))$.

The plots shown in Figure 3.3 are computer generated for the following parameters: $f = 900$ MHz, dipole length $L = 0.5\lambda$, interelement spacing $d = 0.5\lambda$, distance $r = 100$ m, and phase δ is changed. The MATLAB™ listing of the computer program used to generate the radiation patterns is also given in the text.

The listing of the program used for plotting the radiation field patterns follows:

```
% E field pattern
% MATLAB program by T. Naveendra
th = pi/100: pi/100: 2*pi;
ph = pi/100: pi/100: 2*pi;
et = 120*pi; f = 100e6; c = 3e8;
n = input('Enter Antenna Length as Fraction of wavelength:');
k = 2*pi*f/c; l = 3e8/(n*f); i = 10;
r = 1000; % meters
q = et*i/(2*pi*r);
s = k*l;


for i = 1: 200
   for j = 1: 200
      e(i,j) = q*(cos(s*cos(th(i)))-cos(s))/sin(th(i));;
```

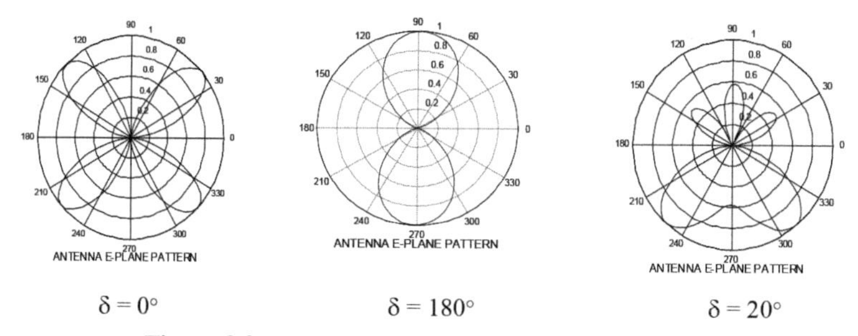

Figure 3.3 Radiation patterns of a two-element array.

```
        x(i,j) = abs(e(i,j))*sin(th(i))*cos(ph(j));
        y(i,j) = abs(e(i,j))*sin(th(i))*sin(ph(j));
        z(i,j) = abs(e(i,j))*cos(th(i));
    end
end

surf(x,y,z);
```

From the resultant field patterns, we observe two general classes of beams emerging. These are 1) the *end-fire array* and 2) the *broadside array.* In the end-fire array, the major part of the radiation pattern is directed along the axis y, which is the axis on which the elements are arranged. There is very little radiation in the z- and x-directions. In the broadside array, the radiation pattern is directed at right angles to the y-axis, the axis along which the elements are arranged. The radiation is in the z-direction, and there is very little radiation in the x- and y-directions. An end-fire array will be useful, for instance, when transmission may be required only in one direction, as in the case of communication systems to transmit to ships in the sea lane. Such antennas should avoid transmitting back into the land area. The end-fire array is used for television reception as well; the Yagi–Uda antennas mounted on most rooftops are end-fire arrays with the beam pointed toward the television broadcasting tower antenna. Indeed, such an antenna is attractive for mobile telephones, where it is desirable to radiate signals only away toward the base station and not into the head of the user. Here again a directive antenna like the end-fire array is useful.

Another important feature about array antennas, which form the radiation patterns for changing phase angle δ, is that the direction of the beam may be controlled by changing the electrical phase difference δ between the two elements. This means that the array antenna beam may be electronically steered. Without using any mechanical device like a motor to direct the antenna in a particular direction, by changing δ, we may make the antenna beam look in a particular direction. This is useful, for instance, in radar systems, where one may require the antenna to be continuously searching for targets in all directions. In mobile communications, smart antennas which exploit this feature may transmit power out to mobile users who may be moving around and are clustered in a certain area. A mobile base station which transmits power out in all directions (omni-directional antenna) is wasteful in power since the mobile stations may not be spread out over the whole area. Therefore, a base station which tracks all the mobile users in its cell may use an electronically steerable antenna to focus its transmission on to the specific area in which the mobile users are located at any instant of time.

Phase shifters are low power modules placed at the input of the transmit amplifier (i.e. the output of the receive amplifier in a two-way communication array antenna). If the phase shifters are placed at the output of the transmit amplifier, there will need to be high power devices that are less accurate and slower. We shall later see that in adaptive arrays, we not only control the phase (δ) of each element but also the amplitude (I) of each element; control of amplitudes allows us to better control the sidelobes and null points of the radiation pattern. Analog type narrow-band phase shifters may use switched lines, where transmission lines of different lengths are switched into the path at the input of the amplifier. Switching is done by PIN diodes or field-effect transistors (FETs) and is digitally controlled. Although these are easily implemented, losses of the order of 0.5 dB per section are present. If the number of line sections or bits is M, then the phase shifter resolution is $\Delta\theta = 2\pi/2^M$. For an M-bit phase shifter, the usable fractional signal bandwidth is given by $\Delta f/f_c = 1/2^M$, where f_c is the carrier frequency. For the extreme case where the phase shift is 2π, the accuracy is $2\pi(\Delta f/f_c)$ rad. For broadband antennas, all-pass filter networks must be used instead of line lengths. These 2–5 bit phase shifters have an insertion loss of about 1 dB.

3.2.2 Two-Element Half-Wave Dipole Array Antenna

In three-dimensional spherical coordinates (r, θ, ϕ), by defining

$$\psi = kd \sin\theta \cos\phi + \delta \tag{3.9}$$

the resultant field based on Equation(3.6),

$$|E| = \frac{2E_{\mathrm{m}}}{R_0} |F(\theta, \phi)| \; |\cos \; \psi/2| \,, \tag{3.10}$$

where $F(\theta,\phi)$ = element factor and $|\cos \; \psi/2|$ = normalized array factor.

Although the array factor and the normalized array factor will have similar patterns, the normalized array factor is always less than or equal to 1. Observe an important property of the normalized array factor: it is symmetrical about the axis of the array. In other words, with the antenna elements arranged along the x-axis, the array factor pattern is symmetrical about the x-axis. This may be verified by sketching the array factor for the zx-plane and the xy-plane. The array pattern on both planes will be identical since in the zx-plane ($\phi = 0$), $\psi = kd \sin \theta + \delta$ and in the xy-plane ($\theta = \pi/2$), $\psi = kd \cos \phi + \delta$. Sketching polar plots of the array factor, or the normalized array factor, for these two ψ forms, we shall get the same array factor pattern. The elemental pattern $F(\theta,\phi)$ may not be symmetrical about the x-axis, and thus the resultant radiation pattern is normally not symmetrical about the x-axis. The elemental pattern $F(\theta,\phi)$ depends on the type of antenna elements used (e.g. dipole antenna elements, horn antenna elements, parabolic aperture antenna elements, etc.), but the array factor for a given d and δ is the same for all types of antenna elements, and it is thus independent of the type of antenna elements we use.

Therefore, the pattern function of an array of identical elements is described by the product of the two directed half-dipoles:

$$|E| = \frac{2E_{\mathrm{m}}}{R_0} \left| \frac{\cos\left((\pi/2) \cos \; \theta\right)}{\sin \; \theta} \right| \; \left| \cos \; \frac{\psi}{2} \right| , \tag{3.11}$$

where ψ is also a function of θ. The E-plane pattern (constant ϕ and θ varies) and the H-plane pattern (ϕ varies, $\theta = \pi/2$) are completely determined by $\cos \psi/2$. When we sketch the electric field radiation pattern of the resultant field, it is important to observe that the radiation pattern in the θ (or ϕ) and ψ coordinates resembles that of a low pass filter. In other words, electric field is peak for a certain value of θ (or ϕ, say $\phi = 0$, or $\phi = \pi/2$) and then it begins to fall off. Then there are sidelobes of much weaker strength. This output signal pattern very much resembles that of an output signal from a low pass filter, although in the low pass filter electronic circuit, filtering occurs in the time or frequency domains. Therefore, an antenna may be considered as a spatial filter, that is, its output is a filtered version in the spherical space coordinates θ and ϕ. When we discuss the aperture antenna, we shall see that the radiation

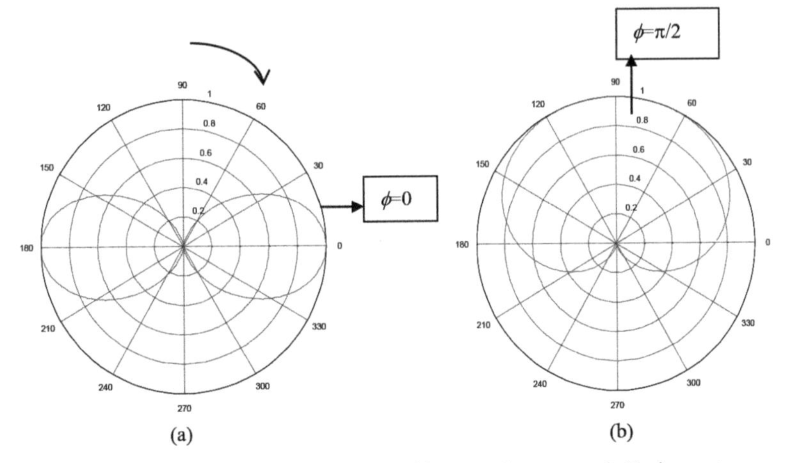

Figure 3.4 Broadside and end-fire beams with two-element and N-element arrays. (a) Broadside array. (b) End-fire array.

pattern is both θ and ϕ dependent. This observation is important in that, when we do antenna beamforming and synthesis, we shall use the same techniques as those used in electronic filter circuit synthesis to design our antenna system to provide a user specified radiation pattern or beam.

Example 1. Obtain the H-plane pattern of two parallel dipoles for 1) $d = \lambda/2$, $\delta = 0$, and 2) $d = \lambda/4$, $\delta = -\pi/2$. In the H-plane, each dipole is omni-directional. Hence, normalized pattern function = normalized array factor

$$|\mathrm{AF}(\phi)| = \left|\cos\frac{\psi}{2}\right| = \left|\cos\left(\frac{1}{2}(kd\ \cos\ \phi + \delta)\right)\right|.$$

1) $d = \lambda/2$, $kd = \pi$, $\delta = 0$ (Figure 3.4(a))

$$|\mathrm{AF}(\phi)| = \left|\cos\left(\frac{\pi}{2}\cos\ \phi\right)\right|,$$

maximum at $\phi = \pm\,\pi/2$ in the broadside direction. It is possible to obtain this broadside beam with an N-element array, as we shall show in the next section.

2) $d = \lambda/4$, $kd = \pi/2$, $\delta = -\pi/2$ (Figure 3.4(b))

$$|\mathrm{AF}(\phi)| = \left|\cos\left(\frac{\pi}{4}(\cos\ \phi - 1)\right)\right|,$$

maximum at $\phi = 0$ resulting in an end-fire array like the Yagi–Uda antenna.

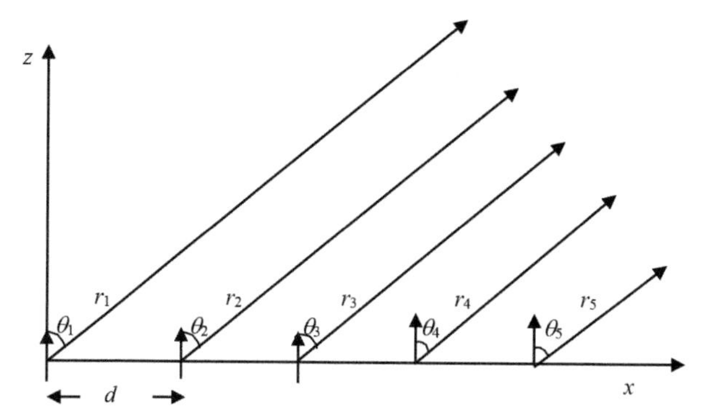

Figure 3.5 Uniform N-element linear array antenna.

3.3 General N-Element Uniform Linear Array

We now consider an array of N antenna elements placed along the x-axis. The array is uniform in that the distance between adjacent elements is d and each element carries a current of I. The array is linear in that the phase angles of the currents in the elements $n = 0, 1, 2, \ldots, (N - 1)$ are $0, \delta, 2\delta, 3\delta, \ldots, (N - 1)\delta$. Microstrip lines or microprocessors may provide these phase shifts. Antenna elements equipped with electronic phased shifters are called phased arrays. These can scan in θ (elevation) and ϕ (azimuth) directions. Widely used in radar and radio-astronomy systems, time delay circuits are used to provide phase shifts. By changing source frequency, we may change time delays. These are called frequency scanning arrays. N antenna elements are equally spaced and are excited with equal magnitude currents, with a phase displacement of $n\delta$ (see Figure 3.5). We shall consider a radiation field pattern in the $\theta = \pi/2$ plane; hence, the array factor is only a function of ϕ, and $\psi = kd \cos \phi + \delta$ since $\sin \theta = 1$.

At far field, the observation point P in Figure 3.5 is very far from the antenna array and the static field and the induction field is much smaller than the radiating field. As a result, we could approximate the angles as

$$\theta 1 \approx \theta 2 \approx \theta 3 \approx \theta 4 \approx \theta 5 \approx \theta.$$

Further, to determine the amplitude of the electric field signal, we could set

$$r_1 \approx r_2 \approx r_3 \approx r_4 \approx r_5 \approx r,$$

whereas the phase of the signal is allowed to have the following distance shifts with respect to the reference distance to allow for the important phase shifts between the elements:

$$r_2 = r - d\sin\theta\cos\phi,$$
$$r_3 = r - 2\,d\sin\theta\cos\phi,$$
$$r_4 = r - 3\,d\sin\theta\cos\phi,$$
$$r_5 = r - 4\,d\sin\theta\cos\phi.$$

In the following discussion, for the sake of simplicity, we shall set $\phi = 0$, or $\cos\phi = 1$. Assuming all the elements have the same current amplitude and dimension, the resultant field for an N-element antenna becomes

$$\begin{aligned} E_{\theta\text{total at point}p} &= E_{\theta 1} + E_{\theta 2} + E_{\theta 3} + E_{\theta 4} + ... + E_{\theta\ N} \\ &= \frac{\mathrm{j}\eta kIL\sin\theta\ \mathrm{e}^{-\mathrm{j}kr}}{4\pi r} \\ &\times \left[1 + \mathrm{e}^{\mathrm{j}(kd\sin\theta + \delta)} + \mathrm{e}^{\mathrm{j}2(kd\sin\theta + \delta)} + \cdots + \mathrm{e}^{\mathrm{j}(N-1)(kd\sin\theta + \delta)}\right], \end{aligned} \tag{3.12}$$

$$\begin{aligned} E_{r\text{total at point p}} &= E_{r1} + E_{r2} + E_{r3} + E_{r4} + E_{r5} \\ &= \frac{J\eta IL\cos\theta\mathrm{e}^{-\mathrm{j}kr}}{2\pi r^2} \\ &\times \left[1 + \mathrm{e}^{\mathrm{j}(kd\sin\theta+\delta)} + e^{\mathrm{j}2(kd\sin\theta+\delta)} + \cdots + e^{\mathrm{j}(N-1)(kd\sin\theta+\delta)}\right] \\ &\approx 0. \end{aligned} \tag{3.13}$$

It is apparent from Equations (3.12) and (3.13) that the total field of the array is equal to the field of a single element positioned at the origin multiplied by a factor which is widely referred to as the array factor. Thus, in our case, for a five-element array of constant amplitude, the array factor is given by

$$\mathrm{AF} = 1 + \mathrm{e}^{\mathrm{j}(kd\sin\theta + \delta)} + \mathrm{e}^{\mathrm{j}2(kd\sin\theta + \delta)} + \cdots + \mathrm{e}^{\mathrm{j}(N-1)(kd\sin\theta + \delta)}. \tag{3.14}$$

The array factor is a function of the geometry of the array and the excitation phase. Thus, by varying the separation d and/or the phase δ between elements, the characteristics of the array factor and of the total field of the array can be controlled. In general, we may state that the total electric field radiated by an array antenna is given by

$$E(\text{total}) = E(\text{single element}) \times \text{array factor(AF)}. \tag{3.15}$$

This is referred to as *pattern multiplication* for arrays of identical elements. Thus, for an N-element array, the array factor can be written as

$$\text{AF} = 1 + e^{j(kd\sin\theta+\delta)} + e^{j2(kd\sin\theta+\delta)} + \cdots + e^{j(N-1)(kd\sin\theta+\delta)} \tag{3.16}$$

$$= \sum_{n=1}^{N} e^{j(n-1)(kd\sin\theta+\delta)} \tag{3.17}$$

$$= \sum_{n=1}^{N} e^{j(n-1)}\psi \tag{3.18}$$

where $\psi = kd\sin\theta + \delta$.

In general, $\psi = kd\sin\theta\cos\phi + \delta$. By multiplying both sides of Equation(3.20) by $e^{j\psi}$, it can be written as

$$\text{AF}\left(e^{j\psi}\right) = e^{j\psi} + e^{j2\psi} + \ldots + e^{j(N-1)\psi} + e^{jN\psi} \tag{3.19}$$

Subtracting Equation (3.14) from Equation (3.19) reduces to

$$\text{AF}\left(e^{j\psi} - 1\right) = \left(-1 + e^{jN\psi}\right) \tag{3.20}$$

which can be written as

$$\text{AF} = \frac{e^{jN\psi} - 1}{e^{j\psi} - 1} e^{j[(N-1)/2]\psi} \frac{e^{j(N/2)\psi} - e^{-j(N/2)\psi}}{e^{j(1/2)\psi} - e^{-j(1/2)\psi}}$$

$$= e^{j(N-1)2]\psi} \left[\frac{\sin(N/2)\psi}{\sin(1/2)\psi}\right] \tag{3.21}$$

Expressing the foregoing analysis in another form, the total electric field E is the electric field E_0 due to a single element multiplied by the array factor. The total field is

$$|E_t| = |E_0| \left|1 + e^{j\psi} + e^{j2\psi} + \cdots + e^{j(N-1)\psi}\right|$$

$$= |E_0| \left|\frac{1 - e^{jN\psi}}{1 - e^{j\psi}}\right| = |E_0| \left|\frac{e^{j(N/2)\psi}}{e^{j\psi/2}}\right| \frac{\sin N\psi/2}{\sin \psi/2}$$

$$= |E_0| \frac{\sin N\psi/2}{\sin \psi/2} \tag{3.22}$$

where $\psi = kd\cos\phi + \delta$ the magnitude of

$$\left|\frac{e^{j(N/2)\psi}}{e^{j\psi/2}}\right|$$

is equal to 1.

We define the normalized array factor

$$|\mathrm{AF}\ (\psi)\ | = \frac{1}{N}\left|\frac{\sin\ N\psi/2}{\sin\ \psi/2}\right|. \tag{3.23}$$

The advantage in handling the normalized array factor instead of the array factor is that the normalized array factor is always less than or equal to 1, irrespective of the number of antenna elements making up the array antenna. Other than the difference in amplitudes, the patterns obtained for both the normalized array factor and the proper array factor are identical. The actual radiation pattern as a function of ϕ depends on kd (=$2\pi d/\lambda$) and δ. As ϕ: $0 \rightarrow 2\pi$, we have ψ: $kd + \delta \rightarrow -kd + \delta$ covering a range of $2kd$ (=$4\pi d/\lambda$) angle of scan. This is the visible range of the array antenna. The significant properties of $|\mathrm{AF}\ (\psi)|$ are as follows:

1) The main beam direction ϕ_0 is obtained from the relation AF(ψ) = 1.0, which is what we get when ψ tends toward zero and the angles in the numerator and denominator of Equation (3.23) become very small. Now $\psi = 0$ when $kd\cos\phi_0 + \delta = 0$ or $\cos\phi_0 = -\delta/kd$.

   (a) Broadside array $\phi_0 = \pm\pi/2$, i.e. $\delta = 0$, when in phase excitation.
   (b) End-fire array $\phi_0 = 0$ for maximum radiation, i.e. $\delta = -kd\cos\phi_0 = -kd = -2\pi\ d/\lambda$.

   Note that if we do a linear sketch of the array factor against ψ, we will get the peak always appearing at $\psi = 0$. However, if we do a linear sketch of the array factor against ϕ, the peak will appear at the ϕ angle toward which the maximum of the array factor is pointed. As we change the electronic angle δ, the peak of the array factor will change in the ϕ direction, while at the same time the array factor will be symmetrical about the x-axis.

2) *Null location.* For nulls, we have $|\mathrm{AF}(\psi)| = 0$, i.e. when

$$\frac{N\psi}{2} = \pm n\pi, n = 1, 2, 3, \cdots. \tag{3.24}$$

3) *Width of main beam*. The first-null beam width (FNBW) is the angular width between first nulls for large *N*. The first nulls occur when

$$\frac{N\psi_{01}}{2} = \pm\pi \text{ or } \psi_{01} = \pm 2\pi/N. \tag{3.25}$$

(a) Broadside array ($\delta = 0$, $\phi_0 = \pi/2$):

$$\psi = kd \cos\phi.$$

First null at ϕ_{01} = width of main beam. $2\Delta\phi = 2(\phi_{01} - \phi_0)$. At ϕ_{01},

$$\cos\phi_{01} = \cos(\phi_0 + \Delta\phi) = \psi_{01}/kd$$

For $\phi_0 = \pi/2$,

$$\cos\left(\frac{\pi}{2} + \Delta\phi\right) = -\sin\Delta\phi = -2\pi/Nkd,$$

$$\Delta\phi = \sin^{-1}(\lambda/Nd) \approx \frac{\lambda}{Nd} \text{ for} Nd >> \lambda. \tag{3.26}$$

With the broadside beam, there will be two main lobes, one with a maximum in the $\phi = \pi/2$ direction and the other with a maximum in the $\phi = -\pi/2$ direction. If we do not want one of the maximum beams (say in the $\phi = -\pi/2$ direction) by placing a reflecting metal plane of wire grid behind the antenna array (i.e. just below the *x*-axis), we could fold the beam in the $\phi = -\pi/2$ direction to point also in the $\phi = \pi/2$ direction. Thus, the strength of the beam in the $\phi = \pi/2$ direction will be almost doubled.

(b) End-fire array ($\delta = -kd$, $\phi_0 = 0$):

$$\psi = kd(\cos\phi - 1)\text{or } \cos\phi_{01} - 1 = \frac{\psi_{01}}{kd} = -\frac{2\pi}{Nkd} = -\frac{\lambda}{Nd},$$

$$\cos\phi_{01} = \cos\Delta\phi \approx 1 - (\Delta\phi)^2/2\text{forsmall}\Delta\phi,$$

$$(\Delta\phi)^2/2 = \lambda/Nd\text{or}\Delta\phi = \sqrt{\frac{2\lambda}{Nd}}, \tag{3.27}$$

since $Nd > \lambda/2$, i.e. the FNBW of the end-fire main beam is greater than the FNBW of the broadside main beam. The end-fire beam could be designed to have one main beam in one direction only (e.g. in the $\phi = 0$ direction) without the use of a reflector.

In wireless mobile communications, base station antennas (normally mounted on top of buildings) use 12-element arrays to get a beam width of about 30° and 24 elements to get 15° beam width with a 120° reflector placed behind the array. The directivity D, or the gain G, of the array antennas can now be obtained from Equations (1.11) and (1.12) of Chapter 1, noting that the HPBW θ_B is approximately equal to FNBW/2, and hence $\theta_B = \Delta\phi$.

4) *Sidelobe locations.* Sidelobes are minor maxima, side is a maximum. That is when

$$\left|\sin\left(\frac{N\psi}{2}\right)\right| = 1 \text{ or } N\psi/2 = \pm(2m+1)\frac{\pi}{2}, \quad m = 1, 2, \ldots. \tag{3.28}$$

The first sidelobes occur when

$$\frac{N\psi}{2} = \pm\frac{3}{2}\pi(m = 1). \tag{3.29}$$

5) First sidelobe level

$$\frac{1}{N}\left|\frac{1}{\sin(3\pi/2N)}\right| \approx \frac{1}{N}\left|\frac{1}{3\pi/2N}\right| = \frac{2}{3\pi} = 0.212 \tag{3.30}$$

or 20 $\log_{10}(1/0.212)$ = 13.5 dB. These may be reduced by making the excitation amplitudes (i.e. peak currents) in the center elements higher than those in the end elements. The directivity for the N-element broadside is $D_b = 2N(d/\lambda) = 2(L/\lambda)$, where L is the length of the array antenna. The directivity of an N-element end-fire array is $D_e = 4N(d/\lambda) = 4(L/\lambda)$. The directivity of an end-fire array antenna is, in general, greater than that of a broadside array antenna. For a given directivity (or gain), we can determine the number of elements required in a linear array. The distance d between two adjacent elements is often limited by the presence of *grating* (or *ambiguity*) lobes periodically located at a distance inversely proportional to the distance d. They give rise to directional ambiguity and could be eliminated by keeping d small. If the interelement distance d is reduced to too small a value, then problems related to mutual coupling, increased number of elements, and increased cost of the antenna arise. Thus, keeping $d = \lambda/2$ has often been found to be a good compromise.

In general, if the antenna must scan between angles $-\theta_s$ and $+\theta_s$, then the interelement distance d must be kept to the following limit to prevent grating lobes appearing in the scanning region:

$$d<\lambda/(1+\sin\theta_s), \tag{3.31}$$

where λ is the wavelength of the signal.

Linear array antenna main beams (array factor) are symmetrical about the axis on which the arrays of individual antenna elements are arranged. To get rid of one of the two main lobes, a reflector is placed behind the antenna to turn one of the main lobes over to add together with the other main lobe. An alternative way of reducing one of the two main lobes is to reduce the spacing below a half-wavelength, for instance, to $d - 0.45\lambda$. The idea here is that the visible region in the ψ domain is $2kd$, and to eliminate the grating lobe (the unwanted main lobe), the visible region should be reduced below the $d = 0.5\lambda$ value of 2π. Since $2\pi/N$ is the grating lobe half-width (maximum to null), we reduce the visible region by ensuring $2kd < 2\pi - (\pi/N)$, or $d < 0.5\lambda(1 - 0.5/N)$.

A simple method of obtaining the resultant pattern of a collection of antennas is sometimes called pattern multiplication. For the two-element array antenna, for instance, the resultant pattern is obtained by multiplying the radiation pattern of a single element (called the unit pattern), say, that of a half-wave dipole, with the array factor, which is called the group pattern. Thus, we have

$$\text{radiation pattern (RP)} = \text{unitpattern(UP)} \times \text{group pattern (GP)} \tag{3.32}$$

In the case of the two-element array made of half-wave dipoles, UP = radiation pattern of the half-wave dipole and GP = array factor. When we considered the N-element array, we had a simple analytical expression for the array factor. This method may be extended to more complex, non-linear, non-uniform arrangements too. What we must do is to break up a non-linear, non-uniform array into two sets of linear, uniform arrays. The first set would form just a two-element uniform linear array called the unit. The distance between the two elements of the unit is d. The second set is called a group consisting of an $N/2$-element uniform linear array. The distance between two adjacent elements in the group is $2d$. The elemental pattern of the unit and group arrays is the elemental pattern of the original array antenna, that is, the elemental pattern of the actual antenna elements used (e.g. the elemental pattern of a half-wave dipole). Once we find the radiation pattern of the unit (UP) and that of the group (GP) separately, by multiplying the two, we get the radiation pattern of the original array antenna.

3.4 Mutual Coupling Between Elements of The Array Antenna

We have thus far ignored the mutual coupling between elements placed in an array. The signal radiated by element number n, say, will hit the adjacent elements and be reflected back. If we consider dipole antenna elements, these will radiate in all directions and part of these radiated signals will hit and be reflected from the other elements. This means that these reflected signals would also be present due to mutual coupling between elements. So far, we have only considered the direct waves uncontaminated by these coupled or reflected waves. The reflected waves could destructively or constructively interact with the direct waves. This gives rise to the presence of blind angles at which the array factor goes to zero due to reflected waves. We shall here obtain an expression for the reflection coefficient of an array antenna; this reflection coefficient will describe the ratio between the reflected waves and the direct waves incident on the elements of the array antenna.

Let the direct waves be p_i, $i = 1, 2, \ldots, N$, from an N-element array antenna. For a transmitting antenna, these direct waves emerge from each of the N-element arrays. In a receiving antenna, these direct waves are those that are incident on each element of the array. Let the reflected waves due to the p_i waves be q_i, $I = 1, 2, \ldots, N$. If the coupling coefficients are C_{im}, characterizing the matrix of coupling coefficients, we have

$$q_i = \sum_{m=1}^{N} C_{im} p_m. \tag{3.33}$$

For an array with an interelement distance d, and if the signal at the 0th element is p_0, the signal at the mth element is given by

$$p_m = p_0 \exp(-jkmd \sin\theta), \tag{3.34}$$

where the difference in signal path length between the 0th and mth element is md sin θ. Therefore, the reflected signals are given by

$$q_i = p_0 \sum_{m=1}^{N} C_{im} \exp(-jkmd\ \sin\theta). \tag{3.35}$$

Therefore, the direction of arrival dependent reflection coefficient, $\Gamma_i(\theta) = q_i/p_i$, is given by

$$\Gamma i(\theta) = \sum_{m=1}^{N} C_{im} \exp(-\mathrm{j}k(i-m)d \sin \theta). \tag{3.36}$$

The coupling coefficient C_{im}, which is a function of the interelement distance d_{im}, where d_{im} = Modulus$(i-m)d$, may be written as

$$C_{im} = c_0 d_{im} \exp(-jkd_{im}), \tag{3.37}$$

where c_0 is a decreasing function of d_{im}, e.g. $c_0 = \rho \exp(-\kappa(i-m)d)$, where ρ and κ are constants.

Thus, the active reflection coefficient may be rewritten as

$$\Gamma i(\theta) = \sum_{m=1}^{N} c_0 d_{im} \exp[-\mathrm{j}k(d_{im} + (i-m)\, d \sin \theta)]. \tag{3.38}$$

It is of interest to note that the active reflection coefficient approaches unity (i.e. total reflection) for certain values of θ. These values of θ are called blind angles since the receiver antenna will not capture any signal at these angles. The blind angles are close to the grating lobe angles and are given by the relation

$$k(d_{im} + (i-m)d \sin \theta) = 2n\pi \quad (n = 1, 2, \ldots),$$

from which we get, with $n = 1$,

$$\sin \theta = \lambda/d - 1. \tag{3.39}$$

3.5 Polarization

In our presentation of antenna theory so far, we have considered only the wire type of antennas. These antennas radiate linearly polarized signals. Thus, for example, if we consider the half-wave dipole antenna radiation in the far field, the electric field is in the $\mathbf{u}_\theta$ direction, but if we consider the significant electric field strength (in the 3 dB range), then, within the 3 dB cone, unit vector is roughly parallel to u_z and the radiation field will be parallel to the line element. This is the reason why in mobile communication and in television transmission, with the transmission antenna kept vertical, the receiver antenna should also be kept vertical to capture the peak radiated signal strength. In amateur radio, where over-the-horizon communication links are formed by bouncing 6–20 MHz signals off the ionosphere (charged particle layer at heights of about 70 km above the earth), horizontally placed antennas are used to transmit and receive horizontally polarized electric fields. However, in satellite communications, frequencies are reused by sending out two signals (each carrying different voice and video signals) polarized perpendicular to each other. Thus, the two orthonormal signals do not interfere with each other, although both are at the same frequency and travel over the same path simultaneously. Two signals, polarized perpendicularly to each other (say in the $\mathbf{u}_x$ and $\mathbf{u}_y$ directions), will produce a resultant signal, which will rotate in space as it travels over free space. This is called a circularly (or elliptically) polarized signal. To transmit and receive such orthonormal signals, we need antennas which can effectively handle signals that are perpendicular to each other and produce a resultant circularly polarized signal (see Figure 3.6). Consider the resultant signal due to two orthonormal signals

$$\begin{aligned}\mathbf{E}_\theta &= \mathbf{u}_x E_1 \cos(\omega t - kr) + u_y E_2 \cos(\omega t - kr - \pi/2) \\ &= \mathbf{u}_x E_1 \cos(\omega t - kr) + u_y E_2 \sin(\omega t - kr) \\ &= \mathbf{u}_x E_x + u_y E_y. \end{aligned} \tag{3.40}$$

The resultant signal in this case rotates in the counterclockwise direction and is called a right-hand-polarized wave (RPW) since its motion is similar to rotating one's outstretched right hand into the body, over the head. At $r = 0$, we have

$$\begin{aligned}\cos \omega t &= \tfrac{E_x}{E_1}, \\ \sin \omega t &= \tfrac{E_y}{E_2} = \sqrt{1 - \left(\tfrac{E_x}{E_y}\right)^2}.\end{aligned}$$

Hence, the trajectory of the resultant electric field is given by

$$\left(\frac{E_y}{E_2}\right)^2 + \left(\frac{E_x}{E_1}\right)^2 = 1.$$

If $E_1 = E_2$, the resultant signal rotates in a circle at a velocity equal to the electrical frequency as it propagates through space at the velocity of light. The direction of rotation is in the anticlockwise direction (right hand circularly polarized signal) or clockwise direction (left hand circularly polarized signal),

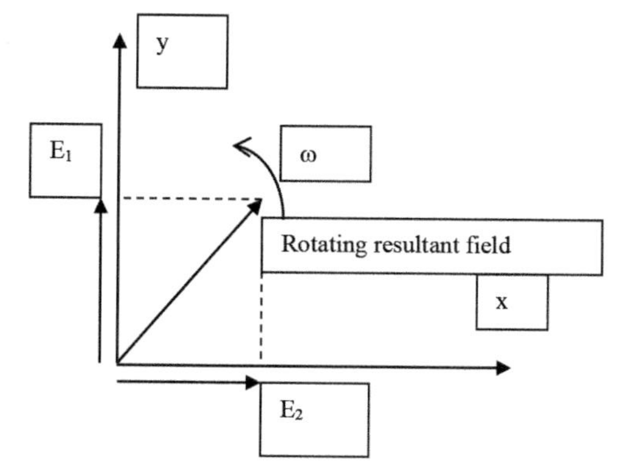

Figure 3.6 Circularly polarized signal.

depending on which signal is leading in phase. If the magnitudes are not equal, then we get the resultant signal rotating over an ellipse, and it is called an elliptically polarized signal. To transmit and receive such circularly polarized signals, we use aperture antennas, which can also produce very narrow beams.

3.6 Aperture Antennas

Aperture antennas are antennas in which the electromagnetic signals are radiated not from conduction currents flowing along wires, as in the case of wire antennas, but from displacement currents, which appear at the opening of a waveguide. Waveguides are hollow conductors, which transmit signals not by conduction currents, as in coaxial cables, but through the displacement currents existing in the empty space inside a hollow tube, normally shaped, in a rectangular hollow conductor. The waveguide is left open and the signals which travel along inside are launched out into space, thus the opening forming an aperture antenna. The shape of the opening differs; one popular shape is that of a horn so that the waveguide is opened out to provide a wider aperture to radiate from.

The aperture antennas work in the microwave frequency spectrum (see Figure 3.7), roughly stretching from 0.1 GHz to about 300 GHz. Above this frequency of 0.3 trillion Hz, the electromagnetic spectrum is termed infrared, mostly associated with heat emission. Close to the visible light spectrum of 430×10^{12} to 750×10^{12} Hz, we have optical waveguides (largely made of glass) which transmit optical electromagnetic signals.

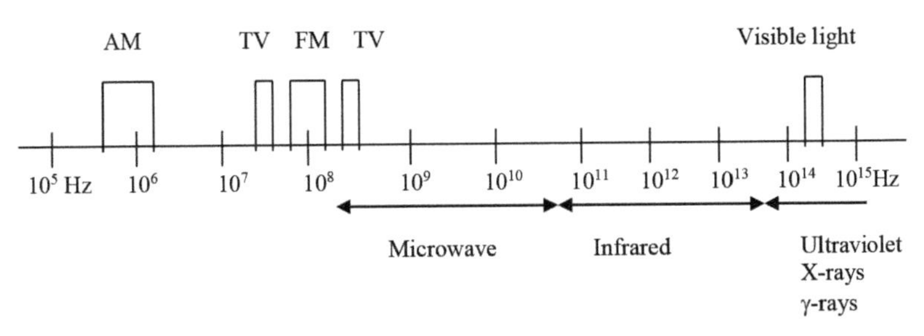

Figure 3.7 Microwave spectrum in which aperture antennas are largely used.

Optical transmission is a major competitor to microwave transmission, with more and more undersea optical fiber cables interconnecting continents for telecommunication purposes. Above the visible light spectrum, we have the ultraviolet electromagnetic signals, which carry enormous amounts of energy (given by Planck's constant $h \times$ frequency f) and are highly life threatening. A limited amount of wireless transmission may be achieved with laser beams traveling over free space; but their use will be highly localized, unlike a microwave satellite link, which can cover about one-third of the earth with one satellite in geostationary orbit. For wireless communications, the microwave region is the highest we can sensibly achieve, and as frequencies increase, newer types of antennas are required to handle these signals. Almost all antennas used at microwave frequencies are of the aperture type.

A rectangular aperture may simply represent a waveguide that is left open at one end or the aperture of a rectangular horn antenna. The signal comes into the aperture through the waveguide and forms displacement currents (ε dE/dt A/m^2, where ε is the permittivity of air) at the aperture. These displacement currents, like conduction currents along a linear array of dipole antennas, radiate out the signals. Consider for a moment the electric field lines that appear at the aperture of the antenna. The electric field lines can be broken into small lines of fictitious elemental dipoles carrying current ε dE/dt (=j$\varepsilon\omega E$) A/m^2. For each one of these little dipoles, we may determine the electric field radiated as (from Chapter 2)

$$E_\theta = \mathrm{j}\eta \frac{kILe^{-\mathrm{j}kr} \sin\theta}{4\pi r}, \tag{3.41}$$

which, for the fictitious dipoles, may be rewritten as

$$E_\theta = (\mathrm{j}k/2\pi) \frac{\eta ILe^{-\mathrm{j}kr} \sin\theta}{2r}. \tag{3.42}$$

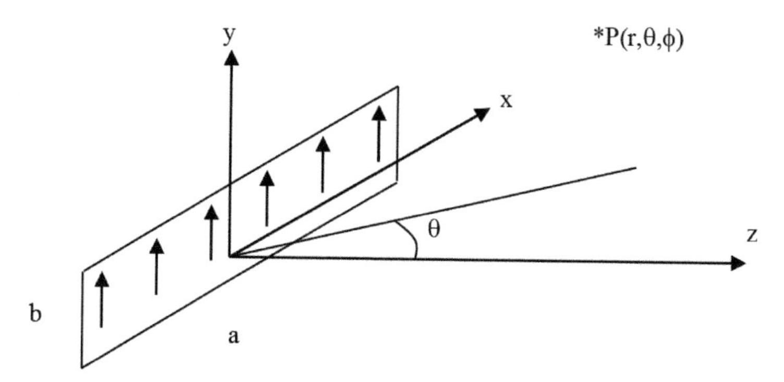

Figure 3.8 A rectangular aperture with uniform electric field.

In the case of the aperture antenna, we replace $\eta I(L/2)$ by the aperture electric field $E_a = E(x',y',0)$ and integrate the electric field over the aperture to get the resultant radiated electric field. The magnetic field $H = E/\eta$.

A general expression relating the radiation field $E(x, y, z)$ to the field at the aperture is

$$E = \frac{\mathrm{j}k}{2\pi}\int \frac{E(x',y',0)\mathrm{e}^{-\mathrm{j}kR}}{R}\,\mathrm{d}s, \tag{3.43}$$

where $E(x', y', 0)$ is the scalar field in the aperture. The surface integration is carried over the aperture.

Although in Equation (3.42), though $r >> x', y'$, even a small difference in distance will make a large difference in the signal phase for a high frequency signal. When we integrate over the aperture, the phase differences will either constructively or destructively sum together. For the R in Equation (3.43), which appears in the denominator, we may assume that $R \approx r$ since we are considering far away observation points, and the R in the denominator impacts the magnitude only.

$$E(x, y, z) = \frac{\mathrm{j}k}{2\pi r}\mathrm{e}^{-\mathrm{j}kr}\int E(x',y',0)\,\mathrm{e}^{\mathrm{j}k\,\sin\theta\,(x'\,\cos\,\phi\,+y'\,\sin\,\phi)}\mathrm{d}x'\,\mathrm{d}y'. \tag{3.44}$$

We shall now evaluate this integral for an $a \times b$ rectangular aperture with a uniform field $E_0\mathbf{u}_y$ appearing in the aperture. This is a slightly idealized case since, in general, the electric fields inside a waveguide have a $\sin(\pi x/a)$ or $\sin(\pi y/b)$ term attached to E_0 (see Figure 3.11).

We have $E(x', y', 0) = E_0\,\mathbf{u}_y$ at $(x', y', 0)$ in the aperture. Hence, the radiation field at point $P(r,\theta,\phi)$ is obtained using Equation (3.44):

$$E = \frac{\mathrm{j}k\mathrm{e}^{-\mathrm{j}kr}}{2\pi r} \int\limits_{x'=-a/2}^{a/2} \int\limits_{y'=-b/2}^{b/2} E_0 \mathrm{e}^{\mathrm{j}k \sin\theta\,(x' \cos\phi + y' \sin\phi)} \mathrm{d}x'\,\mathrm{d}y'$$

$$= \frac{\mathrm{j}kE_0\mathrm{e}^{-\mathrm{j}kr}}{2\pi r} \int\limits_{-a/2}^{a/2} \mathrm{e}^{\mathrm{j}x'k \sin\theta \cos\phi}\,\mathrm{d}x' \int\limits_{-b/2}^{b/2} \mathrm{e}^{\mathrm{j}y' k \sin\theta \sin\phi}\,\mathrm{d}y'$$

$$= \frac{\mathrm{j}kE_0 ab\,\mathrm{e}^{-\mathrm{j}kr}}{2\pi r} \left(\frac{\sin\psi_1}{\psi_1}\right)\left(\frac{\sin\psi_2}{\psi_2}\right), \tag{3.45}$$

where

$$\psi_1 = \frac{ka \sin\theta \cos\phi}{2} \tag{3.46}$$

and

$$\psi_2 = \frac{kb \sin\theta \sin\phi}{2}. \tag{3.47}$$

The terms $\sin\psi/\psi$ determine the radiation pattern of the aperture. Note that it is in some ways similar to the pattern function of a linear array. Thus, the radiation pattern of an aperture is similar to the resultant pattern of two linear arrays arranged along the *x*- and *y*-axes. Indeed this is what we would expect if we place the lines of electric field lines of force at the aperture along which displacement currents flow. These displacement currents are line linear current elements arranged in the *x*- and *y*-directions at $z = 0$.

For $\phi = 0$, we get the *E* radiation pattern in the *zx*-plane as

$$E_{\phi=0} = \frac{kE_0 ab}{2\pi r}\left|\frac{\sin(ka \sin\theta)/2}{(ka \sin\theta)/2}\right|. \tag{3.48}$$

For $\phi = 90°$, we get the E_θ radiation pattern in the *yz*-plane as

$$E_{\phi=90°} = \frac{kE_0 ab}{2\pi r}\left|\frac{\sin[(kb \sin\theta)/2]}{(kb \sin\theta)/2}\right|. \tag{3.49}$$

We may similarly obtain the E_ϕ pattern by setting $\theta = \pi/2$. The three-dimensional radiation pattern will contain both these patterns as well as each pattern for different values of ϕ.

$$Note: \lim_{\psi\to 0} \frac{\sin\psi}{\psi} \to 1.$$

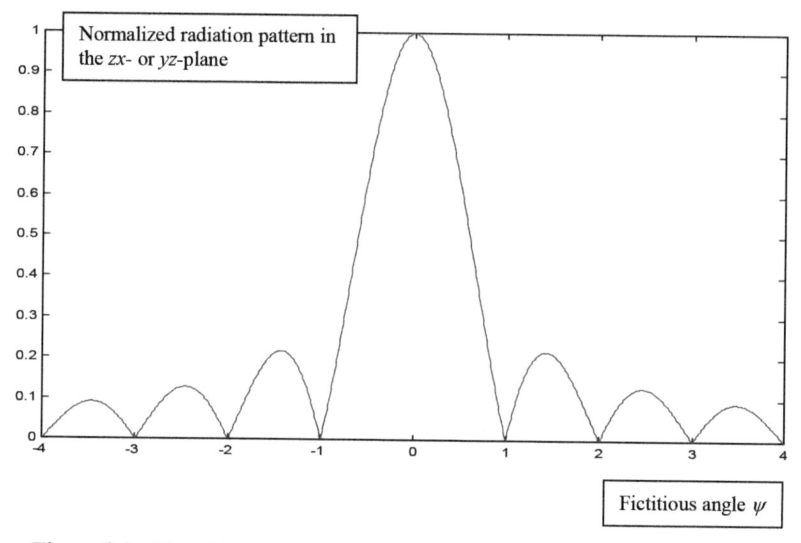

Figure 3.9 Two-dimensional aperture radiation pattern in the *zx*- or *yz*-plane.

We note the identical behavior in both planes (see Figure 3.9) except for dimensions a and b, which will determine the width of the beams. Both E-plane and H-plane patterns will be unsymmetrical for the aperture antenna.

Beam width between first nulls may be obtained from

$$\frac{ka\ \sin\ \theta}{2} = \pi \tag{3.50}$$

or

$$\frac{ka\theta}{2} = \pi,$$

assuming narrow beams. Hence,

$$\theta = \frac{2\pi}{ka} = \frac{\lambda}{a} \text{ or } 2\theta = \frac{2\lambda}{a} = \text{FNBW}_{\phi=0}, \tag{3.51}$$

where FNBW is the first-null beam width. In general, for aperture antennas, the angular beam width is inversely proportional to the aperture size a (or diameter d for circular apertures) normalized to the wavelength λ. An approximate value for the half-HPBW is

$$\theta = \frac{\lambda}{a} = \text{HPBW}_{\phi=0}. \tag{3.52}$$

The directivity of the antenna is given by

$$D = \frac{P_{\mathrm{m}}}{P_{\mathrm{r}}/4\pi r^2}. \tag{3.53}$$

The total radiated power P_{T} is the power density at the aperture multiplied by the area of the aperture for uniformly distributed fields in the aperture. Hence, the total radiated power is given by

$$P_{\mathrm{r}} = \frac{1}{2}\frac{E_0^2}{\eta}(ab) \tag{3.54}$$

To find the maximum power density radiated P_{m}, we must first find the maximum field radiated:

$$E_{\mathrm{max}} = \mathrm{j}\frac{kE_0 ab\, \mathrm{e}^{-\mathrm{j}kr}}{2\pi r}. \tag{3.55}$$

Hence, the maximum power density radiated by the rectangular aperture is given by

$$\begin{aligned} P_{\mathrm{m}} &= \frac{1}{2}\frac{|E_{\mathrm{max}}|^2}{\eta}\ \mathrm{W/m^2} \\ &= \frac{k^2 E_0^2 a^2 b^2}{8\pi^2 r^2 \eta}. \end{aligned} \tag{3.56}$$

Hence, the directivity of the aperture from Equations (3.58), (3.59), and (3.60) is given by

$$D = \frac{k^2 ab}{\pi} = \frac{4\pi}{\lambda^2}(ab). \tag{3.57}$$

Note that the larger the aperture area (*ab*), the greater the directivity. This is one reason why large parabolic reflector dishes are used in microwave communications and in radio-astronomy so as to increase both the directivity (*D*) and gain (*G*) of the antenna. A typical parabolic reflector will have a gain of about 60 dB, a diameter of 12 m for 4 GHz (satellite-to-earth), and 6 GHz (earth-to-satellite) space communications. Thus, the physical size of the antenna is large compared to the millimeter or centimeter wavelengths involved. High gain or high directivity implies that the radiation pattern of the antenna can be synthesized with great accuracy. Space communications require very high gain antennas since distances are very large. The altitude of a geostationary satellite is 36,000 km, the distance between Earth and Moon is 360,000 km, and the mean distance between Earth and Mars is 150 million km. As the frequency spectrum gets filled up, space communications at higher Ku-band frequencies (11–15 GHz) pose further challenges since atmospheric noise increases exponentially with frequency; this means that the antenna must be made to have low noise (cold antennas) and high gain.

For a uniform electric field E_0 at the aperture, the physical aperture (ab) is equal to the effective aperture of the antenna. If the electric field at the aperture was non-uniform and given by $E_0 \cos(\pi y'/b)$, it could be shown that the radiation power will be halved, and directivity (or gain) $D = 32\,(ab)/(\pi\lambda^2)$ and effective aperture $A_{em} = 8(ab)/\pi^2$. Now the effective aperture is less than the physical aperture (ab).

One further point to note is that all along, we have defined D and G for the maximum power density radiated. Therefore, the gain in directions other than the direction of maximum power density radiation will be less than the G and D values we have obtained. If E is proportional to $\sin\theta$, for example, the gain and directivity will be functions of $(\sin\theta)^2$. The effective radiated power (ERP) of the antenna

$$\text{ERP} = P_r = G.P_T, \tag{3.58}$$

where P_T is the power input to the antenna in W. If $G = 1000$ and is 5 W, then ERP is 5000 W.

A widely used aperture antenna is the horn antenna (often used together with the parabolic reflector). The directivity of a horn antenna is given by

$$D = \frac{1}{2}\frac{4\pi(ab)}{\lambda^2}. \tag{3.59}$$

For a pyramidal horn, the gain of the antenna is

$$G = 10A/\lambda \approx 10(ab)/\lambda^2. \tag{3.60}$$

Its vertical –3 dB beam width is given by $\theta_v = 51\lambda/b$ and its horizontal –3 dB beam width $\theta_h = 70\lambda/a$. A 4 × 9 cm pyramidal horn is typically used at a frequency $f = 10.25$ GHz. The vertical –3 dB beam width of a horn radiator is $\theta_v = 51\lambda/b$ and the horizontal –3 dB beam width is $\theta_h = 70\lambda/a$; both angles are given in degrees. Typical gain and beam widths that we obtained with horn antennas at 10 GHz are of the order of 50° and 25°, respectively. The parabolic reflector type of aperture antenna has a gain $G = 6d^2/\lambda^2$ and a –3 dB beam width (HPBW) of $70\lambda/d$ (degrees), where d is the diameter of the dish. The radiation pattern of such an antenna can be synthesized with a sampling step of λ/d. The primary feeds of such parabolic reflectors are horn antennas which direct the waves onto the surface of the parabolic reflectors. Thus, we have two apertures here: that of the horn antenna and that of the reflector. The latter has a much larger aperture area leading to high gain. The typical efficiency of such antennas (i.e. $G/D = P_r/P_T$) is 60%. With structural improvements (e.g. Cassegrain antenna), it is possible to get gains of the order of 80%, but the cost of these antennas is very high. With parabolic dishes, it is possible to get gains of the order of 10^5 and beam widths of the order of 0.1° at operating frequencies of 10 GHz and above. A microwave antenna with a gain of 10^5 and a noise temperature of 55 K (i.e. 17.4 dB with respect to 1 K) will have an antenna gain to total noise temperature ratio (called the *figure of merit*) of 32.6 dB. When a horn antenna is mounted outdoors, it is generally a good idea to cover the aperture with a plastic paper to prevent rainwater getting into the antenna. This is particularly critical with the low noise block converter (LNBC) electronics mounted just adjacent to the horn. The LNBC consists of a frequency downconverter and a low noise amplifier. Only a short strip of waveguide of a few centimeters long was needed to connect the horn antenna to the microwave input terminal of the LNBC. The LNBC is mounted outdoors next to the horn antenna in order to be able to downconvert the received microwave signal to a lower frequency before transmitting it over a coaxial cable into the radio room. In the older installations, the LNBC was placed indoors, which meant that the rather cumbersome waveguides had to be used to transmit the signal from the antenna into the radio room.

An aperture array antenna can be constructed by cutting a series of rectangular slots (apertures) on one of the four walls of a rectangular waveguide. In this case, there will be radiation out of each rectangular slot, and the resultant waveform will be the summation of radiation from each hole or aperture. Such a 7.5 × 1.5 m array with 4000 slots with beam widths of 1° and 4.75° constitutes the airborne warning and control system (AWACS) antenna costing about US$ 300 million with the transmitter/receiver electronics; it is normally mounted on top of a 707 civilian aircraft.

3.7 Patch Microstrip Antennas

One type of aperture antenna, which has become very popular, is the microstrip printed or patch antenna. It is a small, light antenna, made of copper patches mounted on dielectric substrates, as shown in Figure 3.10.

By using dielectric substrates of high permittivity, we make the wavelength short. This means that the antenna can be very small since the size

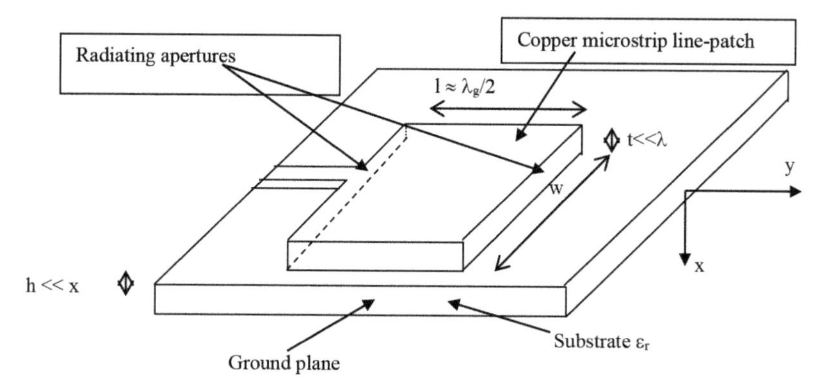

Figure 3.10 A rectangular microstrip patch antenna.

of the antenna depends on the frequency (or half-wavelength) of the signals coming into it. They are particularly attractive candidates for mobile, aircraft, and spacecraft applications. These copper patches are flash mounted, with the feedline, which is also another microstrip line, placed behind the ground plane. These are inefficient radiators and have a narrow frequency bandwidth of about 5% of the center frequency f_0. The relatively narrow bandwidth is due to the patch antenna behaving as a resonator. There are also added disadvantages of low radiation efficiency and poor polarization purity, but the fact that these antennas may be realized with printed circuit technology and are suitable for electronic system miniaturization has made them attractive for telecommunications, radar, mobile communications, space industry, and medical applications. Figure 3.16 shows a rectangular printed antenna, which is the simplest and most widely used geometry. Electromagnetic radiation takes place through the $w \times t$ aperture, with the beam pointing in the y-direction. The radiating elements and the feedlines are usually photoetched on the dielectric substrate.

The antenna has two slots each $w \times h$, placed perpendicular to the feedline. Between the slots is a transmission line of length $l = \lambda_g/2$. Two slots array, each element separated by $\lambda_g/2$. Fields at each slot have opposite polarization ($\lambda_g/2$). The y-components are out of phase and cancel out. Only the TEM mode exists in feedlines.

The far-field region fields from a rectangular microstrip antenna are given by

$$E_r = E = 0, \tag{3.61}$$

$$E_{\phi} = -\mathrm{j}\frac{hwkE_0\ \mathrm{e}^{-\mathrm{j}kr}}{2\pi r}\left\{\sin\ \theta\ \left[\frac{\sin\ \psi_1}{\psi_1}\right]\left[\frac{\sin\ \psi_2}{\psi_2}\right]\right\}, \tag{3.62}$$

where

$$\psi_1 = \frac{kh}{2}\ \sin\ \theta\ \cos\ \phi, \tag{3.63}$$

$$\psi_2 = \frac{kw}{2}\ \cos\ \theta. \tag{3.64}$$

The simplified radiation electric field for the condition $h << \lambda$ is given by

$$E_{\phi} = -\mathrm{j}\ \frac{V_0\mathrm{e}^{-\mathrm{j}kr}}{\pi r}\ \sin\ \theta\frac{\sin\left[(kw/2)\ \cos\ \theta\right]}{\cos\ \theta}. \tag{3.65}$$

The voltage at the aperture is related to the field at the aperture by the following relation:

$$V_0 = hE_0 \tag{3.66}$$

The patch antenna gain is given by

$$G = \begin{cases} \dfrac{1}{90}\left(\dfrac{w}{\lambda}\right)^2 \text{for } w << \lambda, & (3.67) \\ \dfrac{1}{120}\left(\dfrac{w}{\lambda}\right)^2 \text{for } w >> \lambda. & (3.68) \end{cases}$$

The directivity of the antenna is

$$D = \left(\frac{2\pi w}{\lambda}\right)^2 \frac{1}{I} \tag{3.69}$$
$$= \begin{cases} 3 & \text{for } w << \lambda\ (4.77\ \text{dB}), \\ 4 & \text{for } w >> \lambda\ (6.02\ \text{dB}). \end{cases}$$

The bandwidth of the printed or patch antenna may be increased by:

1) increasing the impedance of the line or by increasing h, but increasing h will mean that the antenna is no more low profile;
2) using high ε_r substrate to reduce dimensions of parallel line;
3) increasing L of the microstrip by cutting holes or slots;
4) adding reactive components to reduce the VSWR at the antenna.

Microstrip array antennas are shown in Figures 3.11 and 3.12. These are compact array antennas with high directivity. The compactness is achieved by reducing the wavelength using high permittivity substrate. These antennas may replace wire antennas in mobile communication systems.

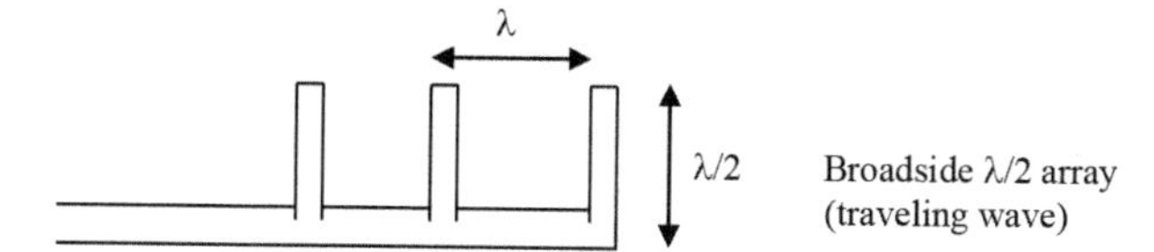

Figure 3.11 Traveling wave microstrip antenna array.

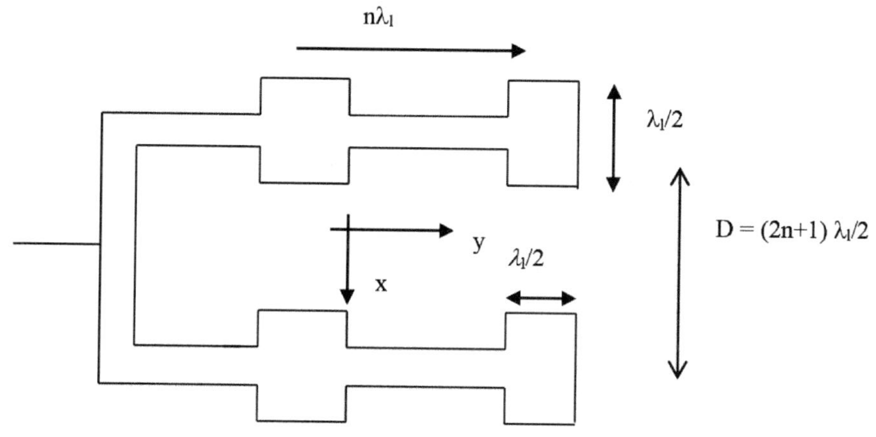

Figure 3.12 A popular patch antenna array.

The array spacing D suppresses radiation in the x-direction and helps to direct the antenna beam along the y-direction. Such an antenna may be used in aircraft-to-satellite communications. Since the antenna is small and relatively flat, it may be placed on top of the aircraft without causing wind friction. The two elements in the top row and the two elements in the bottom row can be electronically phase controlled to steer the beam in the azimuth direction. Such azimuth control will be required as the aircraft flies past the satellite with which it is maintaining communication. When it is flying toward the satellite, the beam will have to look forward. As it flies past the satellite position, the beam will have to be steered so that, after flying past, the beam looks back at the satellite antenna. Thus, azimuth control of the microstrip antenna array is required to keep in contact with the satellite. The two rows of arrays, separated by D, will provide the elevation angle beam since the aircraft cruise altitude (e.g. 10 km) will be much less than the altitude of the satellite (e.g. 36,000 km). The elevation angle beam need not be steered since the altitude difference between the aircraft and satellite will be constant when the aircraft is cruising.

3.8 Corner-Reflector Antenna

The corner-reflector antenna is very useful for UHF reception because of its high gain, large bandwidth, low sidelobes, and high front-to-back ratio. A 90° corner reflector, constructed in grid fashion, is generally used. For a corner reflector, let d be the spacing between the half-wave dipole radiator and the corner. Let the half-wave dipole wire and the reflector wires all be parallel to the z-axis. It is possible to work out the resultant radiation field by considering an image of the dipole on the upper reflector plane, a second image on the lower reflector plane, and a third image at the corner of the reflector. The contribution of the two-element array formed by the feed element and its image at the corner will be $2\cos(kd\cos\phi)$ and the contribution from the two images in the reflector planes will be $2\cos[kd\cos(\pi/2-\phi)]$. The resultant radiation field in the xy- or H-plane in the region $-\pi/4 \leq \phi \leq \pi/4$ is given by

$$\mathrm{AF}(\theta = \pi/2, \phi) = 2[\cos(kd\cos\phi) - \cos(kd\sin\phi), \tag{3.70}$$

and the xz- or E-plane pattern is given by

$$\mathrm{AF}(\theta, \phi = 0) = 2[-1 + \cos(kd\sin\theta)]\left|\frac{\cos((\pi/2)\cos\theta)}{\sin\theta}\right|. \tag{3.71}$$

A wide-band triangular dipole of flare angle 40°, which is bent 90° along its axis so that the dipole is parallel to both sides of the reflector, is used as the feed element. The dipole has a spacing of about (wavelength/2) at mid-band from the vertex of the corner reflector, i.e. $d = \lambda/2$. However, for this choice of d, the input impedance of the antenna is quite high, close to 125 Ω. To bring down the input impedance without too great a drop in the gain, in most practical designs, $d = 0.35\lambda$ is used. Grid length L and grid width W should be kept about $L = 2d$ and $W = 1.5 \times$ length of the antenna feed. Hence, if a half-wave dipole feed is used, $W = 1.5 \times 0.5\lambda = 0.75\lambda$. Keeping $W >$ length of the feed element ensures that there will be no radiation leaking into the back region. The spacing for grid tubing of 0.1λ diameter should be slightly under 0.5λ. An alternative to the dipole feed is the bow-tie antenna element, which has good impedance bandwidth properties. In Figure 3.13 is shown the parabolic reflector and horn antenna arrangement used in satellite communications, space science, and direct broadcasting systems (DBS). The electronic equipment that is located in the control room of an antenna station are also shown.

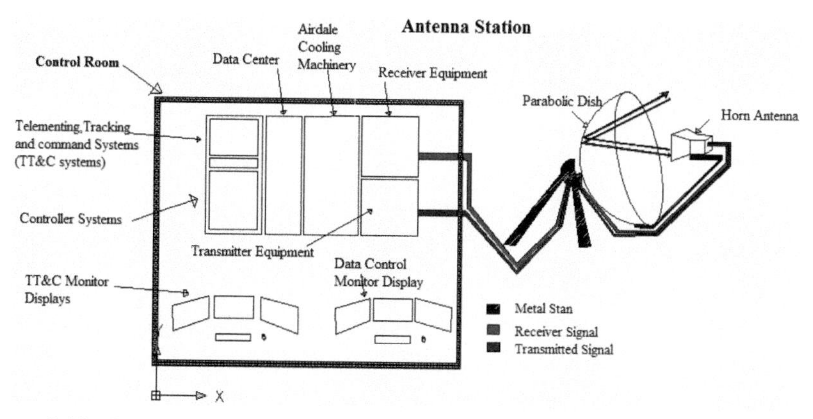

Figure 3.13 Parabolic reflector and horn antenna for DBS (11.7–12.5 GHz), Industry Science Medical ISM Bands (2.4–2.48 GHz), ultra wide band (UWB) (3.1–10.6 GHz), and higher frequency bands including L to Ka bands.

3.9 Finite Length Antenna: A Basic Building Block for Antenna Simulation

We have seen that some of the most complex antennas like array antennas and aperture antennas may be thought of as being made up of discrete line elements. In aperture antennas, for instance, the radiating line elements are the electric field lines that appear at the aperture carrying displacement currents. In electromagnetic image reconstruction too, the line element is expected to yield image reconstruction in shorter time and it inherently contains more information about the region it covers. This includes information about the size of the region and its angle of inclination. The equation of the electric field of a finite line element forms the basis of the imaging model. This model is developed and implemented in this report. This chapter will first set out to derive the equations for the electric field. Studies of the variation of the electric field with the various parameters in the field equation are performed to understand the impact of the model when used in image reconstruction.

Once the radial and tangential fields for a finite length radiator are derived, the structures of the radiation patterns are studied. These radiation patterns have a direct impact on the image quality and also on the relative value of using measurements of the radial component, $\mathbf{E}_r$, tangential components, $\mathbf{E}_\theta$, or magnitude of the electric field, $\sqrt{(\mathbf{E}_r^2 + \mathbf{E}_\theta^2)}$, for imaging in medicine and radar cross-sectional measurements. This is an additional advantage that

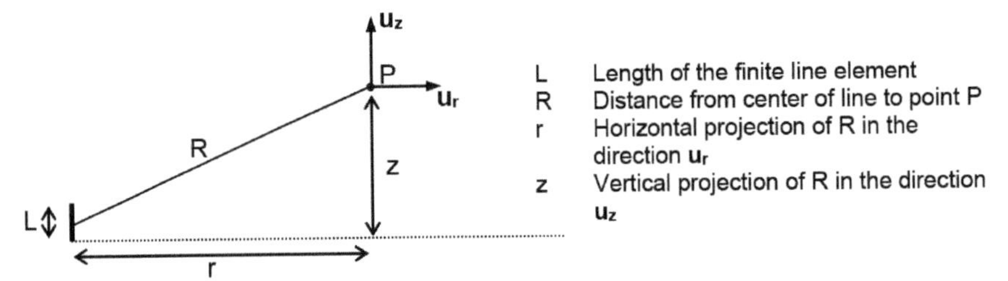

Figure 3.14 Orientation of a finite line element set up for investigation.

the line scatterer model provides over point scatterers, i.e. the use of three different signals for image reconstruction.

Let $[I]$ be the retarded current carried by the element and $[Q]$ be the retarded electric charge. Both quantities are a function of , where t is time, R is the distance from the point P to the center of line element, and c is the speed of light. $[Q]$ is related to $[I]$ by the following equation: $\frac{\mathrm{d}[Q]}{\mathrm{d}t} = [I]$. Based on the line orientation geometry as shown in Figure 3.14, $z_j = z$, $z_1 = 0$, and $z_2 = L$. The radiated electric field is given by

$$\begin{aligned}\mathbf{E}(r,z,t) &= -\mathbf{u}_r\left\{\frac{3r}{8\pi}\sqrt{\frac{\mu_0}{\varepsilon_0}}\left[\frac{1}{r^2+(z-L)^2}-\frac{1}{r^2+z^2}\right][I]\right.\\ &\quad\left.+\frac{\mu_0 r}{4\pi}\left[\frac{1}{\sqrt{r^2+(z-L)^2}}-\frac{1}{\sqrt{r^2+z^2}}\right]\frac{\mathrm{d}I}{\mathrm{d}t}\right\}\\ &\quad-\mathbf{u}_z\left\{\begin{array}{l}\frac{1}{8\pi}\sqrt{\frac{\mu_0}{\varepsilon_0}}\left[\frac{3(z-L)}{r^2+(z-L)^2}-\frac{3z}{r^2+z^2}-\frac{1}{r}\left(\tan^{-1}\left(\frac{z-L}{r}\right)\right.\right.\\ \quad\left.\left.-\tan^{-1}\left(\frac{z}{r}\right)\right)\right][I]+\left(\frac{z}{r}\right)\Big]\\ \frac{\mu_0}{4\pi}\left[\frac{z-L}{\sqrt{r^2+(z-L)^2}}+\frac{z}{\sqrt{r^2+z^2}}\right]\frac{\mathrm{d}I}{\mathrm{d}t}\end{array}\right\}\end{aligned} \tag{3.72}$$

We can write Equation (3.72) in polar coordinates. From Figure (3.15), the equation of the electric field becomes

$$\begin{aligned}\mathbf{E}(R,\theta,t) &= -\mathbf{u}_r\left\{\begin{array}{l}\frac{3R\cos\theta}{8\pi}\sqrt{\frac{\mu_0}{\varepsilon_0}}\left[\frac{1}{R^2+(L/2)^2-RL\sin\theta}-\frac{1}{R^2+(L/2)^2+RL\sin\theta}\right][I]\\ +\frac{\mu_0 R\cos\theta}{4\pi}\left[\frac{1}{\sqrt{R^2+(L/2)^2-RL\sin\theta}}-\frac{1}{\sqrt{R^2+(L/2)^2+RL\sin\theta}}\right]\frac{\mathrm{d}I}{\mathrm{d}t}\end{array}\right\}\end{aligned}$$

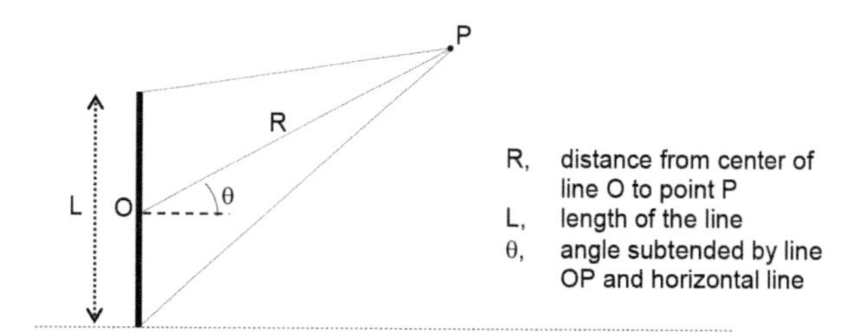

Figure 3.15 Finite line element in polar coordinates.

$$-\mathbf{u}_z \left\{ \begin{array}{l} \frac{1}{8\pi}\sqrt{\frac{\mu_0}{\varepsilon_0}} \left[\begin{array}{l} \frac{3(R\sin\theta - L/2)}{R^2+(L/2)^2-RL\sin\theta} - \frac{3(R\sin\theta+L/2)}{R^2+(L/2)^2+RL\sin\theta} \\ +\frac{1}{R\cos\theta}\left(\tan^{-1}\left(\frac{R\sin\theta-L/2}{R\cos\theta}\right) - \tan^{-1}\left(\frac{R\sin\theta+L/2}{R\cos\theta}\right)\right) \end{array} \right] [I] \\ +\frac{\mu_0}{4\pi}\left[\frac{R\sin\theta-L/2}{\sqrt{R^2+(L/2)^2-RL\sin\theta}} + \frac{R\sin\theta+L/2}{\sqrt{R^2+(L/2)^2+RL\sin\theta}}\right]\frac{dI}{dt} \end{array} \right\} \tag{3.73}$$

From Equation (3.73), it is observed that the electric field of a finite length line element is composed of two components: the $\mathbf{u}_r$ component and the $\mathbf{u}_z$ component. The $\mathbf{u}_r$ component exists both in far-field and near-field regions, while the $\mathbf{u}_z$ component exists only in the near-field region. In this section, the following simulations are carried out to study the variation and structure of the electric field with distance:

1) variation of near-field $\mathbf{E}_z$ component with distance R from the center of the line element;
2) variation of far-field $\mathbf{E}_r$ component with distance R from the center of the line element;
3) variation of total electric field $\mathbf{E} = \sqrt{(\mathbf{E}_r^2 + \mathbf{E}_\theta^2)}$ with distance R from the center of the line element.

Some general observations are as follows:

1. The intensity of the electric field at fixed distance R varies with angular position along the locus of R. Moreover, the pattern for both the far-field and near-field components is different. For the near-field component, the maximum intensity occurs at each quadrant of the locus circle as shown in Figure 3.29. For the far-field component, only two regions of maximum intensity are observed along the perpendicular bisector of the finite length line element.
2. The effect of the E_z component is dominant when the distance R is in the near-field region. The contribution of the E_r component is still apparent in the near-field region. The E_z component broadens the lobes of the

intensity pattern of the E_r component. However, at larger distance R, the contribution of E_z is negligible, and the intensity pattern approximates that of the E_r component.

3. The radiated waves from the finite line element cannot be approximated to a plane wave at small values of R because of the variation of the electric field with angle. However, at larger distances, the propagation of waves can be approximated to plane waves, provided the region covered by the radiation is small.

Acknowledgment

For full discussion of Section 3.9, see Hoole and Hoole (1996) or Hoole (2001).

4

Antenna Beamforming: Basics

P.R.P. Hoole

Abstract

This chapter introduces the essential basics of antenna beamforming. Offline beamforming techniques are presented through the Fourier transform method and the superior Woodward–Lawson sampling method. Both are illustrated using hand calculations to show how from changing the currents in each element of an array antenna relative to other elements, a specified antenna beam or radiation pattern may be obtained. In the final section of the chapter, the adaptive array is presented using the two-element array first, where the weight of the antenna array relative to each other is got using the least mean square (LMS) method. In the final section is described the derivation of the weights for an N-element array antenna using the LMS method.

4.1 Introduction

The following issues are addressed in this chapter:

1) Given a wire antenna, how may the current waveform imposed on the antenna be structured to obtain a desired radiation pattern? In this case, the radiation pattern of the antenna is given and the current on the wire antenna is to be determined. This is termed the inverse problem in some literature. In practically implementing such an antenna, the continuous current waveform obtained may be made up of a number of small antenna elements placed in a straight line, with each element having different current magnitudes to produce the required current waveform.

2) Given an array antenna, how may the currents supplied to each element of the array need to be controlled to obtain a given radiation pattern or array factor? In general, the phase difference and distance between each adjacent element will be fixed; i.e. δ and d will be fixed. The problem is to obtain the current I in each element.
3) Adaptive antennas: how may the above two design solutions be automated such that the beam or radiation pattern of the antenna may be automatically varied to, for instance, receive the maximum power from a mobile transmitter? In this case, the magnitudes of current I and the electronic phase angle δ must be automatically controlled to steer the main beam to be always pointing at the transmitter, and for the nulls to be created in directions from which there are interfering signals (e.g. jammers in military applications) or noise signals are arriving at the antenna elements.

We consider each one of these issues in turn, and in later chapters of this book, we illustrate the implementation of a technique (a) to estimate the state (e.g. position and direction of movement) of a transmitter (Chapter 7), and (b) which achieves beamforming to keep track of a cluster of mobile stations in land or airborne wireless communications (Chapter 8). Section 4.2 of this chapter has been based on Balanis (1997), and Section 4.3 on Compton (1988).

4.2 Antenna Synthesis

While the focus of the first three chapters of this book has been on antenna analysis and forward design, this chapter considers the inverse solution for antennas or antenna synthesis. The analysis problem is one of determining the radiation pattern and impedance of a given antenna structure. Antenna design is the determination of the hardware characteristics (e.g. lengths, antenna geometry, currents, etc.) of a specific antenna to produce a desired radiation pattern and/or gain. Antenna synthesis is similar to antenna design and, in fact, the terms are frequently used interchangeably. However, antenna synthesis, in its broadest sense, is an *inverse problem*. In antenna synthesis or beamforming, we first specify the required radiation pattern and then use a systematic method or combination of methods to arrive at an antenna configuration and also its geometrical dimensions and current excitation distribution, which produce a pattern that acceptably approximates the prescribed radiation pattern.

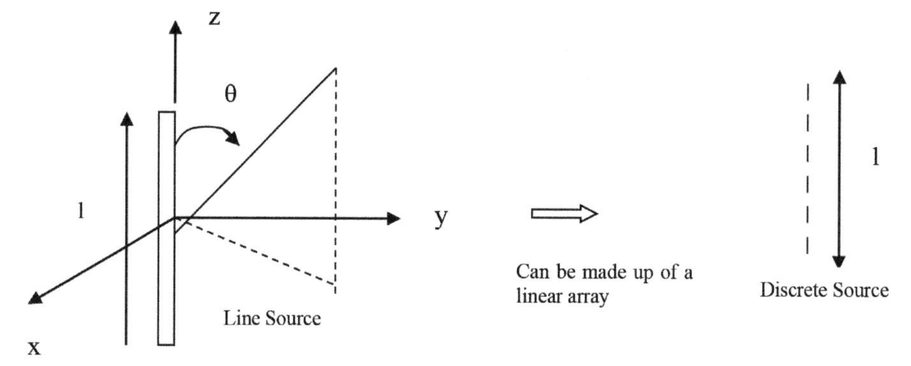

Figure 4.1 Antenna made of a continuous line source or an array of discrete sources.

4.2.1 Line Source

An antenna like a wire antenna is a continuous source in that the current flowing along the conductor is continuous along the wire. Such a continuous source may represent a true continuous source antenna like the half-wave dipole antenna or approximate a discrete source antenna like an array antenna.

Consider the line source of length l placed along the z-axis as shown in Figure 4.1. This may represent an array of discrete sources that have an interelement distance much smaller than the total size or length of the array. The array factor of such a continuous distribution of currents in discrete elements is sometimes called the *space factor*. The space factor (AF_S) of the line source shown in Figure 4.1 is given by

$$\mathrm{AF_S}\left(\theta\right)=\int_{-l/2}^{+l/2} I_n(z')\mathrm{e}^{\mathrm{j}(kz'\ \cos\ \phi+\delta_n(z'))}\,\mathrm{d}z', \tag{4.1}$$

where $I_n(z')$ and $\delta_n(z')$ are the current amplitude and phase distributions along the antenna.

For a constant phase distribution, we have $\delta_n(z') = 0$. The kz' $\cos\phi$ term comes into the equation from the array antenna factor $kd\cos\phi$ (see Equation (3.15) of Chapter 3), where we have divided the continuous source antenna of Figure 4.1 into an array of discrete elements. In general, allowing for the current pulse to travel along the antenna (traveling wave antenna), $\delta_n(z') = k_z z'$; therefore, the current along the antenna is given by

$$I(z') = I\exp(-k_z z'). \tag{4.2}$$

The current travels along the antenna at a phase velocity of ω/k_z, where k_z is the current phase constant of the source. The radiation field of such a traveling wave antenna is obtained by dividing the antenna into little dipole segments and then summing up (integrating) the radiation fields from each small dipole segment. The radiation electric field of a traveling wave antenna of length L is given by

$$\begin{aligned}\mathbf{E}_\theta =& \mathbf{u}_\theta \mathrm{j}kI\eta Le^{-\mathrm{j}kr}((\sin\theta)/4\pi r)(\sin((kL/2)(\cos\theta - k_z/k)))/ \\ &((kL/2)(\cos\theta - k_z/k)).\end{aligned} \tag{4.3}$$

Interestingly, the radiation pattern of the traveling wave antenna is always zero along the line (i.e. z-axis, $\theta = 0$) and symmetrical about the axis of the antenna (z-axis). In our discussion of antenna synthesis, we shall mostly assume that the current does not propagate along the antenna, i.e. $\delta_n(z') = 0$, and, hence, it is a standing wave current pattern that we see on the antenna. However, for traveling wave antenna synthesis, the $\delta_n(z') = k_z z'$ phase delay should be included.

4.2.2 Fourier Transform Method

Given a complete description of the required pattern, this method can be used to determine the excitation of a continuous or a discrete source antenna system which yields, either exactly or approximately, the required antenna pattern. This method is commonly referred to as *beam shaping*.

4.2.2.1 Line Source

Using Equation (4.2), we may rewrite Equation (4.1) for the normalized space factor as follows:

$$\begin{aligned}\mathrm{AF_S}(\theta) &= \int_{-l/2}^{+l/2} I(z')\mathrm{e}^{\mathrm{j}(k\cos\phi - k_z)z'}\,\mathrm{d}z' \\ &= \int_{-l/2}^{+l/2} I(z')\mathrm{e}^{\mathrm{j}\beta z'}\,\mathrm{d}z',\end{aligned} \tag{4.4}$$

where

$$\beta = k\cos\phi - k_z \text{ or } \phi = \cos^{-1}\left(\frac{\beta + k_z}{k}\right). \tag{4.5}$$

For uniform current distribution along the line, we have $I(z') = I_0/l$, and Equation (4.4) reduces to

$$\mathrm{AF_S}\,(\phi) = I_0 \frac{\sin\left[(kl/2)(\cos\,\phi - k_z/k)\right]}{(kl/2)(\cos\,\phi - k_z/k)}. \tag{4.6}$$

For observation angle ϕ to have real values, the following conditions must be satisfied:

$$-\,(k + k_z) \leq \beta \leq (k - k_z). \tag{4.7}$$

The antenna is of finite length l, and beyond this length, the current distribution $I(z')$ is obviously zero. Thus, the limits of the integral in Equation (4.4) may be extended to infinity without losing the accuracy of our formulation:

$$\mathrm{AF_S''}\,(\phi) = \mathrm{AF_S}\,(\beta) = \int_{-\infty}^{+\infty} I(z')\mathrm{e}^{\mathrm{j}\beta z'}\,\mathrm{d}z'. \tag{4.8}$$

Equation (4.8) is recognized as a Fourier transform in the spatial domain (SDFT). The corresponding inverse Fourier transform of Equation (4.8) is

$$\begin{aligned} I(z') &= \frac{1}{2\pi}\int_{-\infty}^{+\infty} \mathrm{AF_S}(\beta)\mathrm{e}^{-\mathrm{j}z'\beta}\,\mathrm{d}\beta \\ &= \frac{1}{2\pi}\int_{-\infty}^{+\infty} \mathrm{AF_S}(\phi)\mathrm{e}^{-\mathrm{j}z'\beta}\,\mathrm{d}\beta. \end{aligned} \tag{4.9}$$

Equation (4.9) is the key equation for a synthesis procedure. We note that Equation (4.9) indicates that if SF(θ) represents the required pattern, the excitation distribution $I(z')$ that will yield the exact required pattern must, in general, exist for all values of z'. With a finite length antenna, we get an approximate $I_\mathrm{a}(z')$ from

$$I_a(z') = \begin{cases} \frac{1}{2\pi}\int \mathrm{AF_S}(\beta)\mathrm{e}^{-\mathrm{j}z'\beta}\,\mathrm{d}\beta, & -l/2 \leq z' \leq l/2, \\ 0 & \text{elsewhere.} \end{cases} \tag{4.10}$$

From Equation (4.10), a very direct, but only approximate, solution for $I_a(z')$ can be obtained by using the truncated excitation distribution. Once we get $I_a(z')$ to double check how accurate our synthesis has been, we may obtain the approximate space factor resulting from this approximate solution from the following integration:

$$\text{Approx}[\text{AF}_\text{S}(\phi)] \approx \int_{-l/2}^{l/2} I_\text{a}(z')\text{e}^{\text{j}\beta z'}\,\text{d}z'. \tag{4.11}$$

The synthesized approximate pattern Approx[AF$_\text{S}(\theta)$] yields the least mean square (LMS) error from the specified or desired pattern AF$_\text{S}(\theta)$ over all values of β. When the values of β are restricted only in the visible region, however, the synthesized pattern will be further distorted.

Example 1. For a desired H-plane radiation pattern which is symmetrical about $\phi = \pi/2$, determine the current distribution and the approximate radiation pattern of a line source placed along the z-axis. The desired space factor is given by

$$\text{AF}_\text{S}(\phi) = \begin{cases} 1, & \pi/3 \le \phi \le 2\pi/3, \\ 0 & \text{elsewhere.} \end{cases} \tag{4.12}$$

This is a sectoral pattern and such patterns are popular for search applications where vehicles are located by establishing communications or by a radar echo in the sector of space occupied by the main beam of the antenna pattern.

Since the pattern is symmetrical, $k_z = 0$. Using Equations (4.5) and (4.7), the values of β are given by $-k/2 \le \beta \le k/2$ and the current distribution can be determined by Equation (4.9)

$$\begin{aligned} I(z') &= \frac{1}{2\pi}\int_{-\infty}^{+\infty} \text{AF}_\text{S}(\beta)\text{e}^{-\text{j}z'\beta}\,\text{d}\beta \\ &= \frac{1}{2\pi}\int_{-k/2}^{k/2} \text{e}^{-\text{j}z'\beta}\,\text{d}\beta = \frac{k}{2\pi}\left[\frac{\sin(kz'/2)}{(kz'/2)}\right]. \end{aligned} \tag{4.13}$$

Although we have solved for a fictitious source that exists in the limits ?$\infty \le z' \le \infty$, a realistic approximation of the current distribution over the finite length of the line source may be written as

$$I_a(z') \approx I(z'), \qquad -l/2 \le z' \le l/2. \tag{4.14}$$

By limiting the length of the line to l, we will only get an approximate array factor AF_S, which will be somewhat different to the specified or desired radiation pattern; the longer the length of the line, the closer the approximate pattern will be to the desired pattern. The approximate pattern obtained from Equation (4.11) using the above truncated current distribution of Equation (4.14) is given by

$$\mathrm{Approx}[\mathrm{AF_S}(\phi)_\mathrm{a}] = \int_{-l/2}^{l/2} I_\mathrm{a}(z')\mathrm{e}^{\mathrm{j}\beta z'}\,\mathrm{d}z'$$

$$= \frac{1}{\pi}\left\{\int_0^{X_1} \frac{\sin\, x}{x}\,\mathrm{d}x - \int_0^{X_2} \frac{\sin\, x}{x}\,\mathrm{d}x\right\}, \tag{4.15}$$

where

$$X_1 = \frac{l}{\lambda}\pi\left(\cos\,\phi + \tfrac{1}{2}\right), \tag{4.16}$$

$$X_2 = \frac{l}{\lambda}\pi\left(\cos\,\phi - \tfrac{1}{2}\right). \tag{4.17}$$

The approximate current distribution of Equation (4.13) is plotted in Figure 4.2(a). This synthesized sector pattern of Equation (4.15) is plotted in Figure 4.2(b) for $l = 5\lambda$. The pattern is plotted in linear form, and in decibels, to emphasize the details of the main beam. Observe the oscillations about the desired pattern on the main beam, called ripple, and the non-zero sidelobes. This appearance of main beam ripple and finite sidelobes is typical of any synthesized pattern.

4.2.2.2 Linear Array

The array factor resulting from an array of identical discrete radiators (elements) is, of course, the sum over the currents for each element weighted by the spatial phase delay from each element to the far-field point. The basic theory of an N-element array was described in Section 3.3. Consider now a linear N-element array antenna, with phase angles $-M\delta$, $-(M-1)\delta, \ldots, -\delta$, 0, $\delta, \ldots, (M-1)\delta$, $M\delta$. The currents in the elements are not equal and are given by $a_{-M}, \ldots, a_{-1}, a_0, a_1, \ldots, a_M$. Since the reference point is taken at the physical center of the array, the array factor for an odd number of elements

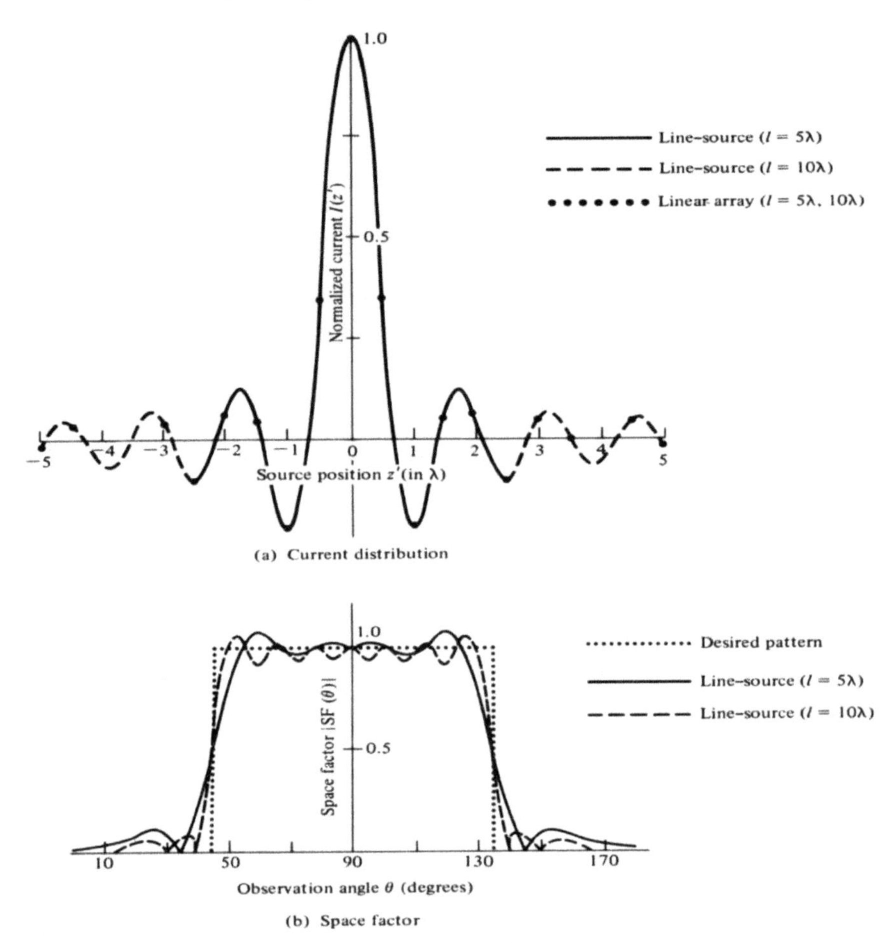

Figure 4.2 Normalized current distribution, desired pattern, and synthesized patterns using the Fourier transform method. (a) Current distribution. (b) Space factor. (Used with permission from Balanis (1997).)

(N = 2M + 1) can be written as (see Section 3.3)

$$\mathrm{AF}(\phi) = \mathrm{AF}(\psi) = \sum_{m=-M}^{M} a_m \mathrm{e}^{\mathrm{j}m\psi}, \tag{4.18}$$

where, considering only the H-plane ($\theta = \pi/2$), we have

$$\psi = kd\cos\phi + \delta. \tag{4.19}$$

Note that ϕ is the angle measured away from the axis on which the array is placed; i.e. it is the angle measured from the x-axis if the array is placed along the x-axis or it is the angle measured from the z-axis if the array is placed along the z-axis, as in the present case. The elements are placed at positions along the z-axis at

$$z_m^{'} = md, m = 0, \pm 1, \pm 2, \ldots, \pm M. \tag{4.20}$$

In general, the array factor of an antenna is a periodic function of ψ, and it must repeat for every 2π radians. To satisfy periodicity requirements for real values of ϕ, we have $2kd = 2\pi$ or $d = \lambda/2$. The excitation coefficients can be determined by the Fourier formula:

$$\begin{aligned} a_m &= \frac{1}{T} \int_{-T/2}^{T/2} \mathrm{AF}(\psi) \mathrm{e}^{-\mathrm{j}m\psi}\, \mathrm{d}\psi \\ &= \frac{1}{2\pi} \int_{-\pi}^{+\pi} \mathrm{AF}(\psi) \mathrm{e}^{-\mathrm{j}m\psi}\, \mathrm{d}\psi, \quad -M \leq m \leq M. \end{aligned} \tag{4.21}$$

The Fourier series synthesis procedure is, then, to use the excitation coefficients a_m calculated from the desired pattern AF(ϕ) to determine the approximate array factor. This Fourier series synthesized pattern provides the least mean squared error over the region

$$\cos^{-1}(-2d/\lambda) \leq \phi \leq \cos^{-1}(2d/\lambda).$$

Example 2. We shall repeat the problem specified in Example 1 using an array antenna to synthesize the radiation pattern. Thus, we need to determine the excitation for a broadside array whose array factor closely approximates the desired pattern. With an interelement spacing of $d = \lambda/2$, the design may be done for 11 elements and repeated for 5 elements. It is obvious that the 21-element array will give us a better radiation pattern than the 5-element array, but the cost of the antenna will increase with the number of elements.

For a broadside array, the progressive phase shift between the elements (δ) is zero. Since the pattern is non-zero only for $\pi/4 \leq \phi \leq 3\pi/4$, the corresponding values of ψ are obtained from Equation (4.19) or

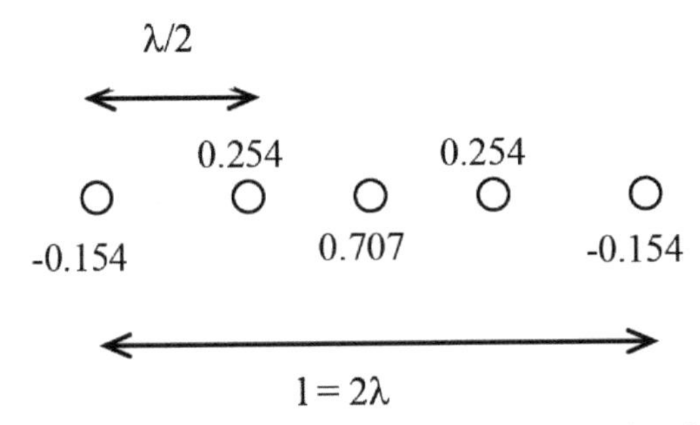

Figure 4.3 Synthesis using a five-element array. Relative amplitudes of currents are shown.

$-\pi/2 \leq \psi \leq \pi/2$. The excitation coefficients are obtained from Equation (4.21) or

$$a_m = \frac{1}{2\pi} \int_{-\pi/2}^{\pi/2} \mathrm{e}^{-jm\psi}\, \mathrm{d}\psi = \frac{1}{2} \left[\frac{\sin(m\pi/2)}{m\pi/2} \right], \tag{4.22}$$

and the excitation coefficients are symmetrical about the physical center (at $z = 0$) of the array (i.e. $a_m(-z_m') = a_m(z_m')$). The excitation coefficients, remember, indicate the relative strength of the currents that must flow in each element of the array antenna to obtain the desired radiation pattern. In general, the synthesized currents or current distribution will be real and symmetric if the desired pattern is real and symmetric, i.e. if AF($-\phi$) = AF(ϕ); in turn, the synthesized pattern will be real and symmetric.

For $N = 21$, $d = \lambda/2$, and $l = 10\lambda$, the maximum values of the excitation coefficients (i.e. leaving out the 2^{-1} factor) are as follows:

$$a_0 = 1.0,\ a_1 = 0.3582,\ a_2 = -0.217,\ a_3 = 0.0558, \ldots, a_{10} = -0.0100.$$

Note that, at the element positions, the line source and linear array excitation values are identical since the two antennas are of the same length (for $N = 11$, $d = \lambda/2 \Rightarrow l = 5\lambda$).

For $N = 5$, $d = \lambda/2$, and $\psi = kd \cos\phi + \delta = \pi \cos\phi$,

$$\mathrm{AF}(\psi) = \begin{cases} 1, & -0.5\pi < \psi < 0.5\pi, \\ 0, & -\pi < \psi < 0.5\pi,\ 0.5\pi < \psi < \pi. \end{cases} \tag{4.23}$$

The excitation currents should have the following ratio to achieve this array pattern: –0.217:0.3582:1.0:0.3582:–0.217. The root mean square quantities are as shown in Figure 4.3.

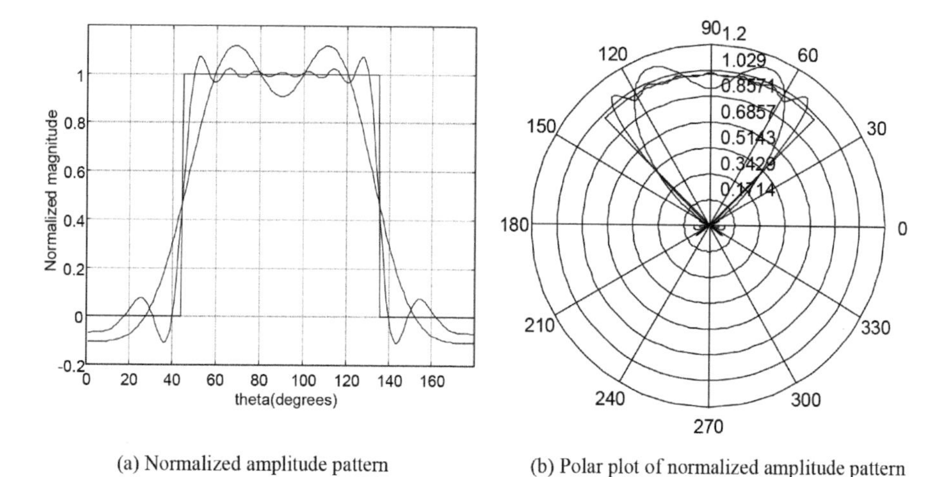

(a) Normalized amplitude pattern (b) Polar plot of normalized amplitude pattern

Figure 4.4 Normalized desired pattern and synthesized pattern for N = 5 and 21 using the Fourier transform method. (a) Normalized amplitude pattern. (b) Polar plot of normalized amplitude pattern.

The approximate array factor that can be determined by Equation (4.18) will be

$$\mathrm{AF_a}(\psi) = 0.707 + \frac{2}{\pi}\sum_{n=1}^{M}\left[\frac{1}{n}\sin\left(0.707n\pi\right) \times \cos\left(n\psi\right)\right], \tag{4.24}$$

where a_0 = 0.707 and a_n = $1/n\pi$ sin($0.707n\pi$).

The normalized array factors are displayed in Figure 4.4. As expected, the larger array (N = 21; d = $\lambda/2$) provides a better reconstruction of the desired pattern.

4.2.3 Woodward–Lawson Sampling Method

A particularly convenient way to synthesize a radiation pattern is to specify values of the pattern at various points, that is, to sample the pattern. Associated with each pattern sample is a harmonic current of uniform amplitude distribution and uniform progressive phase, whose corresponding field is referred to as a *composing function.* For a line source, the composition function is given by

$$\mathrm{CF} = a_m \sin(\psi_m)/\psi_m, \tag{4.25}$$

and for a linear array, the composition function is given by

$$\mathrm{CF} = a_m \sin(N\delta_m)/N\sin(\delta_m). \tag{4.26}$$

The excitation coefficient a_m of each harmonic current is such that the field strength is equal to the amplitude of the desired pattern at its sampled point. The total excitation of the source comprises a finite summation of space harmonics. The corresponding synthesized pattern is represented by a finite summation of composing functions, with each term representing the field of a current harmonic with uniform amplitude distribution and uniform progressive phase.

4.2.3.1 Line Source

Consider a continuous current source of length l placed along the z'-axis with its center at $z' = 0$. We shall decompose this source into a set of small normalized sources. Each discrete source is assumed to have uniform amplitude and linear phase of the form

$$I_m(z') = \frac{a_m}{l} \mathrm{e}^{-\mathrm{j}kz' \cos \phi_m}, \quad -l/2 \le z' \le l/2. \tag{4.27}$$

The angle ϕ_m represents the angles where the radiation pattern is sampled. Assuming an odd number of discrete sources, let the total current $I(z')$ be made of a sum of $2M + 1$ current elements. Each current element has the form indicated in Equation (4.27). Mathematically, therefore, the total current may be written as

$$I(z') = \frac{1}{l} \sum_{m=-M}^{M} a_m \, \mathrm{e}^{-\mathrm{j}kz' \cos \phi_m}, \tag{4.28}$$

where $m = 0, \pm 1, \pm 2, \ldots, \pm M$.

The array factor of each current element of Equation (4.28) is given by Equation (4.6) and may be written as

$$\mathrm{AF}_{\mathrm{S}m}(\phi) = a_m \left\{ \frac{\sin\left[(kl/2)\left(\cos \phi - \cos \phi_m\right)\right]}{(kl/2)\left(\cos \phi - \cos \phi_m\right)} \right\}. \tag{4.29}$$

The maximum value of Equation (4.29) occurs when $\theta = \theta_m$. Therefore, the total radiation pattern of the decomposed current source is obtained by summing the $2M + 1$ radiation pattern in terms of each form given by Equation (4.29). Hence, the approximate space factor of the continuous line source is given by

$$\text{Approx}[\text{AF}_{\text{Sa}}(\phi)] = \sum_{m=-M}^{M} a_m \left\{ \frac{\sin\left[(kl/2)(\cos\,\phi - \cos\,\phi_m)\right]}{(kl/2)(\cos\,\phi - \cos\,\phi_m)} \right\}. \tag{4.30}$$

While determining the values of each term on the right-hand side of Equation (4.30), when one term becomes maximum at $\phi = \phi_m$, all other terms at the other sample points are zero at $\phi = \phi_m$. This means that at each sample point *m*, the resultant field is equal to the single term determined at $\phi = \phi_m$. Therefore, the coefficient at sample point *m* is given by

$$a_m = \text{SF}(\phi = \phi_m). \tag{4.31}$$

Thus, by determining a_m at each sample point and assigning that as the value of the resultant electric field at that point, we may construct the synthesized radiation pattern. This synthesized pattern will be expected to closely resemble the desired or prescribed radiation pattern.

It should be remembered that the sample points should not be arbitrarily chosen, but the selection of the ϕ values must be such that the periodicity requirements of 2π for real values of ϕ are satisfied. Hence, the sample points must be chosen such that $kz'\ s\big|_{|z'|=l} = 2\pi$, and, hence, the step size is given by

$$s = \lambda/l. \tag{4.32}$$

By ensuring that the sample points are separated by step size *s*, we get electric field values at a sample point *m* being determined by the single coefficient a_m only. Therefore, the electric field values at each point *s* are uncorrelated with each other when this condition is satisfied. Hence, the angular location ϕ_m of each sample point must be chosen such that

$$\cos\phi_m = m \cdot s = m(\lambda/l), \tag{4.33}$$

$$\text{i.e. } \phi_m = \cos^{-1}(m\lambda/l), \tag{4.34}$$

with the maximum number of sample points *M* satisfying the condition $M \leq l/\lambda$.

Example 3. Using the Woodward–Lawson synthesis method for an antenna of length $l = 5\lambda$, determine the array coefficients a_m for the problem specified in Example 1. With $l = 5\lambda$ and $M = 5$, the separation of sample points should be $s = 0.2$. Hence, the total number of sampling points for odd-numbered

Table 4.1 Angles and excitation coefficients of the sample points.

M	ϕ_m	a_m = SF(θ_m)	m	ϕ_m	a_m = SF(ϕ_m)
0	90°	1			
1	78.46°	1	?1	101.54°	1
2	66.42°	1	?2	113.58°	1
3	53.13°	0	?3	126.87°	0
4	36.87°	0	?4	146.13°	0
5	0°	0	?5	180°	0

sample points is (M/s) + 1 = 11. The angles where the sampling is performed are given by Equation (4.34)

$$\phi_m = \cos^{-1}\left(\frac{m\lambda}{l}\right) = \cos^{-1}(0.2m), \; m = 0, \pm1, \ldots, \pm5. \tag{4.35}$$

The angles and excitation coefficients of the sample points are given in Table 4.1.

4.2.3.2 Linear Array

The Woodward–Lawson synthesis for a linear array follows the same line of reasoning as that given in Section 4.2.3.1 for a continuous line source. Consider an array of length l = (N?1)d. Each normalized electric field sample may be written as

$$E_m(\phi) = a_m \frac{\sin\left[((N-1)/2)kd\,(\cos\,\phi - \cos\,\phi_m)\right]}{(N-1)\sin\left[\frac{1}{2}kd\,(\cos\,\phi - \cos\,\phi_m)\right]}. \tag{4.36}$$

The total array factor of the entire array of individual antenna elements will be given by the summation of terms represented by Equation (4.36). Hence, the resultant approximate array factor for the array antenna is given by

$$\text{Approx}[\text{AF}(\phi)] = \sum_{m=-M}^{M} a_m \frac{\sin\left[((N-1)/2)kd\,(\cos\,\phi - \cos\,\phi_m)\right]}{(N-1)\sin\left[\frac{1}{2}kd\,(\cos\,\phi - \cos\,\phi_m)\right]}, \tag{4.37}$$

where the coefficients a_m are given by

$$a_m = \text{AF}(\phi = \phi_m). \tag{4.38}$$

In order to ensure that each sample is uncorrelated to neighboring samples, we must sample at sample angles ϕ_m defined by

$$\cos\,\phi_m = m\frac{\lambda}{l} = \frac{m\lambda}{(N-1)d} \quad \Rightarrow \quad \phi_m = \cos^{-1}\left[\frac{m\lambda}{(N-1)d}\right]. \tag{4.39}$$

Therefore, at each element of the odd-numbered or even-numbered array, we must have the following normalized coefficient for currents (or voltages) in order to get the desired (or specified) radiation pattern:

$$a_n(z') = \frac{1}{N}\sum a_m \exp(-jkz_n' \cos \phi_m). \tag{4.40}$$

In Equation (4.40), the z-axis distance z_n' is defined as the distance of the nth element measured from the geometrical center of the array. The center element of an odd-numbered array is placed at the geometrical center of the array.

Program 4.1 contained on disk gives a listing of the computer programs in MATLAB^{TM} for antenna synthesis.

4.3 Adaptive Arrays

In the 1960s, an adaptive antenna was thought of as a self-cohering system which automatically trained its main lobe toward a desired signal of unknown direction. Such systems are easily decoyed by jammers and in any case offer no interference reduction beyond a normal beam pattern. A more recent trend was the development of null-steering systems which automatically steer pattern zeros to jammers and a lobe to the desired signal. The same technique may be used, for instance, in wireless mobile communication systems where a smart antenna constantly keeps adjusting its radiation pattern to get rid of multipath interference. Thus, the basic feature of the adaptive array filter may be described as the ability of the array antenna to change (or adapt) its beam pattern according to the electromagnetic environment surrounding the antenna. The adaptation is done to satisfy a specified optimization criterion which will alter the magnitude and phase of the signal associated with each element of the array. The adaptive algorithm changes element current magnitude and phase to obtain a desired signal from the actual signal impinging on the antenna elements. To do this, a third reference signal may be required. The desired signal may be acquired and its level maintained by a pilot scheme in which a reference signal is compared with the antenna output and the mean square difference is minimized. It maximizes the received power from a signal, of unknown direction, by altering their pattern while receiving by a feedback control process. If interfering signal (e.g. a jammer) or environmental noise are present, adaptive arrays will maximize $S/(N_\mathrm{i} + N_\mathrm{n})$. Another advantage is that the array elements need not be linear or planar.

4.3.1 LMS Adaptive Array

As shown in Figure 4.5, the quadrature hybrid splits the signal into an in-phase signal $X_I(t)$ and a quadrature signal $X_Q(t)$, also denoted here as S_{r1} and S_{r2}:

$$S_r = \frac{S_r}{\sqrt{2}} - j\frac{S_r}{\sqrt{2}} = S_{r1} + S_{r2}, \tag{4.41}$$

$$S_{r1} = \frac{S_r}{\sqrt{2}}, \qquad S_{r2} = \frac{S_r}{\sqrt{2}} \text{ retarded by } 90^\circ. \tag{4.42}$$

Hence, the received signal for each element is given by

$$V_1 = V_{r1} + V_{i1} = \frac{S_1}{\sqrt{2}}(W_{r1} - \mathrm{j}W_{i1}). \tag{4.43}$$

The output of each element is changed in amplitude and phase by multiplying it with the weights and so an antenna pattern to meet desired criteria may be formed.

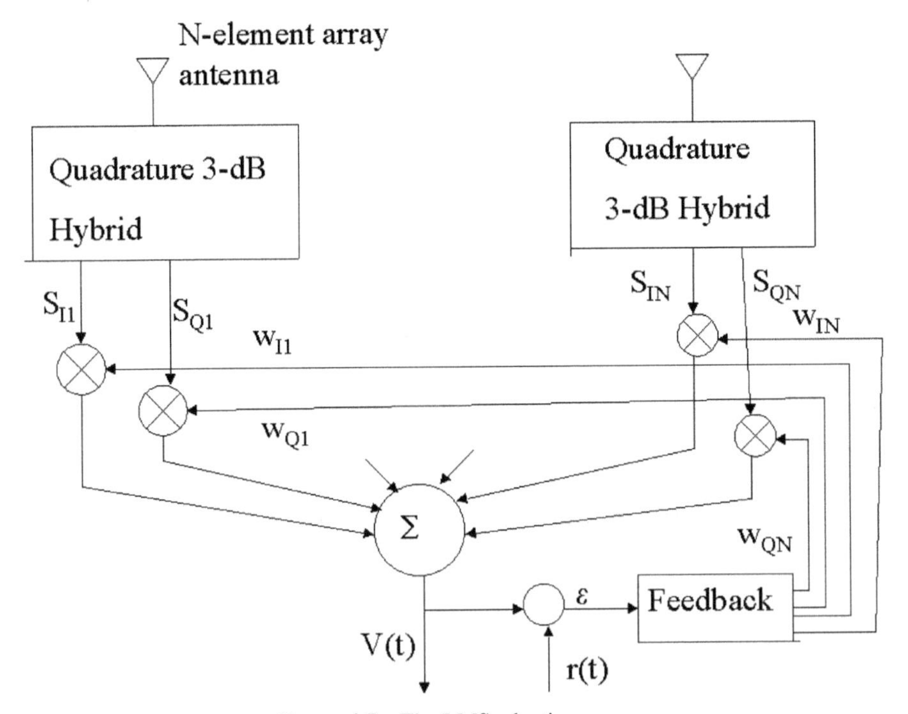

Figure 4.5 The LMS adaptive array.

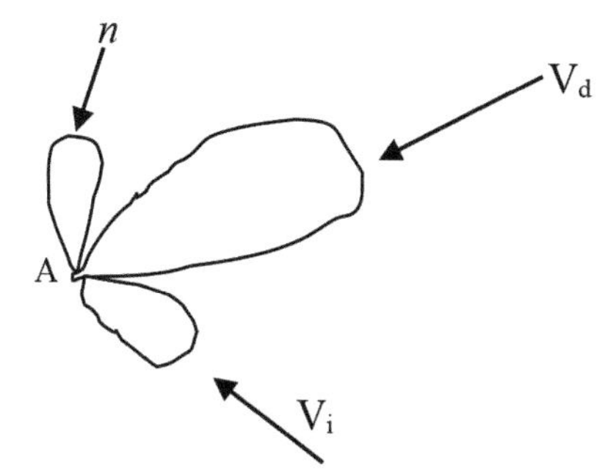

Figure 4.6 An imperfect antenna beam resulting from synthesis.

The LMS array is based on a minimum mean square error (MSE) concept. An error signal $\varepsilon(t)$ is obtained by subtracting the array output $s(t)$, also known as $V(t)$, from another signal called the reference signal $r(t)$

$$\varepsilon(t) = r(t) - V(t), \tag{4.44}$$

where

$$V(t) = V_{\mathrm{d}}(t) + n(t) + V_{\mathrm{i}}(t). \tag{4.45}$$

Here, V_{d} is the desired signal, n is the noise signal, and V_{i} is the interfering signal.

Hence, the error may be written as

$$\varepsilon(t) = r(t) - V_{\mathrm{d}}(t) - n(t) - V_{\mathrm{i}}(t). \tag{4.46}$$

The MSE is

$$E[\varepsilon^2(t)] = E\left\{[r(t) - V_{\mathrm{d}}(t)]^2\right\} + E[n^2(t)] + E[V_{\mathrm{i}}^2(t)]. \tag{4.47}$$

The desired signal, interfering signal, and the noise are all assumed to be uncorrelated zero-mean processes so that all the cross-product terms will be zero. As can be seen from Equation (4.47), $E[\varepsilon^2(t)]$ will be minimum only when both $E[n^2(t)]$ and $E[V_{\mathrm{i}}{}^2(t)]$ are minimum. And the LMS array is able to make $E[V_{\mathrm{i}}{}^2(t)]$ small by forming a pattern with nulls in the direction of V_{i} and n sources, while at the same time maximizing V_{d} by establishing the main beam in that direction. Hence, when $V_{\mathrm{d}}(t) \Rightarrow r(t)$ in amplitude and phase, then $E\{[r(t) - V_{\mathrm{d}}(t)]^2\}$ will be small (see Figure 4.6).

4.3.2 Two-Element Array

We shall first illustrate the inherent strength in using weights to adapt the antenna beam. Consider the two-element array antenna shown in Figure 4.7.

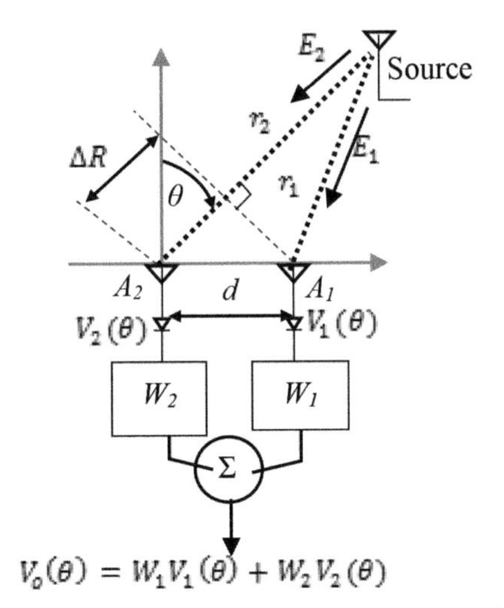

Figure 4.7 A two-element array for beamforming.

The distance between the two elements A_1 and A_2 is d (e.g. $d = \lambda/2$), and, for simplicity, we shall assume that the elements have an isotropic radiation pattern. Thus, the radiation pattern of the antenna is equal to the array factor of the antenna. Consider now a jamming or interfering signal impinging on the array antenna from a direction θ_i. We do not want the antenna to receive signals arriving from the direction θ_i. Thus, the resultant signal V_0 should be equal to zero for $\theta = \theta_\text{i}$. In Figure 4.7, we have V_i = interference or jamming signal, W_2 = complex weight of the second element A_2, and V_0 = sum of the output signals.

We shall consider the case where $W_1 = 1$ To cancel out the signal arriving from the direction θ_I, we must find the value to which the complex weight should be set to, where the complex weight $W_2 = W_0 \exp(-\text{j}\delta)$. If we let $W_0 = 1$, then the radiation beam array factor is minimized in the direction of an interferer atθ_I, that is, $V_0(\theta_\text{i}) = 0$, when

$$\begin{aligned} W_2 &= -\text{Average}\{V_1(t,\theta)V_2^*(t,\theta)\}/(\text{Modulus}\{V_2(t,\theta)\})^2 \\ &= +\exp(\text{j}kd\,\sin\theta_\text{i}). \end{aligned} \tag{4.48}$$

Such adaptive beamforming requires components that will modify the magnitude of the signal by W_0 and phase shift the received signal by δ. Although analog phase shifters like switched-line phase shifters and attenuators like the Quad PIN diode attenuator have been used over many years, in recent times, digital means of achieving this has found many applications. In digital beamforming, the digitized signal is weighted in phase and amplitude by multiplication by a complex weight W_0 exp(–jδ). Then the resultant weighted signal from each element of the array may be digitally summed. Digital beamforming has several advantages, including beamsteering at full clock rate, very accurate phase, and amplitude correction, and sufficient flexibility to minimize power requirement in wireless communication and radar systems. To achieve accurate beamforming, the array antenna may be calibrated by injecting a test signal into each element of the array and measuring the amplitude and phase of the received signal; the weights are adjusted until the correct, expected signal is obtained. The test signal may be injected through wires connected from the source to the point just behind the antenna element. An alternative method is to use another antenna in the far field to transmit a known signal to the antenna to be calibrated.

4.3.3 The LMS Weights

Going back to the general formulation for an N-element adaptive array, let us determine the optimum weights that yield minimum MSE (MSE = $E[\varepsilon^2(t)]$), where the temporal error signal is given by

$$\begin{aligned}\varepsilon(t) &= r(t) - V(t)\\ &= r(t) - \sum_{\substack{j=1\\p=r,i}}^{N} W_{pj} S_{pj}(t).\end{aligned} \tag{4.49}$$

Hence, the MSE is given by

$$\begin{aligned}E[\varepsilon^2(t)] = E[r^2(t)] - 2\sum_{\substack{j=1\\p=r,i}}^{N} W_{pj} E[r(t)S_{pj}(t)]\\ + \sum_{\substack{j=1\\p=r,i}}^{N}\sum_{\substack{k=1\\q=r,i}}^{N} W_{pj}W_{qk}E[S_{pj}(t)S_{qk}(t)].\end{aligned} \tag{4.50}$$

For convenience, we shall define some column matrices. The $2N \times 1$ weight vector

$$W(t) = [W_{r1}(t)W_{i1}(t)W_{r2}(t)W_{i2}(t)\ldots]^{\mathrm{T}}, \tag{4.51}$$

the $2N \times 1$ signal vector

$$S(t)=[S_{r1}(t)S_{i1}(t)S_{r2}(t)S_{i2}(t)\ldots]^{\mathrm{T}}, \tag{4.52}$$

and the cross-correlation vector

$$P = E[S(t)r(t)]. \tag{4.53}$$

The above result may be written more conveniently by using matrix notation

$$E\left[\varepsilon^2(t)\right] = E\left[r^2(t) - 2W^{\mathrm{T}}P + W^{\mathrm{T}}RW\right], \tag{4.54}$$

where the correlation matrix R is given by

$$R = E\left[S(t)S^{\mathrm{T}}(t)\right]$$
$$= E\begin{bmatrix} S_{r1}(t)S_{r1}(t) & S_{r1}(t)S_{i1}(t) & S_{r1}(t)S_{r2}(t) & \cdots \\ S_{i1}(t)S_{r1}(t) & S_{i1}(t)S_{i1}(t) & S_{i1}(t)S_{i2}(t) & \cdots \\ S_{r2}(t)S_{r1}(t) & S_{r2}(t)S_{i1}(t) & S_{r2}(t)S_{r2}(t) & \cdots \\ \cdots & \cdots & \cdots & \cdots \\ \cdots & \cdots & \cdots & \cdots \\ \cdots & \cdots & \cdots & \cdots \end{bmatrix}, \tag{4.55}$$

and the gradient of the MSE in the weight domain is given by

$$\nabla E[\varepsilon^2(t)] = \begin{bmatrix} \frac{\partial E[\varepsilon^2(t)}{\partial W_{r1}} \\ \frac{\partial E[\varepsilon^2(t)}{\partial W_{i1}} \\ \frac{\partial E[\varepsilon^2(t)}{\partial W_{r2}} \\ \cdots \\ \cdots \\ \cdots \end{bmatrix}. \tag{4.56}$$

Hence, the gradient of Equation (4.56) becomes

$$\nabla E[\varepsilon^2(t)] = -2\nabla(W^{\mathrm{T}}P) + \nabla(W^{\mathrm{T}}RW). \tag{4.57}$$

Now the first term on the right-hand side of Equation (4.57) is

$$\nabla(W^{\mathrm{T}}P) = \nabla\left[\begin{matrix} [W_{r1}\ W_{i1}\ W_{r2}\ldots] \end{matrix}\right]\begin{bmatrix} P_{r1} \\ P_{i1} \\ P_{r2} \\ \cdots \\ \cdots \end{bmatrix}$$
$$= \begin{bmatrix} P_{r1} \\ P_{i1} \\ P_{r2} \\ \cdots \\ \cdots \end{bmatrix} = P. \tag{4.58}$$

In order to simplify the matrix relation, consider the following vector:

$$C = A^{\mathrm{T}}BA = \sum_{j=1}^{N}\sum_{k=1}^{N} a_j a_k b_{jk} \tag{4.59}$$

$$\begin{aligned} &= a_1a_1b_{11} + a_1a_2b_{12} + \cdots + a_1a_Nb_{1N} \\ &\quad + a_2a_1b_{21} + a_2a_2b_{22} + \cdots + a_2a_Nb_{2N} \\ &\quad \ldots\ldots\ldots\ldots\ldots\ldots\ldots\ldots\ldots\ldots \\ &\quad \ldots\ldots\ldots\ldots\ldots\ldots\ldots\ldots\ldots\ldots \\ &\quad + a_Na_1b_{N1} + a_Na_2b_{N2} + \cdots + a_Na_Nb_{NN}. \end{aligned} \tag{4.60}$$

Then the first differentiation of C is

$$\begin{aligned} \frac{\partial C}{\partial x_1} &= 2a_1b_{11} + a_2(b_{12} + b_{21}) + \cdots + a_N(b_{1N} + b_{N1}) \\ &= 2\sum_{j=1}^{N} b_{ij}a_j \text{ if } B \text{ is symmetric.} \end{aligned}$$

Therefore,

$$\nabla C = \begin{bmatrix} \frac{\partial C}{\partial a_1} & \frac{\partial C}{\partial a_2} & \cdots \end{bmatrix} = 2\begin{bmatrix} \sum_1^N b_{ij}a_j & \sum_1^N b_{2j}a_j \ldots \end{bmatrix}^{\mathrm{T}}. \tag{4.61}$$

Hence, Equation (4.69) reduces to

$$\nabla E\left[\varepsilon^2(t)\right] = -2P + 2RW. \tag{4.62}$$

The weight vector yielding minimum $E[\varepsilon^2(t)]$, which we denote by W_{opt}, may be found by setting

$$\nabla E\left[\varepsilon^2(t)\right] = 0.$$

From Equation (4.74), we get the optimum weight

$$W_{\mathrm{opt}} = R^{-1}P. \tag{4.63}$$

This is also the steady state weight vector.

We shall now consider how the weights are to be initialized to get the routine working. To ensure that $\varepsilon(t)$ moves toward a minimum, we must consider the time-domain differential

$$\frac{\mathrm{d}\left(E\left[\varepsilon^2(t)\right]\right)}{\mathrm{d}t} = \sum_{\substack{j=1 \\ p=r,i}}^{N} \frac{\partial E\left[\varepsilon^2(t)\right]}{\partial W_{pj}} \cdot \frac{\mathrm{d}W_{pj}}{\mathrm{d}t}. \tag{4.64}$$

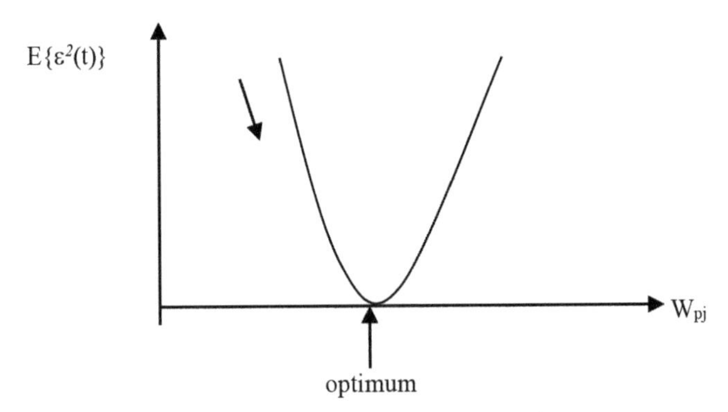

Figure 4.8 Defining the optimum weights.

As shown in Figure 4.8, the weights converge to the desired weights when the slope of the curve becomes zero.

For $E[\varepsilon^2(t)]$ to decrease with time, its derivative has to be negative, which means

$$\frac{\mathrm{d}W_{pj}}{\mathrm{d}t} = -\alpha \frac{\partial E[\varepsilon^2(t)]}{\partial W_{pj}}, \tag{4.65}$$

where α is a positive real constant. Thus,

$$\frac{\mathrm{d}W}{\mathrm{d}t} = -\alpha \nabla \left[E\left(\varepsilon^2(t)\right) \right], \tag{4.66}$$

from which we get

$$\Rightarrow \frac{\mathrm{d}\left[E\left(\varepsilon^2(t)\right) \right]}{\mathrm{d}t} = -\alpha \sum_{\substack{j=1 \\ p=r,i}}^{N} \left\{ \frac{\partial E\left[\varepsilon^2(t)\right]}{\partial W_{pj}} \right\}^2 \tag{4.67}$$

Equation (4.67) yields a value which is always negative. Hence,

$$\frac{\partial E\left[\varepsilon^2(t)\right]}{\partial W_{pj}} = -\frac{\partial}{\partial W_{pj}} \left\{ 2 \sum_{\substack{j=1 \\ p=r,i}}^{N} W_{pj} E\left[r(t)\, S_{pj(t)} \right] \right\}$$
$$+ \frac{\partial}{\partial W_{pj}} \left\{ \sum_{\substack{j=1 \\ p=r,i}}^{N} \sum_{\substack{k=1 \\ q=r,i}}^{N} W_{pj} W_{qk} E\left[S_{pj}(t)\, S_{qk}(t) \right] \right\}$$

$$= -2E\left[r(t)S_{pj}(t)\right] + 2\sum_{\substack{k=1\\q=r,i}}^{N} W_{qk}E\left[S_{pj}(t)S_{qk}(t)\right]$$

$$= -2E\left\{S_{pj}(t)\left[r(t) - \sum_{\substack{k=1\\q=r,i}}^{N} W_{qk}S_{qk}(t)\right]\right\}$$

$$= -2E\left[S_{pj}(t)\,\varepsilon(t)\right]. \tag{4.68}$$

Hence, from Equation (4.67), we get

$$\frac{dW_{pj}}{dt} = 2\alpha E\left[S_{pj}(t)\varepsilon(t)\right]. \tag{4.69}$$

Equation (4.69) in vector form yields

$$\frac{dW}{dt} = 2\alpha E\left[\varepsilon(t)S(t)\right]. \tag{4.70}$$

Since it takes time to obtain the expected value (by a time average), we can instead estimate $E\left[\varepsilon(t)S_{pj}\right]$ by $S_{pj}\varepsilon(t)$. Equation (4.70) yields

$$\frac{dW_{pj}}{dt} = 2\alpha S_{pj}\varepsilon(t), \tag{4.71}$$

which is known as the Widrow LMS algorithm.

As $\varepsilon(t)$ and S_{pj} are stochastic, W_{pj} will vary randomly about their mean values. Hence, α should be chosen to be small enough so that the weights average out the stochastic variations. The implementation of the algorithm is shown in Figure 4.9.

In the time domain, the weight is determined from

$$W_{pj}(t) = 2\alpha\int S_{pj}(t)\,\varepsilon(t)\;dt, \tag{4.72}$$

with each feedback loop forming the product $2\alpha S_{pj}(t)\varepsilon(t)$ and integrating it to give the weights. The flow chart for an adaptive antenna algorithm is given in Figure 4.10. The direction of arrival (DOA) of the received signal is assumed to be known. The estimation of the position of a mobile transmitter, and hence the DOA, is described in Chapter 7.

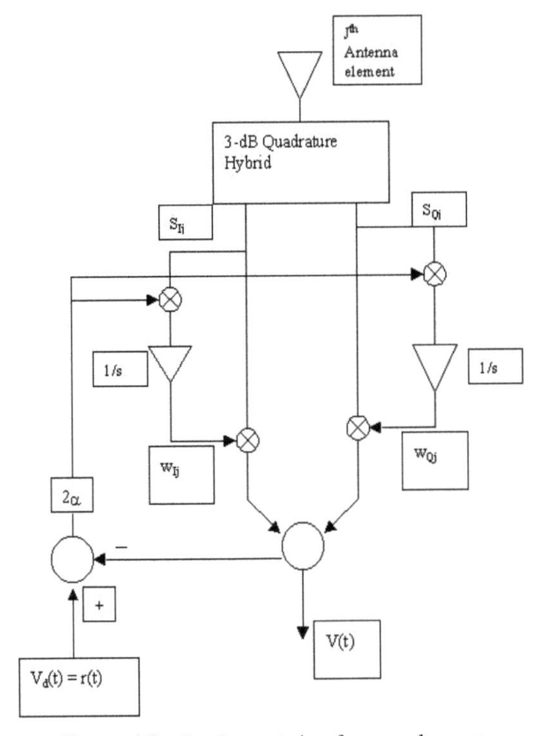

Figure 4.9 Implementation for one element.

4.3.4 Complex Signal Notation

One way of simplifying the analysis of an adaptive array is the use of analytic signal notation with complex weights. We redefine some column matrices as follows:

$$W(t) = [W_1(t)W_2(t)W_3(t)W_4(t)\ldots]^{\mathrm{T}}, \tag{4.73}$$

$$S(t) = [S_1(t)S_2(t)S_3(t)S_4(t)\ldots]^{\mathrm{T}}, \tag{4.74}$$

$$P = E[S(t)^* r(t)], \tag{4.75}$$

$$R = E[S(t)^* S(t)^{\mathrm{T}}], \tag{4.76}$$

where W is the weight factor, S is the signal vector, P is the correlation vector, and R is covariance matrix.

In place of Equation (4.66), i.e. $E[\varepsilon^2(t)] = E[r^2(t) - 2W^{\mathrm{T}}P + W^{\mathrm{T}}RW]$, we have

$$E\left[\varepsilon^2(t)\right] = E\left[r^2(t)\right] - 2Re\left(W^{\mathrm{T}}P\right) + W^{\mathrm{T}}RW. \tag{4.77}$$

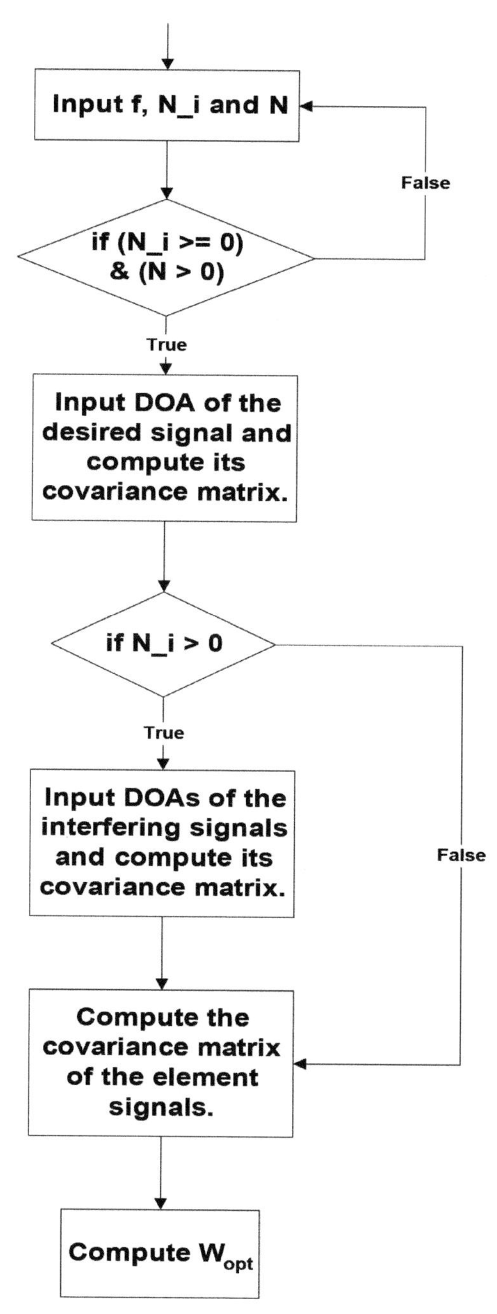

Figure 4.10 Flow chart for an adaptive antenna algorithm.

Defining $W = W_r - \mathrm{j}W_j$ and $P = P_r - \mathrm{j}P_j$, the equation becomes

$$E\left[\varepsilon^2(t)\right] = E\left[r^2(t)\right] - 2W_r^{\mathrm{T}}P_r + 2W_j^{\mathrm{T}}P_j + W^{\mathrm{T}}RW. \tag{4.78}$$

The quadratic form W^TRW is shown to have partial derivatives of

$$\partial\left(W^{\mathrm{T}}RW\right)/\partial W_r^n = 2Re\left(RW\right)^n, \tag{4.79}$$

$$\partial\left(W^{\mathrm{T}}RW\right)/\partial W_j^n = 2Im\left(RW\right)^n. \tag{4.80}$$

Using these results, we have

$$\partial E\left[\varepsilon^2(t)\right]/\partial W_r^n = -2P_r^n + 2Re\left(RW\right)^n, \tag{4.81}$$

$$\partial E\left[\varepsilon^2(t)\right]/\partial W_j^n = 2P_j^n + 2Im\left(RW\right)^n. \tag{4.82}$$

Equating all partial derivatives to zero yields

$$\mathrm{Re}\left(RW\right) = P_r, \tag{4.83}$$

$$\mathrm{Im}\left(RW\right) = -P_j, \tag{4.84}$$

and adding the previous two equations gives the final complex form

$$RW = \mathrm{Re}\left(RW\right) + \mathrm{j}\ \mathrm{Im}\left(RW\right) = P_r - \mathrm{j}P_j = P, \tag{4.85}$$

which yields

$$W = R^{-1}P. \tag{4.86}$$

This is the steady state weight vector. The analytic output signal for the entire array is

$$\tilde{s}(t) = W^{\mathrm{T}}S(t), \tag{4.87}$$

and the voltage pattern of the array is the normalized magnitude of the output signal. The listing of a MATLAB™ program for LMS adaptive beamforming is given in Chapter 12, and the performance of the LMS beamforming technique compared with the perceptron artificial intelligence beamforming technique is found in Chapter 9.

Acknowledgment

We acknowledge the contribution of Mr. Timmy Tan to the development of this chapter.

5

A New Smart Antenna for 5/6G Wireless Systems: Narrow 360° Steerable Beam With No Reflectors

K. Pirapaharan[1], P.R.P Hoole[2], H. Kunsei[3], K.S. Senthilkumar[4], and S.R.H. Hoole[5]

[1]Faculty of Engineering, University of Jaffna, Jaffna, Sri Lanka
[2]Wessex Institute of Technology, Chilworth, Southampton, United Kingdom
[3]Papua New Guinea University of Technology, Lae, Papua New Guinea
[4]Department of Computers and Technology, St. George's University, Grenada, West Indies
[5]FIEEE as Professor of Electrical Engineering (Retired), Michigan State University, USA

Abstract

In this chapter is presented a computationally efficient and fast method to construct smart antennas using an analytical, phase shift technique. A phase shifting method for a nonlinear array antenna is described. The array antenna elements are arranged in a regular polygon geometry, and it is shown that by controlling the electronic signal phase of the array elements, without any extra calculations as in most smart antennas, the antenna beam may be steered through 360°. Moreover, it is only a single beam that is generated, with minimal side lobes and no back lobes so that the wireless system is specifically focused toward a single direction, minimizing interference from other transmitters or multipath.

Keywords: Datacenter design, energy efficiency of data center, energy efficient metrics, datacenter carbon footprint computation.

5.1 Introduction

In antenna technology, a common practice is to place a reflector behind the array of antennas to ensure that the antenna radiates only in a single direction.

A single antenna beam in a prescribed direction ensures that the antenna does not radiate in other undesired directions. It also maximizes radiation toward a desired mobile or base station (BS). Alternatively, nonlinear arrays with repeated, numerical computation of weights are used to get single-beam smart antennas, as proposed in Chapter 9. In this chapter, we discuss the conditions for producing a single-beam antenna without the use of reflectors and no repeated computation of weights, thus reducing the mechanical weight of the antenna, its cost and to enable fast steering of beams using little computational time and memory. The smart antenna plays a critical role in the 5/6G wireless systems, and designing an array antenna that radiates only in a single given direction without a reflector is a highly desirable smart antenna. A single rotatable beam pointing toward a single preferred direction cannot be achieved by placing the antenna array in a straight line or with what is known as the linear antenna. In order to obtain a steerable antenna single beam, the geometrical structure of the antenna must be changed from a linear array to a two-dimensional array where except for the given, say forward looking, beam structure, each array element radiates in other directions so as to cancel each other's radiation in the undesired directions. In order to obtain a single beam that rotates over the entire 360° space surrounding the antenna, the adaptive signal processing algorithm is designed for an antenna array placed in a polygonal geometry. An advantage of this, as we shall see, is that the intensive computational burden of adaptive signal processing is very much simplified such that both computational time and computational burden on the signal processor memory are kept to a minimum. Moreover, in an environment such as in a densely populated area, express ways with large number of vehicles and in underground communication, remote sensing or radar systems, an omni-directional antenna creates a situation where the tunnel walls lead to multipath reflections, where the signals may interact destructively with each other to cancel out radiated signal energy in directions which one cannot predict or control. A single-beam antenna allows us to focus in a narrow direction to minimize the signals contacting the tunnel walls; moreover, having a single steerable beam, it will be possible to place the antenna in a densely populated area or at a junction of roads or tunnels, enabling the transceiver smart antenna beam to be rotated and almost simultaneously communicate to a small cluster of people or along each road/tunnel of the junction. Not only in 5/6G communication systems but also in other applications such as the challenging problem of detecting and tracking small drones, a narrow rotatable beam has significant advantages in radar and remote sensing applications. In the multiple elements, nonlinear array antenna that is discussed in this chapter is where the multiple elements of the array are placed on the circumference of a circle. The numbers of antenna elements for which results are presented include three, four, and six elements placed on the circumference of a circle.

In mobile communication systems, a major problem facing the proper operation of line of sight (LOS) systems is multipath interference. The desired direct ray is usually interfered by reflected, diffracted, or reflected/diffracted rays causing fading and sometimes blockage. The direct ray is highly attenuated or blocked under certain circumstances in mobile communication. Multipath fading has a particularly deleterious effect because it causes deep fades in the signal amplitude that lead to decoding errors at the receiver. The effects of multipath are information data corruption, signal nulling, and increasing or decreasing the signal amplitude. Therefore, multipath environment and null suppression of interference signal is one of the major challenges in mobile communication and wireless systems that are applied in many different fields from safety/security systems to medical and electric power systems. Thus, the desirability of the fast, steerable single beam is presented in the chapter.

One major reason that multiple antenna elements on mobile handsets are prohibitive is the additional battery power needed for the signal processor. Although a two-element array antenna with only a single parameter estimation for beamforming is attractive, its limitations are significant. We present in this paper an antenna radiation field model-based signal processor to make the mobile station (MS) antenna simultaneously increase reception of the desired signal (or increase radiation in the desired direction, when operating as transmitter) and cut off signals from interferers (e.g. other mobile users or reflected signals), or when acting as an MS transmitter, to minimize radiation in directions other than the direction of the receiver. The smart array antenna presented herein may be used in a mobile handset which simultaneously strengthens reception or transmission in the desired direction, while nulling the single interference signal. It is a promising technique that may be used to exploit the diversity principle used in multiple-input and multiple-output (MIMO) where at times the reflected signal (normally considered an interferer) may be used as the main signal when the direct signal, for instance, may be cut down in strength due to shadowing created by buildings [1]. Moreover, such MS-based beam steering may also be the best way to seek to compensate the negative effects of MS velocity.

The 5/6G wireless systems will include a major shift in the use of smart antennas. In the future wireless systems, the BSs may be completely replaced by the MSs taking over the roles of the BSs. In this case, the mobile transceivers (whether handsets or mounted on a vehicle) will need to be adequately equipped to be the relaying points between two other MSs communicating with each other, as well as to intelligently hand over the BS function to another MS when it is no longer in the vicinity to play the role of the BS. This means that the smart antennas need to be able to be agile enough to handle cells that are constantly and dynamically changing as the MSs move away with respect to each other. The small light weight mobile handsets need to functionally sound, while being able to operate with minimum hardware and software to be low in memory use, fast in operations and the cost of power be low. Single-beam smart antennas will not only be used in the BS which more readily accommodate multielement array antennas but also on the mobile handsets. The transmitting and receiving antennas on the handset will need to be very small to be used on the small-sized handsets and will need to operate on low battery power (maximizing battery life) while providing sufficient amount of power for transmission. The gain and directivity of the antenna should be maximized while being able to handle the placement of the hand over the area where the antenna is embedded inside. Spatial diversity of the antenna needs to be maintained to combat multipath signal interference even when the hand is moved over the unit. Impedance compensation for the presence of the hand as well as using the elements that are not covered by the hand helps to improve the performance of the smart antenna used in the MS handset. Smart antennas with steerable beams may also help to reduce the amount of power radiated into the human head by steering the beams away from the head. Moreover, smart antennas on MSs will reduce multipath fading, suppress interference signals, improve call reliability, give better spectral efficiency, increase data rates, improve call reliability, lower absorption rate, and mitigate against dead zones.

5.2 A Narrow Steerable Single-Beam Smart Antenna without a Reflector

In most antenna arrangements, when the array elements are placed in a straight line, a reflector needs to be used to direct the rotatable main beam (single beam) of the smart antenna to an azimuth angle that varies from 0° to 180°. Thus, the unavoidable mirror image of the rotatable main beam that appears in the space between 180° and 360° can be redirected in the direction of the main beam. Therefore, we need two sets of arrays placed side by side

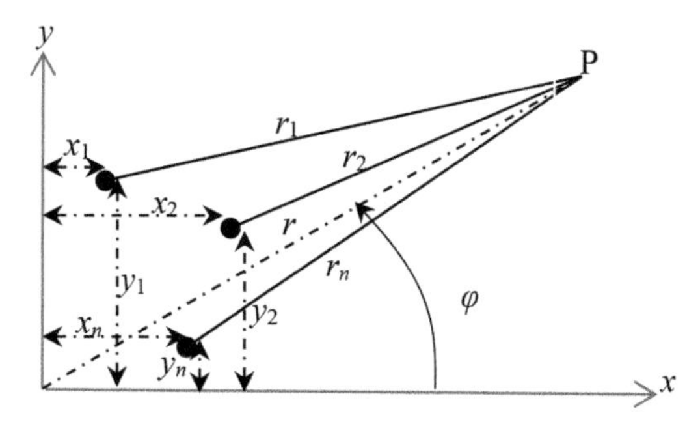

Figure 5.1 Schematic diagram of dipole placement (Pirapaharan et al., 2016).

along with reflectors to have the single-beam coverage of the entire azimuth angle from 0° to 360°. This is tedious to construct and to install, while also being a heavy structure. Consider now the dipole elements placed to form the nonlinear array shown in Figure 5.1.

It should be remembered that though we present the design and analysis in terms of dipole antenna elements, the elements may be made of other antenna elements as well, including the horn element antenna, the patch antenna, the helical antenna, the half-wave dipole, the parabolic dish antenna elements, etc. Any of these antennas may be arranged to form the array antenna discussed herein. The results apply to all array antennas, irrespective of the elements used. The respective complex current phasors of the dipoles I_1, I_2, and I_n yield the following electric far-field at the observation point P:

$$E = A_0 I_1 e^{-j\beta r_1} + A_0 I_2 e^{-j\beta r_2} + \cdots + A_0 I_n e^{-j\beta r_n} \tag{5.1}$$

where A_0 and β are a constant and the angular phase constant, respectively.

Consider first a linear dipole array placed along the x-axis. Substituting for r_1, r_2, $\ldots$, r_n in terms of the distance r of P from the origin, and setting weights $w_n = A_0 I_n e^{-j\beta r}$ and $y_1 = y_2 \ldots = y_n = 0$ for dipoles in a straight line

$$E(\phi) = w_1 e^{j\beta\, x_1 \cos\phi} + w_2 e^{j\beta\, x_2 \cos\phi} + \cdots \ldots + w_n e^{j\beta\, x_n \cos\phi} = E(-\phi) \tag{5.2}$$

Symmetry is observed from $E(\phi) = E(-\phi)$. Hence, weights will not exist for a single E beam on one side of the x x-axis. Solution exists only for weights for a single E beam in the positive/negative direction of the x x-axis (the end fire array) [2]. But this single beam cannot be rotated to other directions. Therefore, rotatable single beam in any and every direction cannot

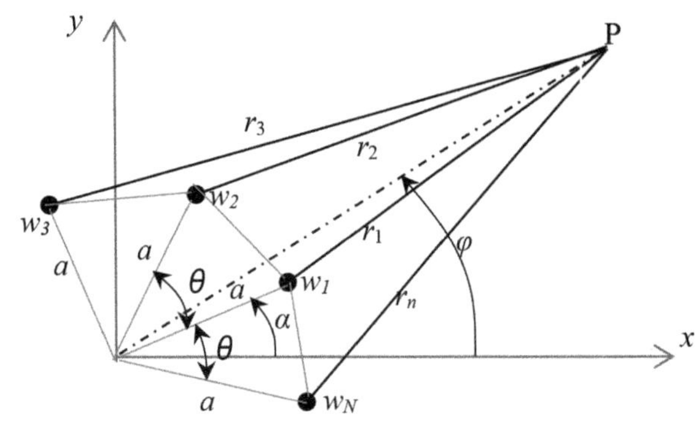

Figure 5.2 Schematic diagram of dipole placement in a regular polygon [3].

be obtained by means of weight optimization when all dipoles are in a straight line (i.e. a linear array antenna). Thus, for instance, in the broad side array, beams will exist in both directions perpendicular to the array axis. Therefore, when a linear array antenna is used, a reflector placed along the x-axis is needed to form the beam on only one side of the x-axis.

When all dipoles are placed in a straight line, the radiation pattern is symmetrical along the axis of dipole placement. This is not the case with a nonlinear array antenna. Consider now the nonlinear array antenna with an array of dipoles placed as shown in Figure 5.2.

When the polygon has N sides, the angle between two vertices from the center of the polygon θ is $2\pi/N$. The electric field at point P for the dipoles given in Figure 5.2 may be rewritten using Equation (5.2) as

$$\begin{aligned} E(\phi) = {} & w_1 e^{j\beta\, a(\cos\alpha\cos\phi+\sin\alpha\sin\phi)} + w_2 e^{j\beta\, a(\cos(\theta+\alpha)\cos\phi+\sin(\theta+\alpha)\sin\phi)} \\ & + w_3 e^{j\beta\, a(\cos(2\theta+\alpha)\cos\phi+\sin(2\theta+\alpha)\sin\phi)} + \cdots \\ & + w_N e^{j\beta\, a(\cos((N-1)\theta+\alpha)\cos\phi+\sin((N-1)\theta+\alpha)\sin\phi)}. \end{aligned} \tag{5.3}$$

The expression in Equation (5.3) may be simplified to

$$\begin{aligned} E(\phi) = {} & w_1 e^{j\beta\, a\ \cos(\alpha-\phi)} + w_2 e^{j\beta\, a\ \cos(\theta+\alpha-\phi)} \\ & + w_3 e^{j\beta\, a\ \cos(2\theta+\alpha-\phi)} + \cdots + w_N e^{j\beta\, a\ \cos((N-1)\theta+\alpha-\phi)}. \end{aligned} \tag{5.4}$$

For the arbitrary weight values, the electric field at the point P may be calculated from the expression given in Equation (5.4). Now we may rotate the weights such that the second element with weight w_1 and the third element with w_2 and the final first element with weight w_N. For the new weight values, the electrical field can be obtained using Equation (5.4) as

$$\begin{aligned}\bar{E}(\phi) &= w_N e^{j\beta\, a\, \cos(\alpha-\phi)} + w_1 e^{j\beta\, a\, \cos(\theta+\alpha-\phi)} \\ &\quad + w_2 e^{j\beta\, a\, \cos(2\theta+\alpha-\phi)} + \cdots + w_{N-1} e^{j\beta\, a\, \cos((N-1)\theta+\alpha-\phi)}.\end{aligned} \tag{5.5}$$

Using the trigonometric identity that

$$\cos(\alpha-\varphi) = \cos(2\pi+\alpha-\varphi) = \cos(N\theta+\alpha-\varphi) \tag{5.6}$$

we may rewrite Equation (5.5) as

$$\begin{aligned}\bar{E}(\phi) &= w_N e^{j\beta\, a\, \cos(N\theta+\alpha-\phi)} + w_1 e^{j\beta\, a\, \cos(\theta+\alpha-\phi)} \\ &\quad + w_2 e^{j\beta\, a\, \cos(2\theta+\alpha-\phi)} + \cdots + w_{N-1} e^{j\beta\, a\, \cos((N-1)\theta+\alpha-\phi)}.\end{aligned} \tag{5.7}$$

Further, rearranging Equation (5.7), we may write

$$\begin{aligned}\bar{E}(\phi) &= w_1 e^{j\beta\, a\, \cos(\theta+\alpha-\phi)} + w_2 e^{j\beta\, a\, \cos(\theta+\theta+\alpha-\phi)} + \cdots \\ &\quad + w_{N-1} e^{j\beta\, a\, \cos(\theta+(N-2)\theta+\alpha-\phi)} + w_N e^{j\beta\, a\, \cos(\theta+(N-1)\theta+\alpha-\phi)} \\ &= E(\theta+\phi).\end{aligned} \tag{5.8}$$

Therefore shifting weights through the adjacent vertices, the entire field pattern could be shifted by the same angle θ. This characteristic will help us to rotate the beam in steps by an angle θ without recalculating weights. This is only possible when the dipoles are placed at the vertices of a regular polygon [3]. It is a computationally efficient arrangement of the array antenna elements, where the beam may be rotated to a new position without any need to recalculate the weights for the new position to which the single beam needs to be rotated to. This will meet the demands on the new generation of 5/6G smart antennas needed for future wireless systems.

Since $E(\varphi) = E(\theta+\varphi)$, we may say that shifting the weights through the adjacent vertices, the entire field pattern could be shifted by the same angle θ. This characteristic will help us to rotate the beam in steps of angle θ without recalculating weights. For example, if the dipoles are placed in an equilateral triangle, the entire beam can be rotated in steps of 120° and 240° (because $\theta = 120^\circ$) without recalculating weights. Moreover, the geometrical center of the array antenna, with the elements placed on the circumference of

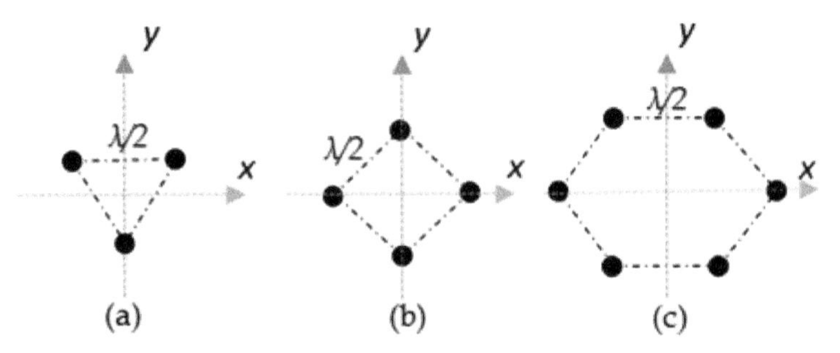

Figure 5.3 Schematic diagram of array models.

a circle, corresponds to the center of the circle. Depending on the number of sides of the regular polygon the step-angle, θ may vary.

We therefore advocate that the individual antenna elements (e.g. dipoles, patch, horn, etc.) of an array antenna be not placed in a straight line (i.e. not as a linear array antenna) but in one of the forms shown in Figure 5.3 (i.e. a nonlinear array antenna). It is evident from the proposed model that the minimum number of dipoles for single-beam beamforming is 3. However, using more elements will ensure a narrower beam shape.

It has been shown that there are advantages in placing the dipoles elements in a regular polygon (that is, the circumference of a circle) as shown in Figure 5.3. Rotating weights will move the resultant beam by an angle of $2\pi/N$, where N is the number of sides of the polygon. We present results for a three-element and six-element single-beam reflector-less array antenna with the antenna elements placed at the vertices of an equilateral triangle and a regular hexagon with optimized current component to rotate the main beam (single beam) from 0° to 360° while retaining a narrow beam width and low levels of side lobes.

We shall implement the above polygonal array design to demonstrate the single beam and its control. Using least mean square (LMS) optimization for a three-element and six-element array models as shown in Figures 5.3(a) and 5.3(c), we get a single beam in the desired angle direction of 60°. Having rotated the coefficients calculated for 60°, we can again rotate the beam to 180° and 300° for triangular array as shown in Figure 5.4(a) and to 120°, 180°, 240°, 300°, and 360° for hexagonal array as shown in Figure 5.4(b), respectively.

Hence, by optimizing the weights for a single angle, we can rotate the beam to multiple angles by simply phase shifting the weights without recalculating them by, for instance, the LMS method. Thus, we have verified

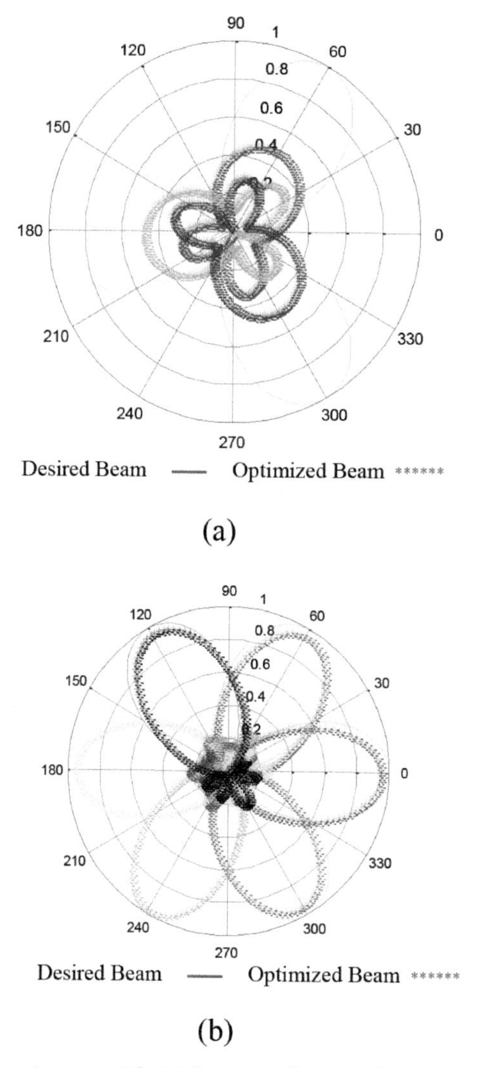

Figure 5.4 (a) Rotated beam to 60°, 180°, and 300° with 60° weights [5]. (b) Rotated beam to 60°, 120°, 180°, 240°, 300°, and 360° with 60° weights [5].

that the set of optimized weights coefficients for a single angle (60°) can be used for specific set of directions without repeating the optimization computation for new sets of desired directions. This is true for any array antenna where the dipole elements are placed at the vertices of a regular polygon.

5.3 Adaptive Array Model and Analytical Beamforming

In this section, we consider the rotation of the single beam using the LMS method, whereby the angle of rotation is not limited by the number of elements in the fast rotational method reported in Section 5.2. By placing array elements in the unconventional geometry presented in Section 5.2 and appropriately adjusting the current component of every element as per the requirement of the main beam direction by using LMS optimization, we are able to rotate the single beam electronically in any direction through $0^{\circ}-360^{\circ}$. The LMS-based adaptive solver matches the beam generated to any desired beam [4]. The beam width and the strength of the beam may be altered by changing the distance between each adaptive antenna element. With a single rotating beam, the system can cover an area of 360° while nullifying all other beams outside the look angle with a minimum of three elements.

The respective complex current phasors of the dipoles are taken as I_1, I_2, and I_n. Hence, the electric field (far-field) at the observation point P as shown in Figure 5.1 is given in Equation (5.1). Substituting for r_1, r_2, and r_n in terms of the distance from the origin, Equation (5.1) can be simplified to

$$E = w_1 e^{j\beta(x_1 \cos\phi + y_1 \sin\phi)} + w_2 e^{j\beta(x_2 \cos\phi + y_2 \sin\phi)} + \cdots + w_n e^{j\beta(x_n \cos\phi + y_n \sin\phi)} \tag{5.9}$$

where w_1, w_2, and w_n are the complex weights that are proportional to the complex current phasors I_1, I_2, and I_n I_n, respectively, as stated earlier. To achieve the objective of forming a resultant single beam, the values of the complex weights w_1, w_2, and w_n need to be optimized such that the resultant field must match a desired single smart antenna beam function $f(\phi)$. Thus, Equation (5.9) may be rewritten as,

$$w_1 e^{j\beta(x_1 \cos\phi + y_1 \sin\phi)} + w_2 e^{j\beta(x_2 \cos\phi + y_2 \sin\phi)} + \cdots + w_n e^{j\beta(x_n \cos\phi + y_n \sin\phi)} = f(\phi). \tag{5.10}$$

The optimization of complex weights w_1, w_2, and w_n, may be done either analytically or iteratively. Since the number of dipole elements will be confined to as few as possible, an analytical optimization method is more appropriate.

The analytical method employed to optimize the complex weights w_1, w_2, and w_n is as follows. Each term in Equation (5.10) is multiplied by its complex conjugate. We begin with the first term multiplying Equation (5.10) with the complex conjugate term respective of w_1, $e^{-j\beta(x_1\cos\phi\ +y_1\sin\phi\ )}$, and integrate by angle ϕ over the limit from 0 to 2π, to get,

$$\begin{aligned} &w_1 \int_0^{2\pi} d\phi + w_2 \int_0^{2\pi} e^{j\beta[(x_2\cos\phi+y_2\sin\phi)-(x_1\cos\phi+y_1\sin\phi)]} d\phi + \cdots \\ &+ w_n \int_0^{2\pi} e^{j\beta[(x_n\cos\phi+y_n\sin\phi)-(x_1\cos\phi+y_1\sin\phi)]} d\phi \\ &= \int_0^{2\pi} f(\phi)\ e^{-j\beta(x_1\cos\phi+y_1\sin\phi)} d\phi. \end{aligned} \tag{5.11a}$$

And for the *n*th term,

$$\begin{aligned} &w_1 \int_0^{2\pi} e^{j\beta[(x_1\cos\phi+y_1\sin\phi)-(x_n\cos\phi+y_n\sin\phi)]} d\phi \\ &+ w_2 \int_0^{2\pi} e^{j\beta[(x_2\cos\phi+y_2\sin\phi)-(x_n\cos\phi+y_n\sin\phi)]} d\phi + \cdots + w_n \int_0^{2\pi} d\phi \\ &= \int_0^{2\pi} f(\phi)\ e^{-j\beta(x_n\cos\phi+y_n\sin\phi)} d\phi. \end{aligned} \tag{5.11b}$$

We can obtain *n* number of such equations, if we consider *n* dipole elements. Hence, the above *n* such equations may be written in matrix form as,

$$\begin{bmatrix} A_{11} & A_{12} & \ldots & A_{1n} \\ A_{21} & A_{22} & \ldots & A_{2n} \\ \ldots & \ldots & \ldots & \ldots \\ A_{n1} & A_{n2} & \ldots & A_{nn} \end{bmatrix} \begin{bmatrix} w_1 \\ w_2 \\ \ldots \\ w_n \end{bmatrix} = \begin{bmatrix} b_1 \\ b_2 \\ \ldots \\ b_n \end{bmatrix} \tag{5.12}$$

where

$$A_{ij} = \int_0^{2\pi} e^{j\beta[(x_j\cos\varphi+y_j\sin\varphi)-(x_i\cos\varphi+y_i\sin\varphi)]} d\varphi \tag{5.13}$$

and

$$b_i = \int_0^{2\pi} f(\phi)\ e^{-j\beta(x_i\cos\phi+y_i\sin\phi)} d\phi.. \tag{5.14}$$

Thus, the optimized coefficients can be obtained by taking inverse operation of Equation (5.12)

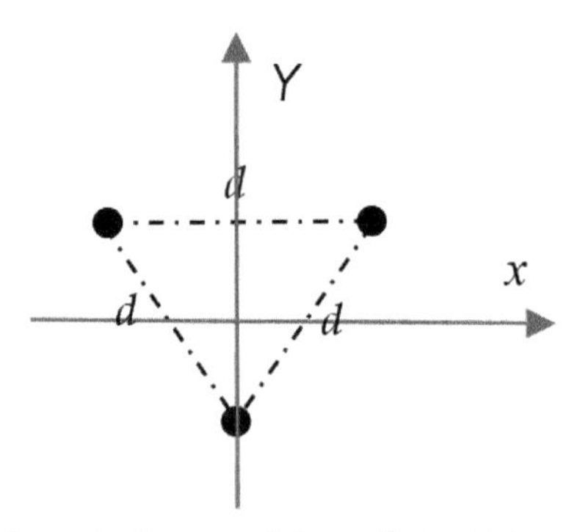

Figure 5.5 Schematic diagram of the equilateral triangular array model.

$$\begin{bmatrix} w_1 \\ w_2 \\ \ldots \\ w_n \end{bmatrix} = \begin{bmatrix} A_{11} & A_{12} & \ldots & A_{1n} \\ A_{21} & A_{22} & \ldots & A_{2n} \\ \ldots & \ldots & \ldots & \ldots \\ A_{n1} & A_{n2} & \ldots & A_{nn} \end{bmatrix}^{-1} \begin{bmatrix} b_1 \\ b_2 \\ \ldots \\ b_n \end{bmatrix} \tag{5.15}$$

where the matrix elements $A_{ij}$and b_i are numerically calculated.

We have considered three-, four-, and six six-element arrays as shown in Figure 5.3. In Figure 5.5 is shown the three three-element antenna with a general distance of inter-element separation *d*. The desired function selected is the sinc function as defined in the following equation (5.16).:

$$f(\phi) = \text{sinc}(\phi - \phi_0) \tag{5.16}$$

where ϕ_0 is the desired angle of maximum antenna transmission and reception.

With these simulation parameters, we turn to MATLAB^{TM} to verify our analytical results. The radiation pattern in Figure 5.6 shows the results obtained from our MATLAB^{TM} simulations when the distance between the dipoles is half-wavelength while the desired angles are $60^0 60^\circ$, 180^0 180^{0°, and $300^0 300^\circ$. It is seen that the optimized pattern falls in line with the desired pattern, except that the signal strength is smaller. The results shown in Figure 5.6 also demonstrate that when the desired beam is rotated over the entire space, the optimized beam using the weights calculated from Equation. (5.15) is able to match and to steer the beam to follow the geographical cluster of users. In cognitive radio systems, thus the spectrum may be efficiently used for both stationary and mobile users, enabling the unused spectrum which itself may be time, and space, dependent in its characteristics.

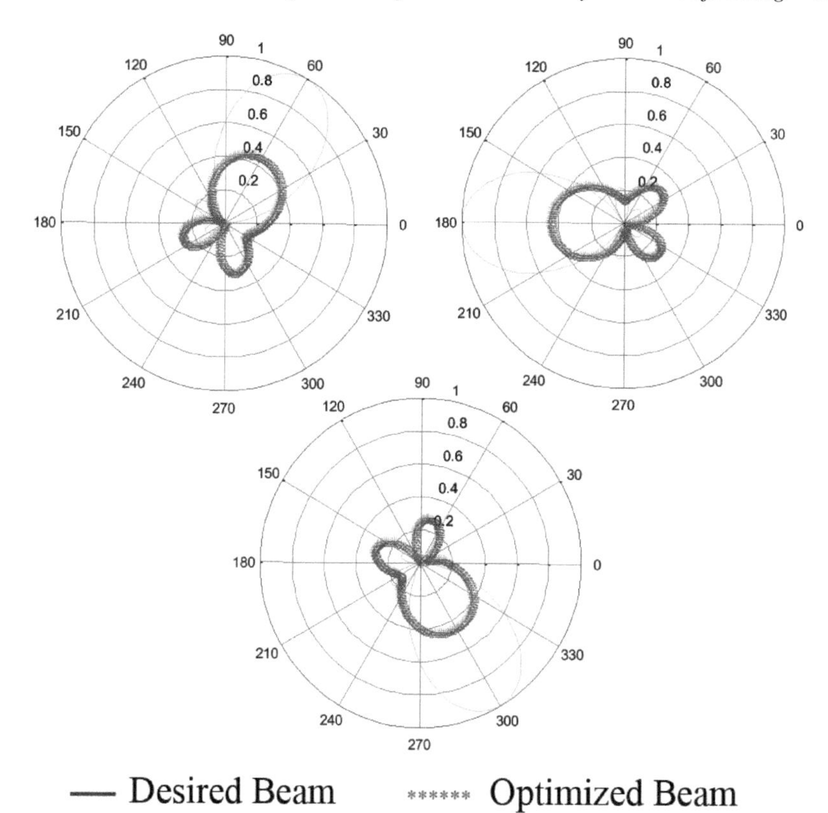

Figure 5.6 Beam rotation of 3 three-Eelement Aarray [3].

In order to study the beam- width and side-lobe variation with the change of the distance between dipoles, the distance between two dipoles are selected as quarter, half, and full wavelengths. The comparison between the radiation patterns and the distance between two dipoles are obtained as shown in Figure 5.7. As expected, when the distance between two elements is smaller, the beam- width is larger while the number of side-lobes reduces. The inter inter-element distance dependence is due tobecause the delay between the two field components at the receiver point is smaller since the delay depends on the angle of arrival and the distance between two elements. Hence, it is expected that there is a large angle of arrival (beam- width) between peak and null points. On the other hand, when the distance between two elements increases, the angle of arrival (beam- width) is small between the peak and null points while the number of side lobes increases. . In this case, we optimized the beam

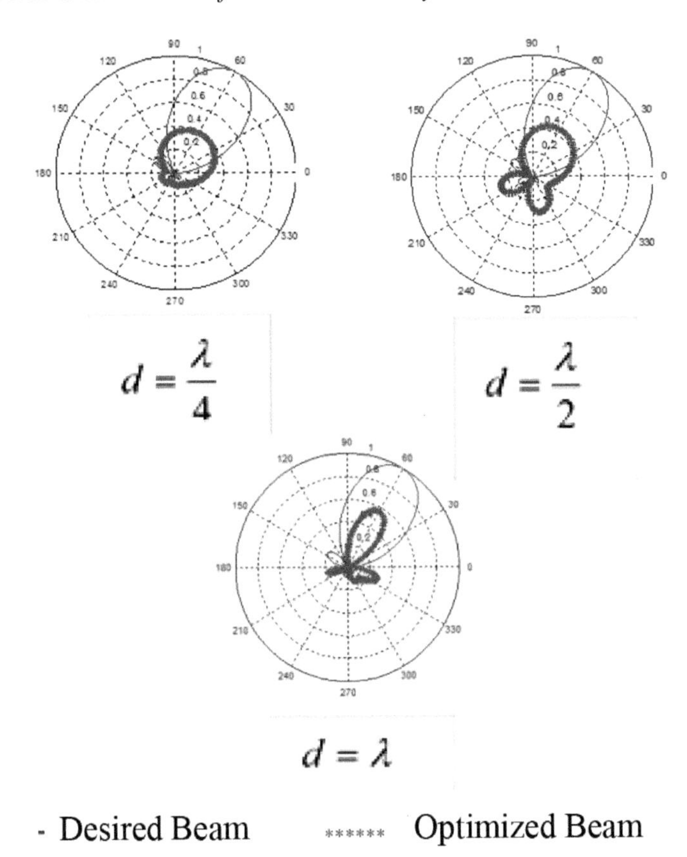

Figure 5.7 Beam width and side lobes with varying distance between antenna elements for 3 three-element array [3].

using three elements. Therefore, the peak direction is single but the number of side lobes increases with increasing distance between two adjacent elements.

Finally, for half-wave dipole elements, the three three-dimensional radiation patterns were obtained and shown in Figure 5.8. The three-dimensional beam patterns are shown for different observation directions. In case the radiation pattern of the beam needs to be changed in the vertical plane, a different type of dipole elements is to be selected [5]. The entire beam characteristics in the three three-dimensional space depend on the type of dipole and the distance between two adjacent dipole elements. The overall results obtained ensure that a single rotatable beam is possible with three dipole elements placed not in straight line, but at the vertices of an equilateral triangle.

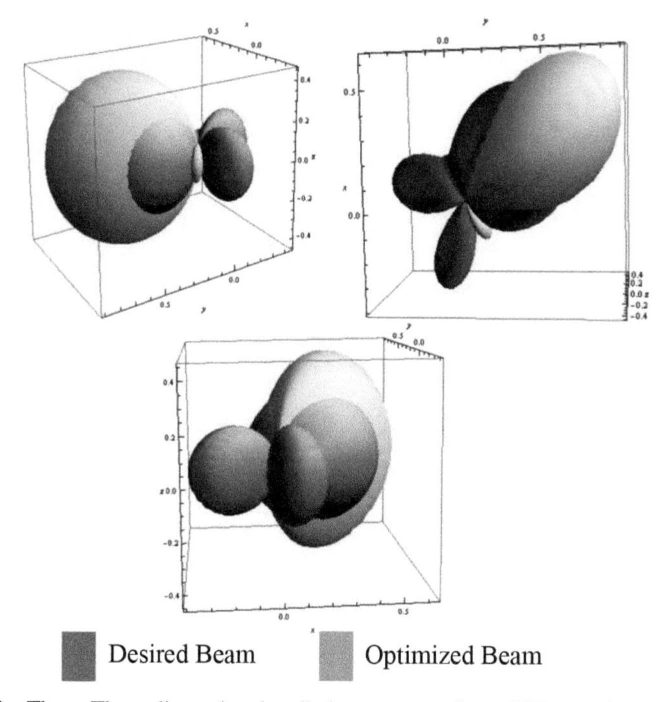

Figure 5.8 Three Three-dimensional radiation patterns from different view points when half-wave dipoles are placed in half-wavelength distance of an equilateral triangle. [3].

The radiation patterns, shown in Figures 5.9–5.11 show significant matching between the desired pattern and the optimized pattern. It may be observed from Figure 5.9 that as the number of elements is increased, the optimized beam pattern develops to an almost perfect match to the desired beam from three to four elements. In addition to beam pattern matching, the beam width is also reduced. A narrow beam would have a greater coverage while utilizing less power.

Figures 5.10 and 5.11 demonstrate our ability to rotate the beam to cover effectively an area covering a 360^0 360° angle in the azimuth. The results in Figures 5.10 and 5.11 are for 4four-element array and 6six-element array smart antennas, respectively. Irrespective of the number of elements, it is shown here that whatever the desired look- angle, we may electronically steer the beam to that desired direction.

In Figure 5.4 was shown that it is possible to generate a single beam in the desired direction with no image in the opposite direction, thus conserving power and reducing interference and reflection by radiating only

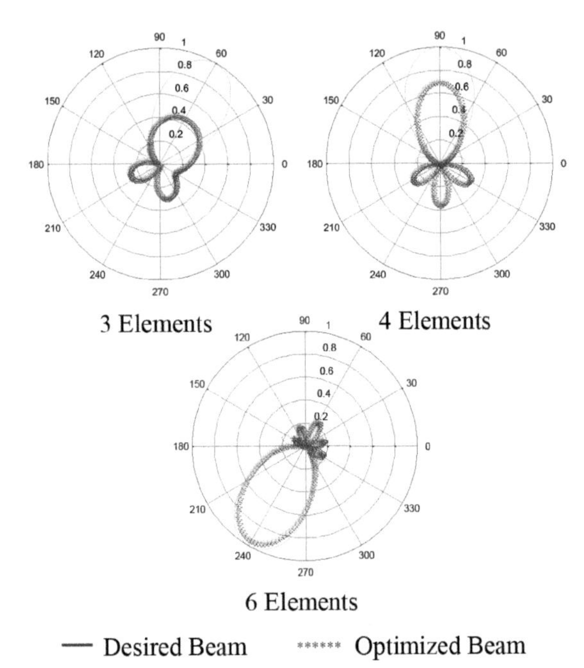

Figure 5.9 Comparison of Radiation radiation Patterns patterns of 3three, 4 four, and 6 six Elements elements [3].

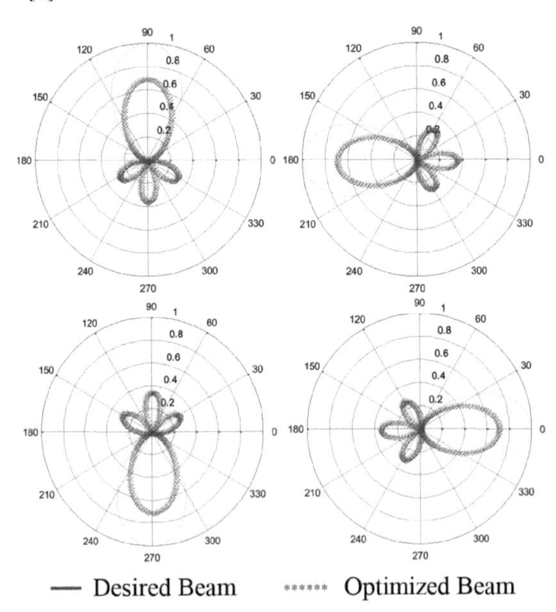

Figure 5.10 Beam Rotation rotation of 4 Efour-element Array array [3].

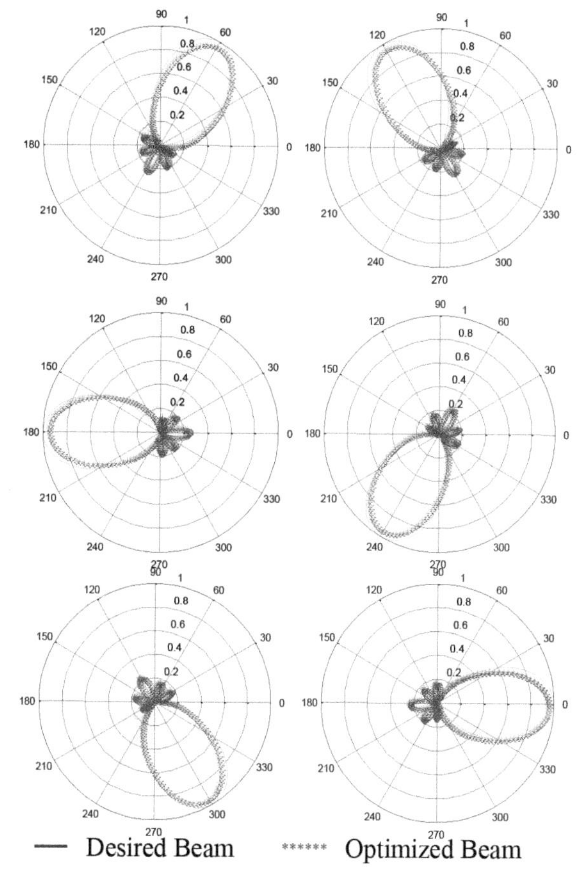

Figure 5.11 Beam Rotation rotation of 6 Esix-element Array array [3].

in the desired direction. Figure. 5.9 shows that with increasing number of elements will certainly improve power utilization which dissipates only in the desired direction extending the range. We have shown that three elements are sufficient for 5/8G smart antenna transcievers transceivers where space, power, and improved performance are vital in systems design.

5.4 Conclusions

This chapter described the development of an array antenna with a non-conventional geometrical shape driven by the need to get a single single-beam antenna without the use of reflectors. Although with three elements placed at the vertices of an equilateral triangle it is possible to get a highly directive single single-beam antenna, we have shown that with an increase in number of elements narrower, more focused beams may be obtained. Furthermore, we have shown that if the elements are placed in the circumference of a circle, where the geometrical center of the array should correspond to the center of the circle, the need for recalculating the weights is avoided since all that is needed is to phase shift the weights by known angles in order to continuously rotate the beams from 0 to 360°. Thus, a rotatable beam that is narrow is obtained with minimum computational burden, where once the weights are calculated for a specific direction, to rotate the beam on to other desired directions, no additional computation of the weights is required.

References

[1] Mamta Agiwal, Abhishek Roy and Navrati Saxana, "Next Generation 5G Wireless Networks: A comprehensive Survey," IEEE Communications Surveys & Tutorials, Vol. 18, No. 3, pp. 1617-1655, 2016.

[2] K.S.Senthilkumar, K. Pirapaharan, G.A.Lakshmanan, P.R.P. Hoole and S.R.H. Hoole, "Accuracy of Perceptron Based Beamforming for Embedded Smart and MIMO Antennas," in *Proceedings of International Symposium on Fundamentals of Electrical Engineering 2016 (ISFEE2016)*, Bucharest, Romania, June 30-July 2, 2016.

[3] K. Pirapaharan, H. Kunsei, K.S. Senthilkumar, P.R.P. Hoole and S.R.H. Hoole, "A Single Beam Smart Antenna for Wireless Communication in Highly Reflective and Narrow Environment," in *Proceedings of International Symposium on Fundamentals of Electrical Engineering 2016 (ISFEE2016),* Bucharest, Romania, June 30-July 2, 2016.

[4] L.J. Griffiths, "A Simple Adaptive Algorithm for Real-Time Processing in Antenna Arrays," *Proc. IEEE*, vol. 57, no. 10, pp. 1696-1704, 1969.

[5] K. Pirapaharan, H. Kunsei, K.S. Senthilkumar, Paul R. P. Hoole, and Samuel R. H. Hoole, "A Robust, 3-Element Triangular, Reflector-less, Single Beam Adaptive Array Antenna for Cognitive Radio Network: Inter-element Distance Dependent Beam," *Journal of Telecommunication, Electronic and Computer Engineering*, vol.8, no.12, pp79-82, 2017.

5.5 Appendix 5.1. The MATLAB™ code for calculating the optimized weights of a three-element array

The separation between adjacent vertices of the unilateral triangular dipole placement is $?/2$ while the desired function of the beamforming is $\text{sinc}\,(f - p/3)$.

```
x=0:pi/100:2*pi;
a11=2*pi; a12=quad(exp(i*2*pi/2*cos(x)),0,2*pi);
a13=quad(exp(i*pi/2*(cos(x)-sqrt(3)*sin(x))),0,2*pi);
a21=quad(exp(-i*2*pi/2*cos(x)),0,2*pi); a22=2*pi;
a23=quad(exp(-i*pi/2*(sqrt(3)*sin(x)+cos(x))),0,2*pi);
a31=quad(exp(-i*pi/2*(cos(x)-sqrt(3)*sin(x))),0,2*pi);
a32=quad(exp(i*pi/2*(cos(x)+sqrt(3)*sin(x))),0,2*pi); a33=2*pi;
b1=quad(sinc(x-pi/3).*exp(i*pi/2*cos(x)),0,2*pi);
b2=quad(sinc(x-pi/3).*exp(-i*pi/2*cos(x)),0,2*pi);
b3=quad(sinc(x-pi/3).*exp(i*pi/2*sqrt(3)*sin(x)),0,2*pi);
A= [a11 a12 a13
    a21 a22 a23
    a31 a32 a33];
b= [b1
   b2
   b3];
w=inv(A)*b
```

6

Synthetic Aperture Antennas and Imaging

Tan Pek Hua, Dennis Goh, P.R.P. Hoole, and U.R. Abeyratne

Abstract

This chapter presents the basic technique for single dimension electromagnetic signal processing of radar signals to obtain two-dimensional images. The synthetic aperture radar (SAR) is a powerful technology for getting pictures whether in the day time or in the night time when a normal camera or the naked eye cannot be used to get a visual image of the scene, the SAR, where a moving radar antenna looking at an object (e.g. a spy plane radar flying over a military installation on ground) may be used with suitable electromagnetic signal processing of the reflected electromagnetic signals picked up the radar antennas. In the inverse synthetic aperture radar (ISAR), although the radar antenna itself is a stationary single element antenna, the moving target, such as an aircraft, is like a set of antenna elements as it moves and, from each point, reflects a signal back to the radar antenna. This set of signals makes the moving target a synthetic array antenna. The technique presented here, illustrated for three different points on the target, may be used to get a picture of the target (e.g. an aircraft). The same technique is used with an array of stationary antenna elements on earth pointing skyward to obtain images of galaxies.

6.1 Basic Principles of Radar Signal Processing

6.1.1 Introduction

Moving a single element antenna over space generates a synthetic aperture antenna. An example of this is an aircraft-mounted antenna that operates (i.e. transmits and receives signals) as the aircraft moves at a velocity of 300 m/s, say. In such a case, over a time period of 1 s, the antenna will be at 30 discrete points of 10 m apart. We now have an artificial array antenna of 30 elements with an interelement distance of 10 m. This is called a synthetic array or synthetic aperture antenna. We get such a situation in mobile communication systems too, where the mobile phone moves in space as it communicates with the base station. Hence, the mobile phone antenna is a synthetic aperture antenna. In this chapter, we shall illustrate the principles of synthetic aperture antennas by looking at its performance in a radar system. A radar system is a system in which an antenna transmits a pulsed signal and then pauses before transmitting another pulse to collect the reflected return pulse of the first signal. The return pulse signals are processed to obtain details like distance and velocity of the target, as well as an image of the target. The synthetic aperture antenna allows us to perform such powerful processing to do remote sensing.

A radar system is made up of a transmitter (to generate the pulse to be radiated out), a receiver (to pick up the return signal), an antenna which is physically moved (in addition, its beam may be electronically steered in some cases), and a signal processor to get one-dimensional and two-dimensional (image) details of the target that was observed. The transmitter generates a sequence of pulses (or bursts) which is launched out toward the target by a narrow beam antenna. The pulsed signals form electromagnetic waves in space and travel at the velocity of light ($c = 3 \times 10^8$ m/s) in free space. In media other than free space with a relative permittivity of ε_r (e.g. 80 for seawater), the velocity of the waves will be reduced to $u = c/(\varepsilon_r)^{1/2}$. The waves hit the target, and part of the wave energy is reflected back toward the radar antenna. These reflected waves are the return signals picked up by the radar antenna and processed by the receiver. The time delay, τ, between the transmitted and return pulse is a measure of the two-way distance between the radar antenna and the target. This two-way distance is twice the range of the target, where range is the line-of-sight (LOS) distance between the radar antenna and the target given by

$$\text{Range} = \frac{u\tau}{2}, \tag{6.1}$$

where u is the speed of the electromagnetic wave and τ is the time delay or echo of the return signal.

Since the beam of the antenna is narrow and pointed in a known direction, the amount of wave energy reflected back from the target will depend on the direction and the reflectivity (reflection coefficient) of the target in that direction. This variation may be used to determine the direction in which the target is situated. Now using the data on the range and angular direction of the target, we may reconstruct or synthesize the position and reflectivity map of the target. The issue becomes more complicated when there are two or more targets falling into the antenna beam at the same time. In such a case, the reflected signal carries energy of waves reflected by both objects, and depending on the distance between the two targets, the time difference between the two reflected signals may not be sufficiently large to separate the two targets. This raises the issue of ambiguity since it is not clear which of the two targets or reflecting points is represented by the return signals. Thus, a central issue becomes spatial resolution that could be achieved from the return signals; the return signals should be processed to achieve minute range and azimuth resolutions.

It can be shown that range resolution is inversely proportional to the bandwidth of the transmitted signal. Hence, a signal with a wide bandwidth will help to achieve finer range resolution. One way to achieve resolution in the azimuth direction is to use an antenna with a narrow azimuth beam width. Narrow antenna beam width means larger antenna aperture size or large radar frequency. One practical constraint on large antenna length is the difficulty of mounting a physically large antenna on an aircraft. Increasing transmitted frequency means greater adverse effects of the mediums: as frequency is increased above, say, 20 GHz, the transmitted and return signal will be significantly attenuated by air or any other medium through which the signals travel. Hence, to improve azimuth resolution, we have to resort to synthetic array or aperture.

6.1.2 Synthetic Aperture Radar

In synthetic aperture radar (SAR), a moving radar collects many pulses of return signals, and by comparing them achieves better azimuth resolution. The azimuth resolution obtained is much finer than that of a stationary antenna with the same real beam width; the movement of the antenna artificially extends the length of the antenna and hence narrows the overall azimuth beam width. By illuminating the target at different angles and different locations, an image (two-dimensional or three-dimensional) of the target could be obtained.

Figure 6.1 illustrates the three common SAR imaging modes, namely the spotlight, stripmap, and scan modes.

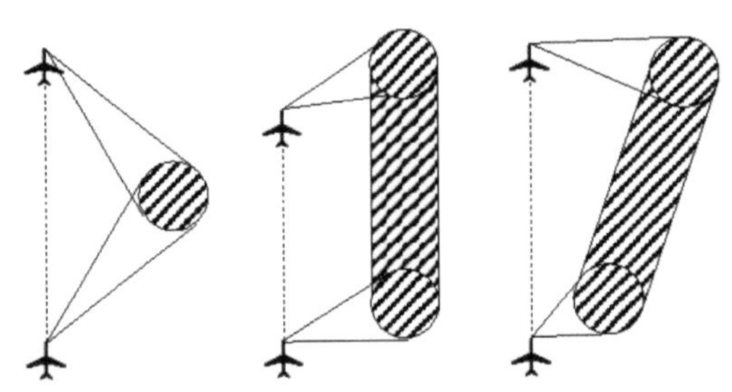

Figure 6.1 Three common SAR imaging modes: spotlight, stripmap, and scan, respectively. (Reprinted with permission from Spotlight Synthetic Aperture Radar Signal Processing Algorithms, by W.G. Carrara, R.S. Goodman, and A. Majewski's. Artech House, Inc., Norwood, MA, USA. www.artechhouse.com (Carrara et al., 1995 in Bibliography).)

6.2 Inverse Synthetic Aperture Radar

The inverse SAR (ISAR) mode is a fourth mode of synthetic aperture antenna imaging; here the antenna is kept stationary at one point and the object is moved around the antenna. As the object moves around the antenna, different views of it are presented to the antenna so that the return pulses carry back a complete range of information on the entire target. The principles governing SAR and ISAR are the same since what is basic is the relative motion between the antenna and the target. In the case of ISAR, the moving target re-radiating back return signals acts as a synthetic array antenna. In this chapter, the ISAR scenario is described. The case study considered is that of imaging an aircraft in flight by a stationary antenna.

6.3 One-Dimensional Imaging with Point Scattering

6.3.1 Overview

This section is intended to give the reader an overview of the one-dimensional imaging process before going on to the mathematical description. Consider M point targets or M number of scattering points on a single target, as shown in Figure 6.2. We assume that they all lie in a straight line. The distances separating two adjacent points are unequal.

An antenna transmitter located at $x = 0$ radiates a pulsed signal $s(t)$ of finite duration T. The leading edge of the $s(t)$ reaches the first scattering point at time $t_1 = x_1/u$, where u is the speed of wave propagation in the medium where the targets reside. The first scattering point reflects back a portion of the signal s(t) back to the antenna receiver reaching $x = 0$ in t_1 s later. Assuming

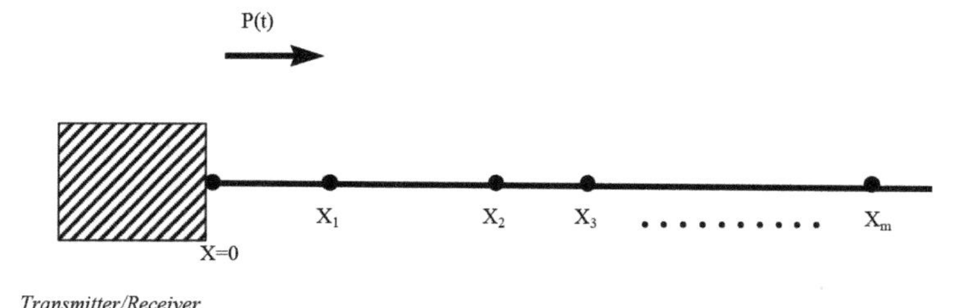

Figure 6.2 Geometry for one-dimensional imaging.

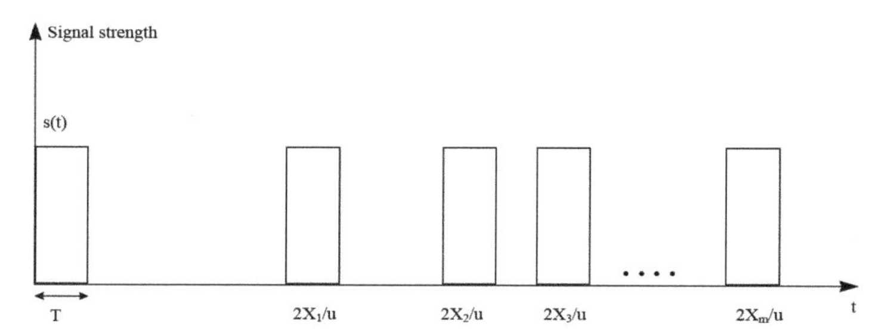

Figure 6.3 Transmitted and returned pulses.

that the reflected signal is very small in comparison to the strength of the transmitted signal $s(t)$, the same signal strength $s(t)$ now hits scattering point 2, and later 3 and so on until it hits point m. Each scattering point will reflect back a portion of $s(t)$ depending on the reflection coefficients of the points. Figure 6.3 shows the transmitted pulse and the m number of reflected pulses. We have ignored the attenuation of the signal due to the transmission medium (e.g. air) and beam spreading.

If the scattering points are close together, the returned pulses may overlap each other as shown in Figure 6.4, in which case it will be difficult to resolve between the scattering points at x_2 and x_3.

It is seen from Figure 6.4 that if we shorten the pulse duration T, then we can get better resolution by avoiding overlapping return pulses. Of course, shortening the pulse has the disadvantage of reducing the energy in the signal and hence the distance up to which we can still detect the return signal. The requirement is that closely spaced points can be resolved if they are separated in time delay by at least the width of the transmitted pulse, T. If the distance of separation between two adjacent scatterers is less than $uT/2$, then the return pulses will overlap, making it impossible to determine where one point ends and the other begins. The distance $uT/2$ is called the

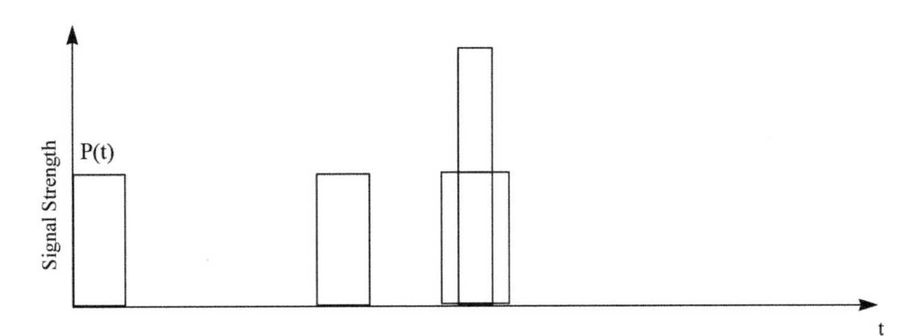

Figure 6.4 Overlapping returned pulses.

range resolution of the radar. As noted, reducing T cannot be done without the overall consideration of the maximum distance to which we require the radar to perform. The transmitted radar pulse should carry more energy if the target is very far so that the little energy reflected back can be processed by the receiver electronics without getting swamped out by the electronic noise. Hence, an alternative is to design a pulse shape that has sufficiently short time duration while having the required energy and may be processed to distinguish different scatterers. The pulse should be designed such that overlapping returns from different scatterers can be separated. This means that the correlation between the pulses should be small everywhere except when the time delay between signals is zero. From the autocorrelation of several functions shown in Figure 6.5, we note that one of the best functions to use as waveforms inside each pulse is the linear frequency modulation (FM) pulse, also known as the chirp pulse.

Since the linear FM or chirp pulse can be compressed into a sharp pulse, even when there is an overall overlap between two return pulses, processing the return pulses by a matched filter will enable us to separate two overlapping returns as they each become compressed into sharp spikes on autocorrelation. Thus, the range resolution can be further enhanced by the compression of the chirp pulse through a matched filter. Then each transmitted pulse is a pulse of linear FM signal, which is a kind of a wavelet. The reader would now realize that designing an antenna system to perform complex activity like imaging requires not only a synthetic array or synthetic aperture antenna but also a waveform that is well shaped and suited for the particular application. In telecommunication systems, we use other kinds of signal shaping or modulation to improve on the system performance. Hence, antenna design cannot be considered independently of the best suited signal for complex systems.

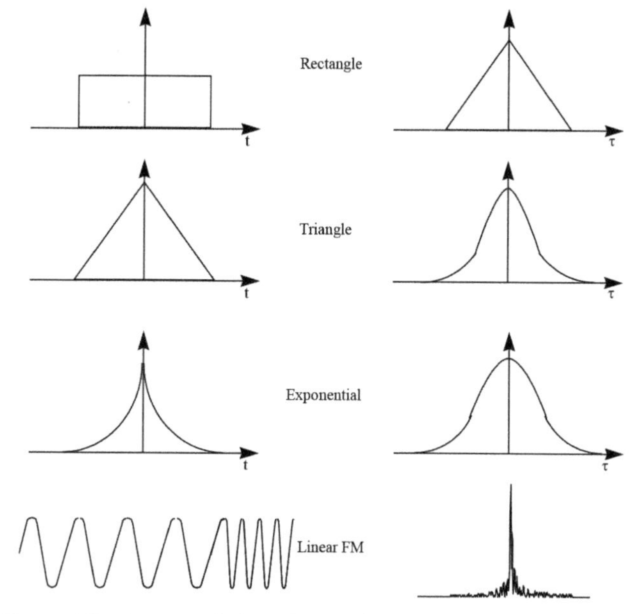

Figure 6.5 Autocorrelation of several functions (from Fitch, 1988).

Linear FM (chirp) waveform is used as the transmitted radar pulse. As mentioned in the previous section, linear FM is used so as to reduce the superposition effect of the two return pulses or echoes which differs by a small time delay. The linear FM waveform is characterized by the equation

$$S'(t) = 2\pi(ft + 0.5at^2), \tag{6.2}$$

where f is the initial frequency of the linear FM and a is the chirp rate or the rate of change of frequency.

This basic signal is up-converted to a higher frequency known as the carrier frequency f_c by mixing before transmission. The higher frequency is required to improve transmission and reception of the wave. Hence, the final transmitted wave is given by

$$s(t) = \cos[2\pi(f_c t + 0.5at^2)]\ \mathrm{rect}((t - T/2)/T), \tag{6.3}$$

and it is shown in Figure 6.6.

The definition of the rect( ) function in Equation (6.3) is as follows:

$$\mathrm{rect}((t - T/2)/T) = \begin{cases} 1, & 0 \le t \le T, \\ 0, & \text{otherwise.} \end{cases}$$

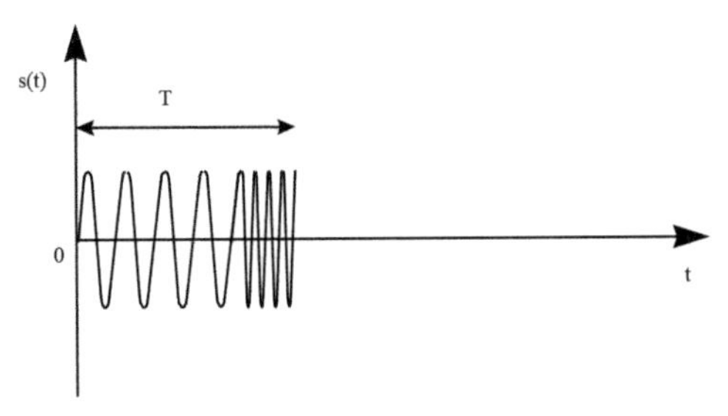

Figure 6.6 Chirp waveform.

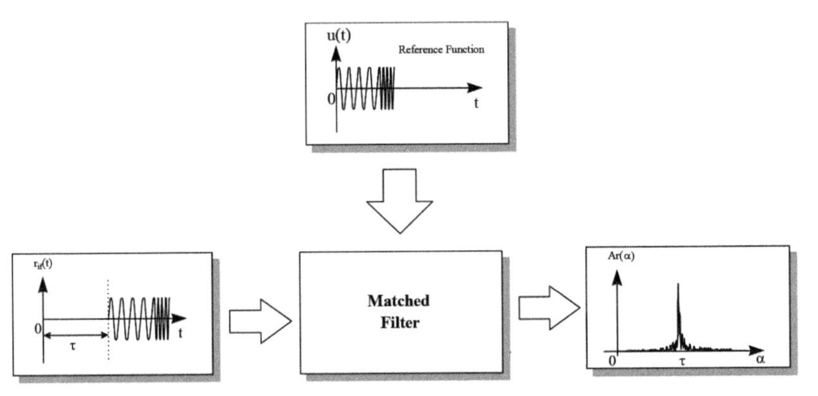

Figure 6.7 Range compression by matched filter.

Now consider the reflection of the transmitted signal of an object at distance R from the antenna. The received return pulse is the result of $s(t)$ and its reflected portion having traveled a distance $2R$ from transmission to reception. This two-way distance corresponds to a time delay of $\tau = 2R/u$ and the received signal is of the form $\sigma.s(t-\tau)$, where the factor σ (reflection coefficient) is a function of the target geometry and material property at carrier frequency f_c.

By mixing (down-conversion) the return signal picked up by the antenna, the down-converted signal r_{if} is given by

$$r_{\text{if}}(t) = \sigma.\cos[2\pi(-f_c\tau + f_{\text{if}}t + 0.5a(t-\tau)^2)]\text{rect}((t - T/2 - \tau)/T). \quad (6.4)$$

The return signal $r_{\text{if}}(t)$ is processed by a matched filter, and the compressed output will be scaled by some σ and located at a time delay of τ as shown in Figure 6.7.

The matched filter correlates two signals. It correlates the return signal with a reference signal. The reference signal used here is a replica of the transmitted linear FM signal $s(t)$. Based on the mixed down signal of Equation (6.3), the appropriate reference pulse for the matched filter is therefore given by

$$s_{\mathrm{ref}}(t) = \cos[2\pi(f_{\mathrm{if}}t + 0.5at^2)]\mathrm{rect}((t - T/2)/T). \tag{6.5}$$

Thus, the output $A(\alpha)$ of the matched filter is the correlation of two signals:

$$\begin{aligned} A(\alpha) &= \int r_{\mathrm{if}}^*(t)\, s_{\mathrm{ref}}(\mathrm{t} + \alpha)\ \mathrm{dt} \\ &= \int \mathrm{e}^{-\mathrm{j}2\pi(-f_c\tau + f_{\mathrm{if}}t + 0.5a(t-\tau)^2)}\ \mathrm{e}^{\mathrm{j}2\pi(f_{\mathrm{if}}(t+\alpha) + 0.5a(t+\alpha)^2)}\ \mathrm{d}t. \end{aligned} \tag{6.6}$$

Consider $\tau = 0$ first to get the mathematical equation of the output of the matched filter.

$$\begin{aligned} A(\alpha) &= \mathrm{e}^{\mathrm{j}2\pi\left(f\alpha + 0.5a\alpha^2\right)} \int \mathrm{e}^{\mathrm{j}2\pi a\alpha t}\ \mathrm{d}t \\ &= \mathrm{e}^{\mathrm{j}2\pi\alpha[f + a(\alpha + 0.5t)]} \frac{\sin\left[\pi a\alpha\left(T - |\alpha|\right)\right]}{\pi a\alpha} \quad \text{for } T \le \alpha \le T. \end{aligned} \tag{6.7}$$

The general shape of this function is easier to see when Equation (6.7) is rewritten as

$$A(\alpha) = U(T - |\alpha|)\frac{\sin[\pi a\alpha(T - |\alpha|)]}{\pi a\alpha(T - |\alpha|)}, \tag{6.8}$$

where U is the unit magnitude phase contribution $\mathrm{e}^{\mathrm{j}2\pi\alpha[f + a(\alpha + 0.5T)]}$. The $U(T-|\alpha|)$ term is a triangle function weighting the $\sin(x)$ over x or the sinc function which follows. The match filtering process is shown in Figure 6.8.

Similarly, if the returned signal is delayed by τ, then by imposing the time shifting property in correlation, the output of the matched filter will be a shifted version of $A(\alpha)$; the shift is by a factor of τ as shown in Figure 6.9.

6.3.2 Range Resolution

The stationary (conventional) SAR or ISAR radar systems all resolve targets in the range dimension in the same way. It is the way in which azimuth resolution is achieved that they differ. Range resolution is achieved through the use of pulsing and time delay sorting as illustrated in Figure 6.3.

In Figure 6.3, if T is the pulse width and u is the speed of the pulsed electromagnetic wave, then the range resolution before compression is given by

$$\Delta R = \frac{uT}{2}. \tag{6.9}$$

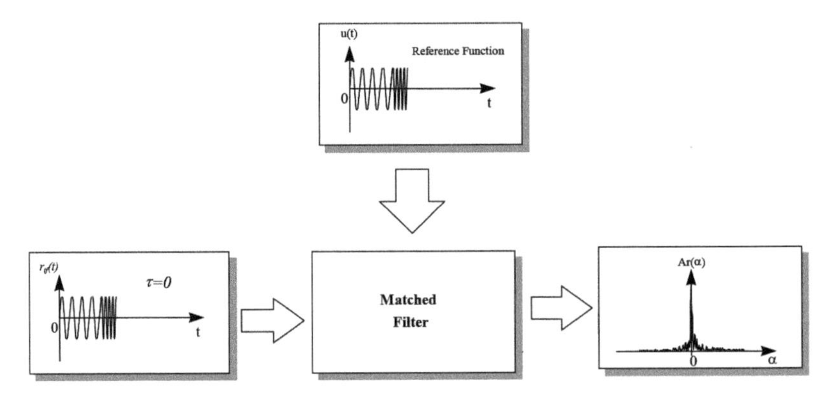

Figure 6.8 Autocorrelation of chirp pulse.

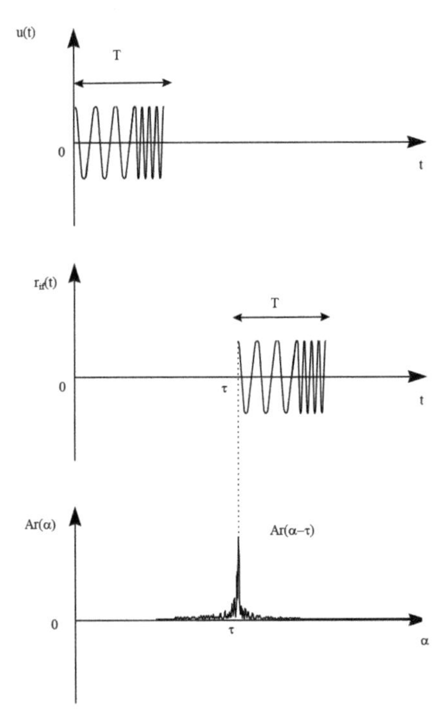

Figure 6.9 Correlated output of a target's echo, $r_{\mathrm{if}}(t)$, shifted by τ.

This is derived from the fact that to ensure non-overlapping reflections, targets must be separated in time delay by at least the width T of the transmitted pulse. In a conventional pulsed radar, the generation of a pulse of duration T

requires a transmitter bandwidth of the order of

$$B \approx 1/T. \tag{6.10}$$

Hence, expressing range resolution in terms of bandwidth, we have

$$\Delta R = \frac{u}{2B}. \tag{6.11}$$

Hence, a wide-band radar transmitter and receiver are important to achieve good range resolution. For a linear FM (chirp) pulse of duration T, the bandwidth B is given by

$$B = aT, \tag{6.12}$$

where a is the chirp rate.

Hence, the range resolution after compression is given by

$$\begin{aligned} \Delta R &= \frac{u}{2aT} \\ &= \frac{uT}{2aT^2}. \end{aligned} \tag{6.13}$$

Comparing Equations (6.11) and (6.13), it is seen that range resolution after chirp compression improves by a factor of aT^2. In other words, range resolution is further enhanced by the compression of the chirp pulse through the matched filter, which is what we expect. In other words, a chirp pulse gives a range resolution that is improved by a factor of aT^2.

This can be illustrated by a simple simulation: For a rectangular pulse of duration T = 0.5 μs, the range resolution is 75 m as given by Equation (6.11). But if chirp pulse of the same pulse duration T and chirp rate $a = 3 \times 10^{14}$ is used, this resolution can be improved by 75 times. Thus, the new theoretical range resolution is 1 m (refer to Equation (6.13)). Figure 6.10 shows two targets separated by 1 m barely resolved. When the targets are 7 m apart, they can be distinguished clearly as shown in Figure 6.11. This suggests that the range resolution is about 1–2 m, which is quite close to the theoretical resolution (1 m).

6.3.3 Effect of Pulse Width Variation

Figure 6.12 shows the comparison of the compressed amplitude response to the same point target with different pulse width T.

The dotted line is simulated with T = 0.2 μs, while the solid line is T = 0.5 μs. The result shows that the main lobe is narrower as the pulse width

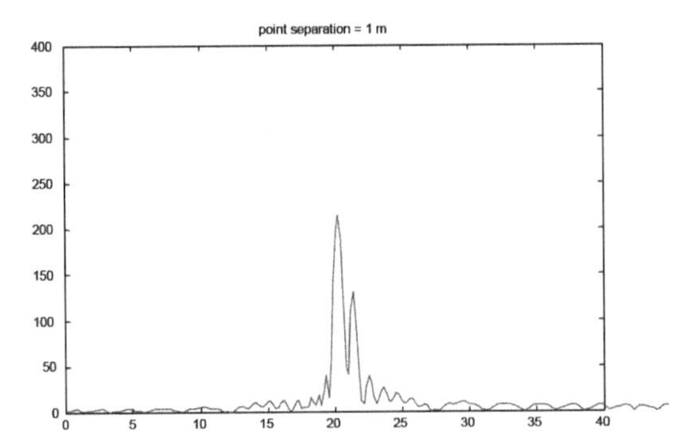

Figure 6.10 Two targets indistinguishable (1 m apart).

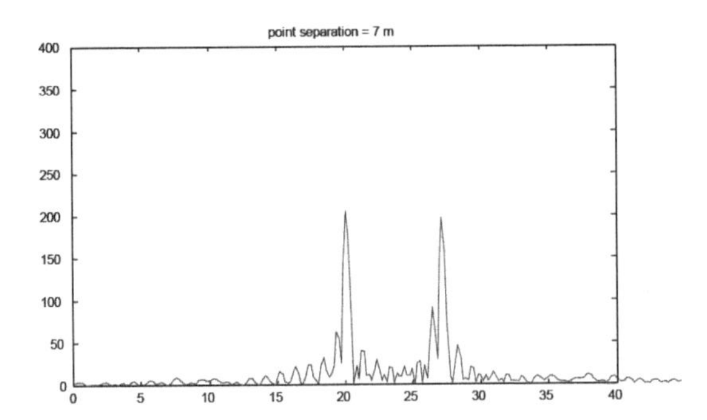

Figure 6.11 Two targets separated by 7 m.

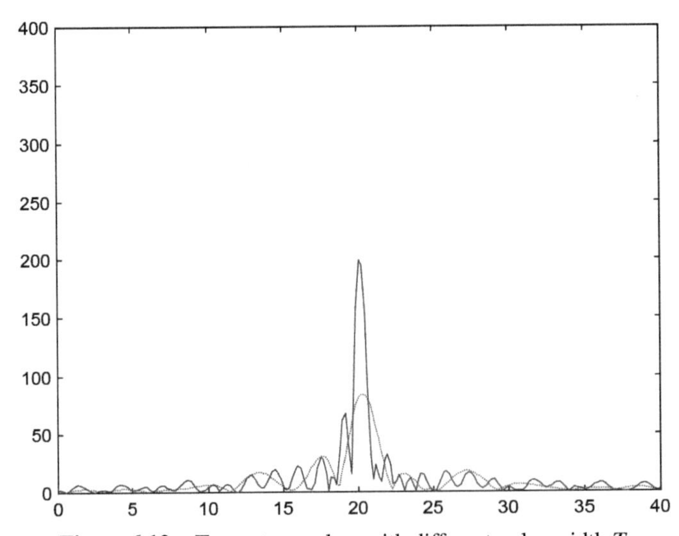

Figure 6.12 Two return pulses with different pulse width T.

T becomes larger. It implies that the resolution is finer. This is true because by increasing the pulse duration T in chirp signaling, the bandwidth of the transmitted pulse also increases as shown in Equation (6.12). As discussed previously, the range resolution increases as the bandwidth of the system increases (see Figure 6.13). It should also be noted that increasing pulse duration increases the overall transmitted power or, equivalently, improves the output signal-to-noise power ratio. This accounts for the difference in the amplitude of the pulse as shown in Figure 6.13. This is not true in rectangular pulse signaling, where increasing T improves signal-to-noise power ratio but results in a poorer range resolution.

6.3.4 Effect of a Chirp Rate Variation

In Equation (6.13), it is evident that the range resolution after compression is dependent on the chirp rate, a. This is because by increasing the chirp rate, the bandwidth of the signal also increases (refer to Equation (6.12)). Since range resolution is inversely proportional to the bandwidth of the transmitted signal, with increasing bandwidth, the range resolution improves as shown in Figure 6.14. It can be seen that the main lobe of the impulse is narrower as the chirp rate a is increased, and hence resulted in a better resolution.

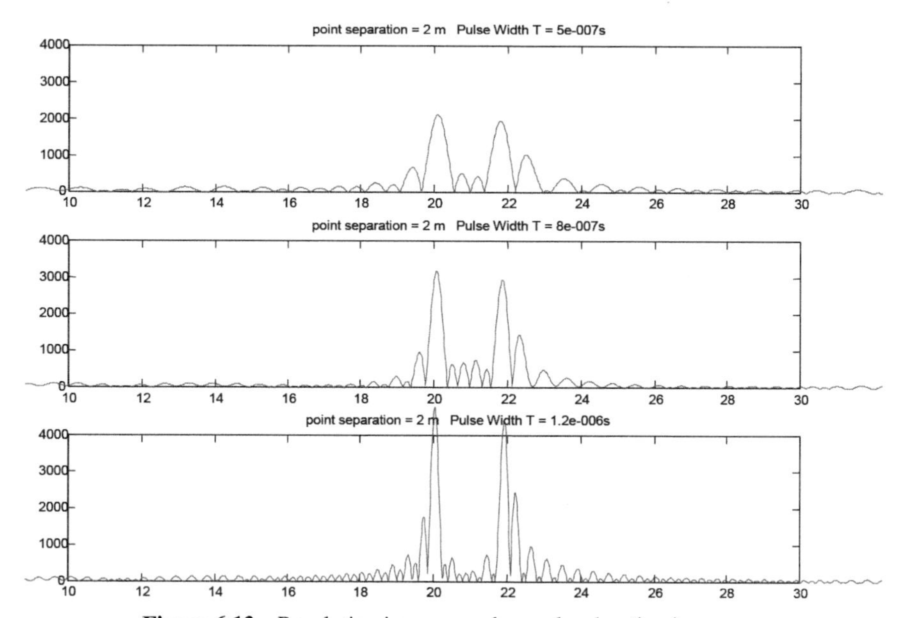

Figure 6.13 Resolution improves when pulse duration increases.

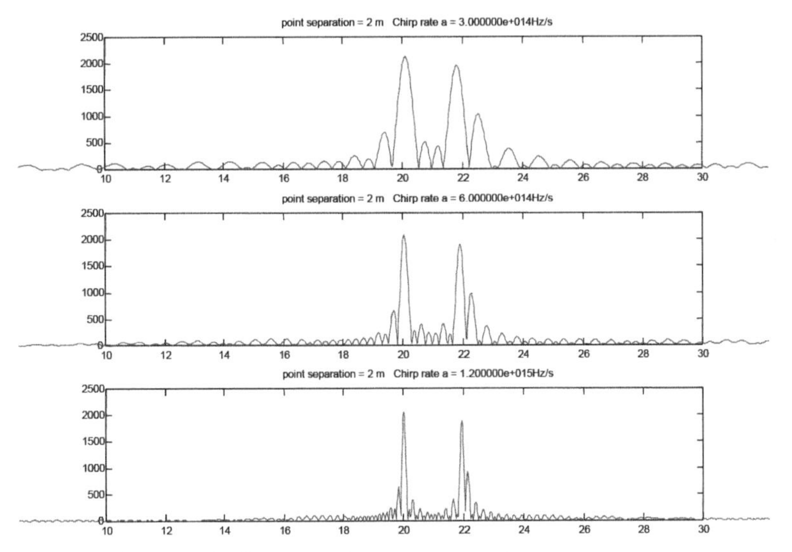

Figure 6.14 Resolution improves when chirp rate increases.

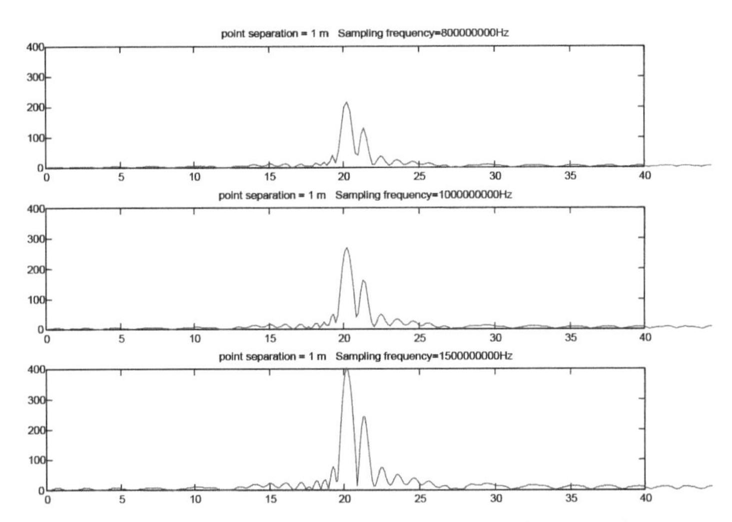

Figure 6.15 Resolution unchanged when sampling frequency increases.

6.3.5 Effect of Sampling Frequency Variation

When faced with a lack of resolution in the image, it is natural to attempt to increase the size of the frequency-sampled array by sampling it at a faster rate. As shown in Figure 6.15, however, this is fruitless: although there are more sampling points, they still cover the same bandwidth. As such, and according to Equation (6.12), the range resolution remains unchanged. Thus, the addition of more sampling points over the existing bandwidth does not improve resolution.

6.4 Two-Dimensional Imaging with Point Scattering

In the one-dimensional imaging discussed earlier, it is assumed that the scattering points are all arranged in a straight line. In general, some points could be displaced in the azimuth direction.

For instance, in Figure 6.16, point 1 and point 3 have the same relative azimuth position, but point 2 is displaced at a distance of $A_{12,\,32}$ to the right of point 1 and point 3.

In this case, range imaging alone does not bring out the complete picture (see Range View in Figure 6.16). It only presents a one-sided restricted view of the positions of the point scatterers and it may be mistaken for three colinear points. Therefore, a two-dimensional picture of the target will give more information in terms of the relative range and azimuth positions of the point scatterers on the target, and the characteristic of the target can be seen clearly.

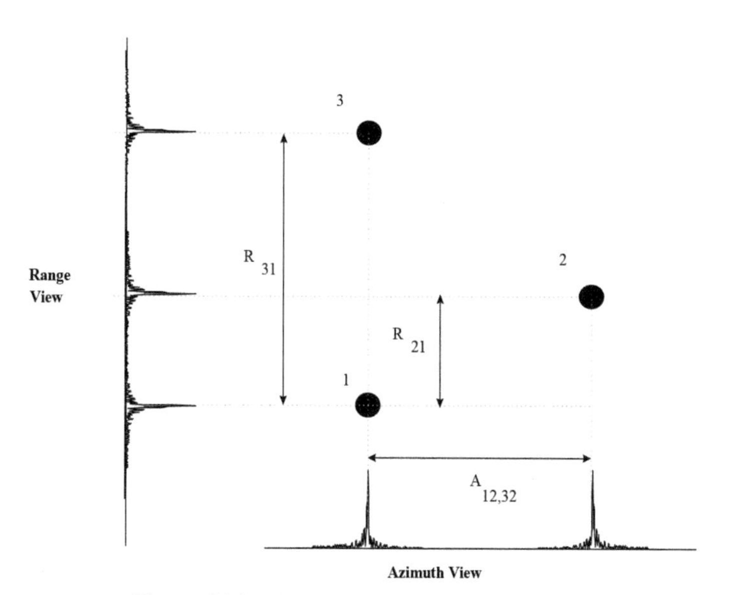

Figure 6.16 Plan view of three point scatterers.

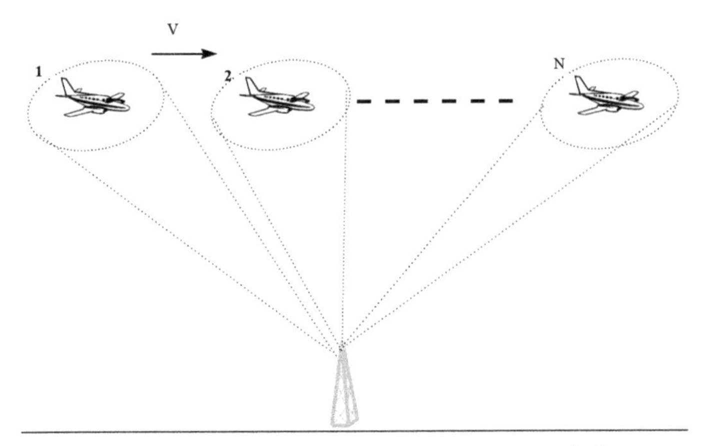

Figure 6.17 Data collection geometry for linear ISAR.

6.4.1 Overview

The creation of the two-dimensional image is extended from one-dimensional imaging. In order to extract the azimuth information, the target is required to move around the radar. In the imaging process, the target is sequentially illuminated at different positions by the radar beam as the latter travels with uniform velocity in a straight line as shown in Figure 6.17.

For each position, the one-dimensional range image of the target (after compression by the matched filter) can be easily obtained. This is repeated at every time interval of $T_{\text{p-p}}$, where $T_{\text{p-p}}$ is the time interval between two consecutive transmitted pulses. For different positions *n*, the one-dimensional range image of the target (which is actually an array of absolute numbers Ar_{n1}, Ar_{n2}, Ar_{n3}, ..., Ar_{nm}) is collected and stored into a two-dimensional array as shown in Figure 6.18.

By a proper processing of this collection of range echoes, the final two-dimensional image of the target can be generated. The detailed procedures of plotting the image of the target are presented in the next section.

6.4.2 Procedures for Two-Dimensional Imaging

6.4.2.1 Data Collection

Consider a target moving with uniform velocity *V* horizontally over the antenna with an aperture *L* as shown in Figure 6.19. Assume that there are three point scatterers (a, b, and c) on the target and they are arranged in a triangular formation.

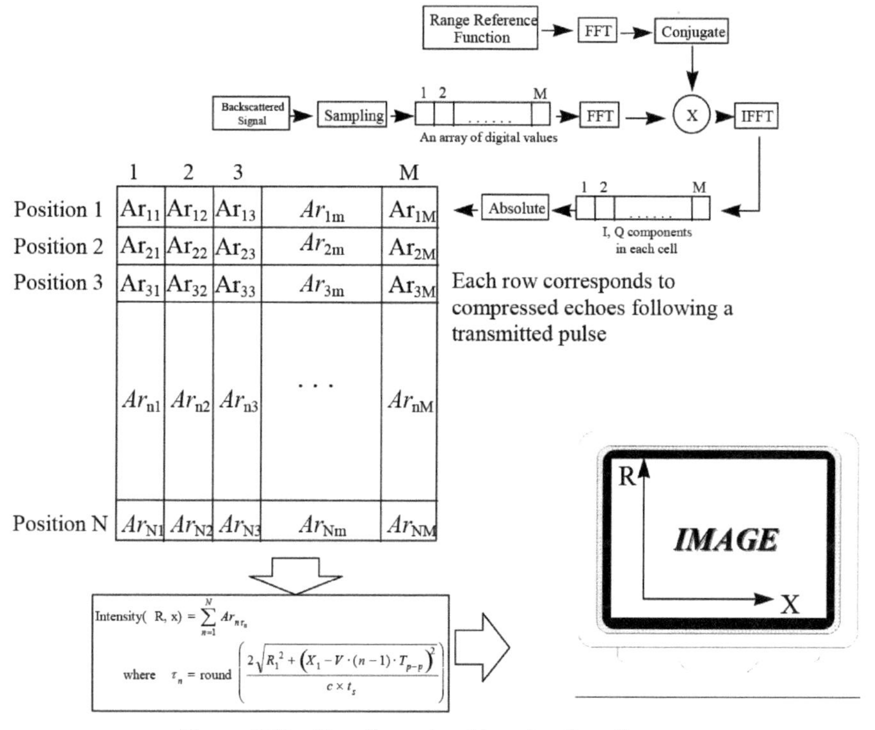

Figure 6.18 Two-dimensional imaging flow diagram.

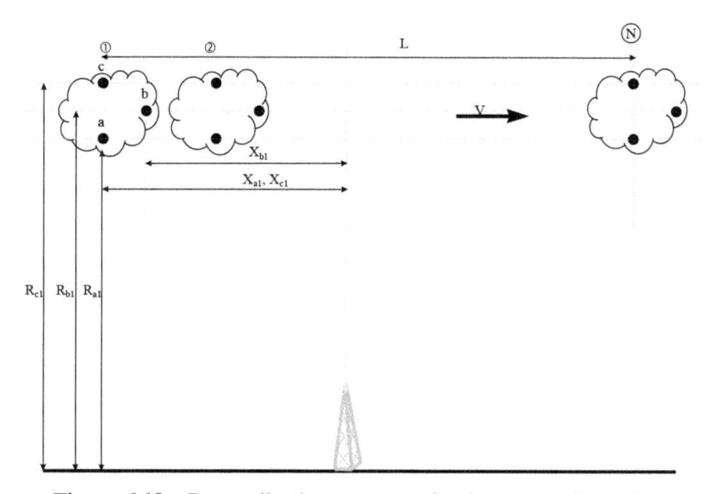

Figure 6.19 Data collection geometry for three scattering points.

When the target is first at position 1, it is illuminated by the radar beam and the return chirp pulse from the three scatterers is compressed into three narrow pulses by matched filtering as shown in Figure 6.21 (top figure).

The time delays for each point scatterer at position 1 are given below:

$$\tau_{\text{a1}} = \frac{2\sqrt{R_{\text{a1}}^2 + X_{\text{a1}}^2}}{c},$$

$$\tau_{\text{b1}} = \frac{2\sqrt{R_{\text{b1}}^2 + X_{\text{b1}}^2}}{c},$$

$$\tau_{\text{c1}} = \frac{2\sqrt{R_{\text{c1}}^2 + X_{\text{c1}}^2}}{c}.$$

As the antenna would be emitting chirp pulses at regular intervals $T_{\text{p-p}}$, during this time interval, the target has actually traveled a distance of $VT_{\text{p-p}}$ to reach position 2. Similarly, another one-dimensional range image of the target at the new position can be obtained (see Figure 6.20, middle figure). Note

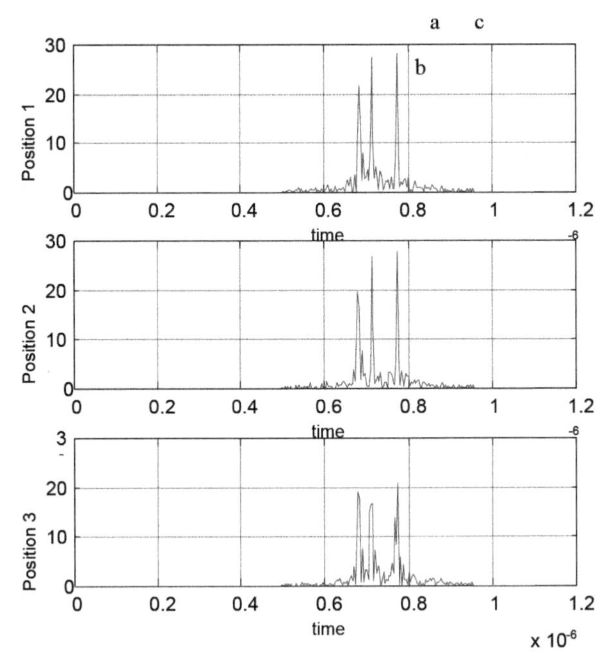

Figure 6.20 One-dimensional range imaging for the first three positions (generated by program Fig.6_21.m).

that the output of the matched filter is actually a $1 \times M$ array of complex numbers which are then converted to absolute numbers and plotted to give the one-dimensional range image. This process is carried out for N positions. The array of real values for all the N positions is collected and stored into a table as shown in Figure 6.18. This $N \times M$ array can be plotted and shown in Figure 6.21. The plot shows the intensity variations accumulated around three arcs corresponding to each of the three scatterers.

6.4.2.2 Concept for Two-Dimensional Imaging

For a target moving in a straight path with constant velocity V, the general relationship for the time delay of each point scatterer on the target at position n is

$$\tau_n = \frac{2\sqrt{R_1^2 + \left(X_1 - V(n-1)T_{\text{p-p}}\right)^2}}{c} \quad \text{for } n = 1, 2, 3, \ldots, N. \tag{6.14}$$

This expression has two arguments R_1 and X_1 which represent range and azimuth length at position 1, respectively (see Figure 6.19). Referring to Figure 6.19, if the point scatterer "a" located at (40, 100) (refer to Table 6.1)

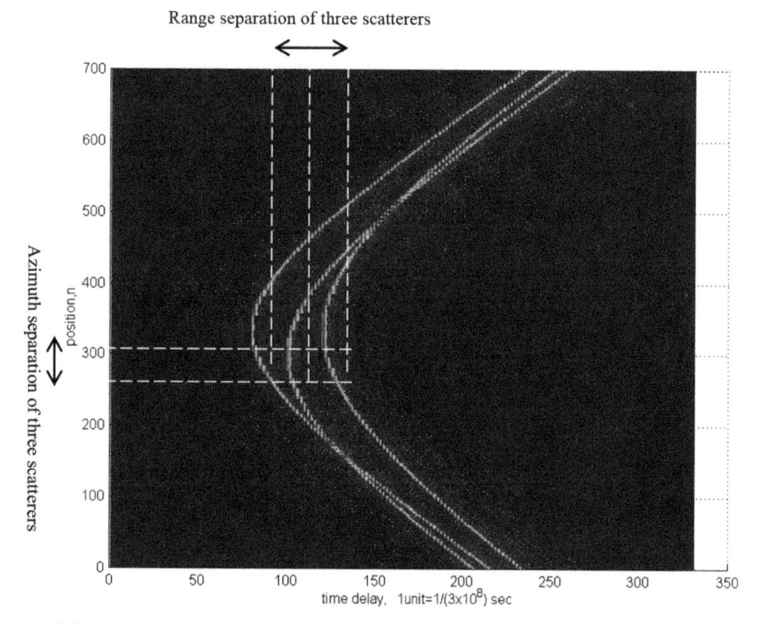

Figure 6.21 Range migration of three scattering points (generated from fig5_22.m).

Table 6.1 Parameters used in the simulation.

Parameters	Descriptions	Values	
System			
	f_c	Carrier frequency	1275 MHz
	f_{if}	Intermediate frequency	2 MHz
	T	Chirp pulse duration	0.18 s
	A	Chirp rate	0.8×10^{15} Hz/s
	$T_{p\text{-}p}$	Pulse to pulse interval	1 ms
	N	Number of positions	700
	t_s	Sampling interval	$1/(3 \times 10^8)$ s
	M	Number of sampling points	330
Target			
	V	Target velocity	300 m/s
Three point scatterers Considered			
Point "a"	R_{a1}	Range	40 m
	X_{a1}	Azimuth	100 m
Point "b"	R_{b1}	Range	50 m
	X_{b1}	Azimuth	90 m
Point "c"	R_{c1}	Range	60 m
	X_{c1}	Azimuth	100 m

is allowed to move from position 1 to position N, then the corresponding time delay at each position can be calculated using Equation (6.14). This is shown in Figure 6.22, where the horizontal axis is the time delay and the vertical axis represents the position number. The simulation parameters provide a useful illustration only; they do not represent any specific radar system.

Note that the location (R_{a1} = 40, X_{a1} = 100) of the point scatterer is not known. The task now is to substitute all possible combinations of (R_1, X_1) into the time delay expression in Equation (6.14) to match the same range migration curve in Figure 6.22, which is traced out uniquely by scatterer "a." This is shown in Figure 6.23, where four combinations of R_1 and X_1 values are tried and the corresponding traces are obtained. Hence, this shows that for a particular point scatterer located at (R_1, X_1), the time delay curve traced out is unique and it will only match when the correct value of (R_1, X_1) is used. With this, the location of the point scatterer can be obtained.

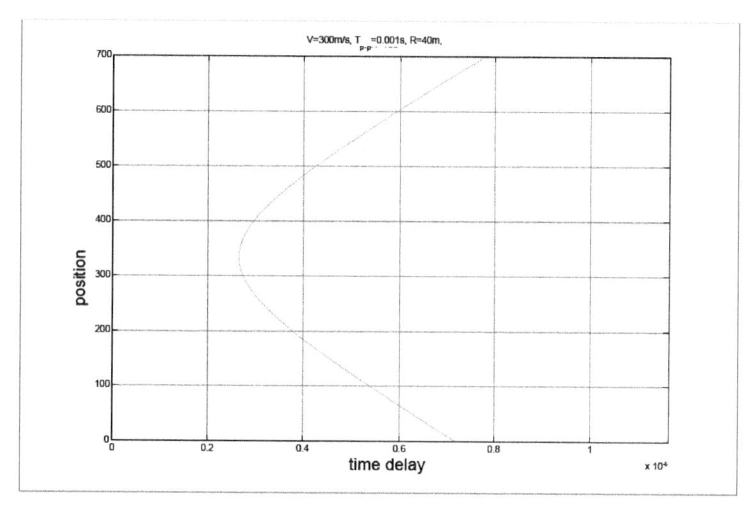

Figure 6.22 Plot of time delay of one point scatterer.

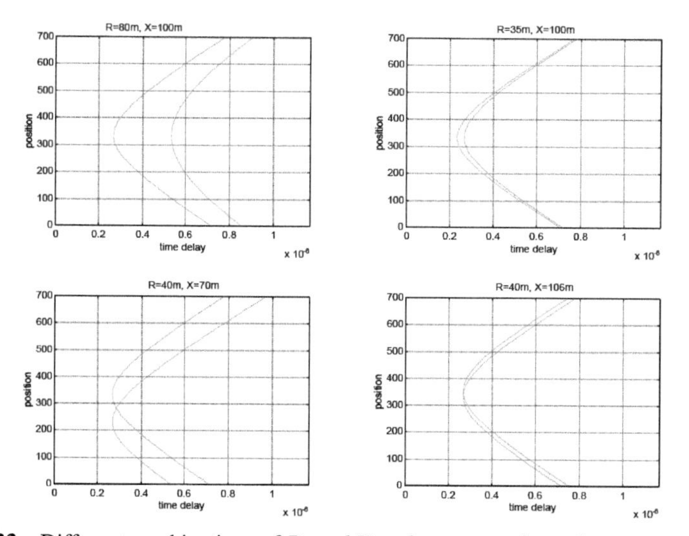

Figure 6.23 Different combinations of R_1 and X_1 values attempting to locate correct values.

6.4.2.3 Development and Implementation

This section will show how to apply and implement the concept described above to obtain a two-dimensional image (with range and azimuth) with the $N \times M$ array of data gathered during the data collection stage.

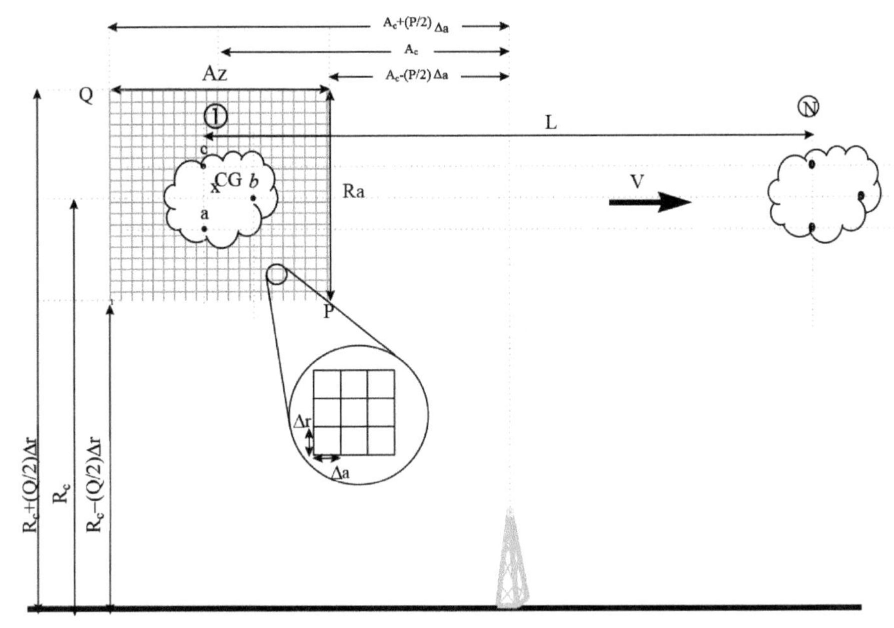

Figure 6.24 Parameters used to customize the plot.

Figure 6.24 shows the geometry and the parameters that will be used in the discussion. In order to produce a two-dimensional image plot of the target, the following information must be provided to customize the plot:

1) Az and Ra – the values of Az and Ra define the size of the plot and they are chosen to cover the target completely. The approximate size of the target of interest is usually known.
2) Δr and Δa – range and azimuth resolutions, respectively.
3) A_{c} and R_{c} – they define the centroid of the plot. A_{c} and R_{c} are given by the position of the centroid of the target where it is first illuminated by the beam (see Figure 6.25).

Therefore, the plot will contain $Q \times P$ pixels, where

$$P = \frac{\mathrm{Az}}{\Delta a}, \qquad Q = \frac{\mathrm{Ra}}{\Delta r}.$$

Suppose the parameters shown in Figure 6.24 assumed the values given in Table 6.2.

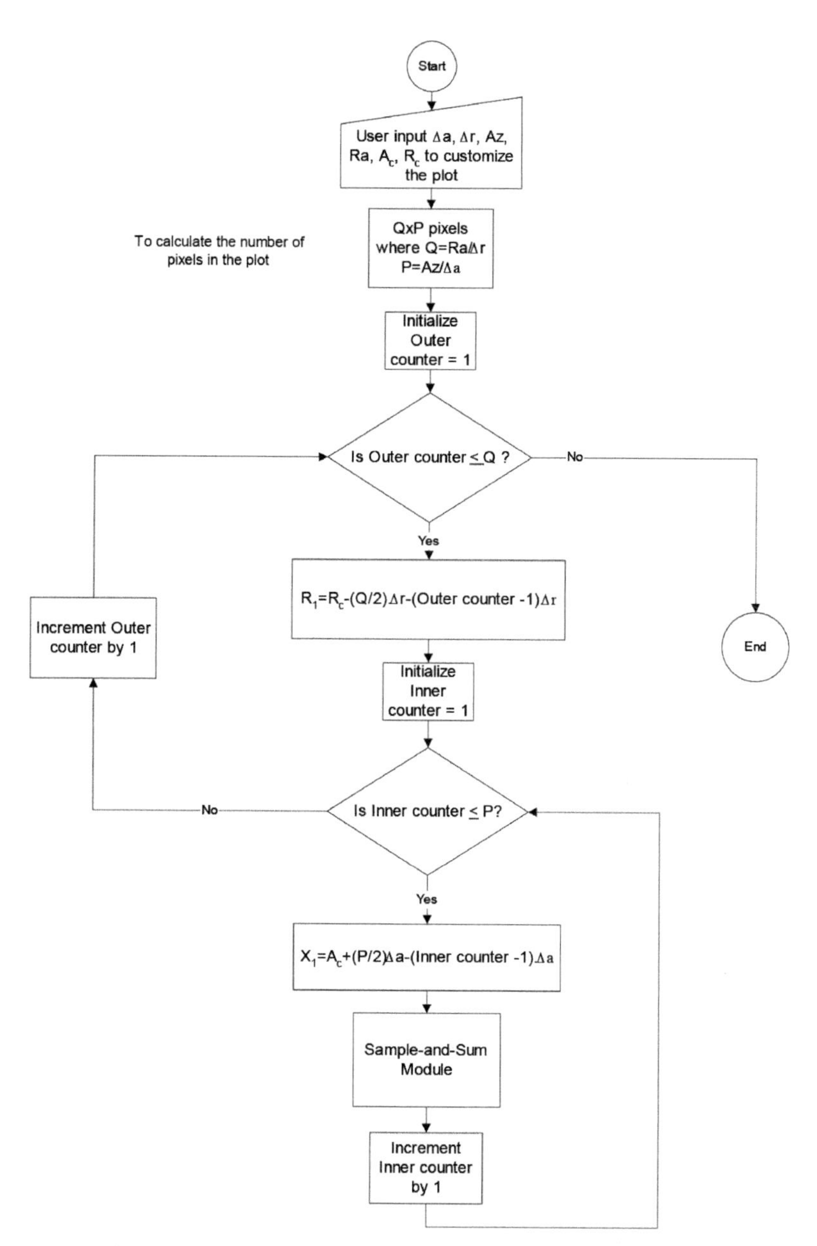

Figure 6.25 Flow chart showing the sequential testing of pixels.

Table 6.2 Values of the parameters used in Figure 6.24.

Parameters	Description	Values
R_c	Range scene center	50 m
A_c	Azimuth scene center	95 m
Δr	Range resolution	0.5 m
Δa	Azimuth resolution	0.5 m
Ra	Range scene size	50 m
Az	Azimuth scene size	50 m
Q	Number of pixels in range direction	100
P	Number of pixels in azimuth direction	100
V	Velocity of target	300 m/s
$T_{\text{p-p}}$	Pulse to pulse interval	1 ms
L	Synthetic aperture length ($L = VNT_{\text{p-p}}$)	210 m

Recall that during the data collection stage, a matrix of data is obtained as shown in Figure 6.18. With these data and applying the curve matching concept discussed earlier, the level of intensity at each pixel can be given by

$$\text{Intensity}(R, x) = \sum_{n=1}^{N} \text{Ar}_{n\tau_n},$$

$$\text{where } \tau_n = \text{round}\left(\frac{2\sqrt{R_1{}^2 + (X_1 - V(n-1)T_{\text{p-p}})^2}}{c \times t_s}\right). \quad (6.15)$$

The process of sampling and summing is carried out sequentially and systematically for every pixel as shown in Figure 6.25. The algorithm for implementing the sample and sum operation is also provided in Figure 6.26.

6.4.3 Simulation Results

Based on the parameter values given in Tables 6.1 and 6.2, the two-dimensional plot of the target is then obtained as shown in Figure 6.27.

The three small patches indicate the location of the scattering points. With the two-dimensional plot above, the relative range and azimuth positions of the point scatterers on the target can be easily seen. The arrangement of the point scatterers used in the simulation is also shown in Figure 6.30 for comparison.

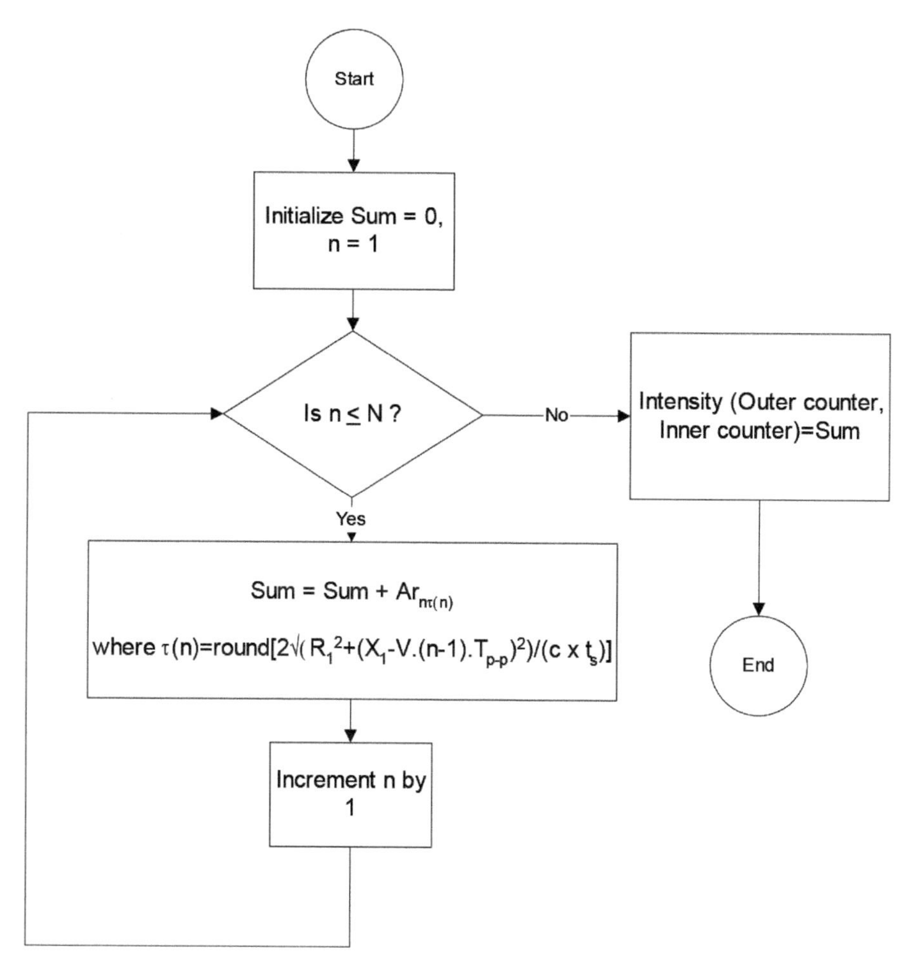

Figure 6.26 Flow chart for range compression.

As shown in Figure 6.27, the locations of three points reconstructed via the procedure outlined in Section 6.4.2 match quite well with their actual position. The shape of the targets, however, has been distorted due to the sidelobes associated with the procedure. The white streaks surrounding each point target result from these sidelobes.

Next, a 16-point-scatterer model of an aircraft is considered. Each wing of the aircraft was considered to be made from two point scatter centers. Each of the two drop tanks were replaced by a scatter center. The rest of the aircraft

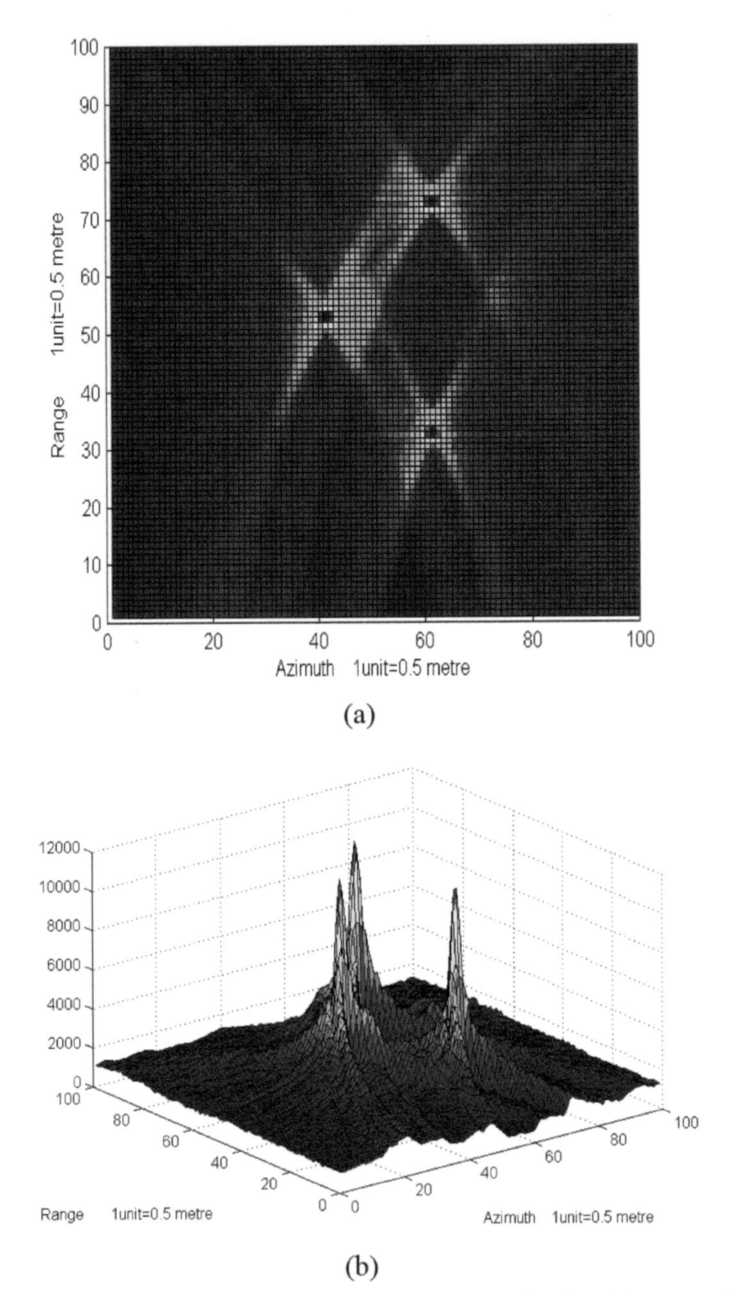

Figure 6.27 Reconstructed two-dimensional image. (a) Intensity plot of the target with three scattering points. (b) Three-dimensional plot of the target with three scattering points.

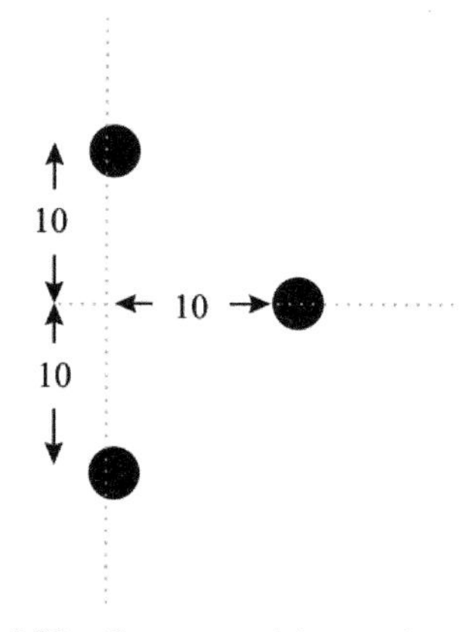

Figure 6.28 Geometry of three point scatterers.

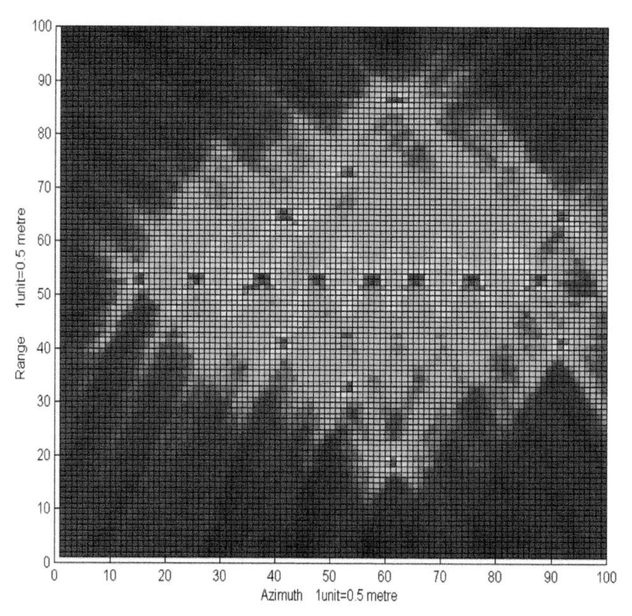

Figure 6.29 Reconstruction of an aircraft image; image of an aircraft using 16 point scatterers.

was considered to consist of 10 point scatter centers. The ISAR reconstruction of Next, the three-point model is extended to a 16-point-scatterer model of an aircraft image, is shown in Figure 6.29.

7

Smart Antennas: Mobile Station Antenna Location

Stetson Oh Kok Leong, Ng Kim Chong, P.R.P. Hoole, and E. Gunawan

Abstract

In order to have a base station (BS) smart antenna beam pointing always toward a desired mobile station (MS; e.g. wireless communication with a desired customer) or a mobile cluster of transceivers (e.g. a cluster of soldiers communicating with their commander), it is necessary to be able to dynamically track the MS in order to automatically form the BS beam toward the MS. This is also the case when an MS antenna should always be turned toward the BS with which it desires to be in contact. This chapter presents the computer technique to automatically track the MS using electromagnetic signal processing based on a model of the signal transmitted from the MS antenna. The chapter illustrates how the position of the MS may be accurately tracked using this model-based electromagnetic signal processor as the MS moves form one BS cell to another BS cell of a wireless systems.

7.1 Mobile Radio Environment

Objects surrounding the base station (BS) and mobile station (MS) severely affect the propagation characteristics of the uplink and downlink channels of cellular systems. This propagation path loss, including reflection and shadowing, tends to degrade system capacity. The height of the MS antenna (e.g. 2 m) is normally much lower than that of the surrounding buildings and natural features. Furthermore, the carrier frequency wavelength is also much less than the size of the surrounding structures. Due to this, an MS will experience significant changes of its received signal strength as it moves. The mobile receiver is characterized by "multipath reception." Its received signal contains a number of electromagnetic waves from the same source (transmitter) arriving at the receiver antenna along different paths. Even when line-of-sight (LOS) is available (this is rare in urban areas), the additional electromagnetic waves beside the direct wave result from the reflection, refraction, scattering, and diffraction of the transmitted signal off objects along the propagation path. These extra waves arrive at the receiving antenna displaced with respect to each other in time and space. Due to this phenomenon, the resultant received signal that appears at the receiver amplifier could be much weaker or stronger than the direct wave. As shown in Section 2.11, every half wavelength in space, nulls of fluctuation at the baseband can be visualized, although the nulls do not occur at the same level. If the MS moves at a faster speed, the rate of fluctuation observed by the receiver becomes faster.

Fading due to multipaths and shadowing, Doppler spread (due to the motion of one of the antennas) and delay spread are some of the main channel effects that arise from these phenomena (see Figure 7.1).

7.1.1 Fading

When an MS moves over small distances of a fraction of a wavelength, the instantaneous field strength at the receiver antenna may rapidly fluctuate as much as four or five orders of magnitude. This is known as signal fading or small-scale fading. There are two approximately separable effects known as fast and slow fading. Fast fading is characterized by deep fades that occur within fractions of a wavelength and is caused by multipath scattering of the signal off objects in the vicinity of the mobile. It is most severe in heavily built-up areas where the number of waves arriving from different directions with different amplitudes and phases cause the signal amplitude to follow a

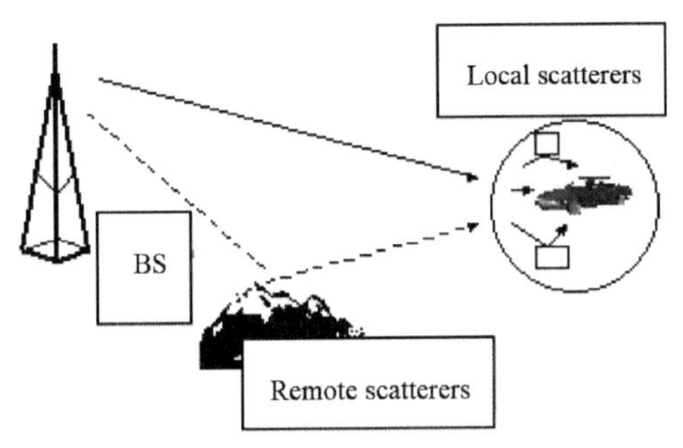

Figure 7.1 Multipath.

Rayleigh distribution. However, fast (Rayleigh) fading also occurs to a certain extent in suburban areas.

The overall signal with multipath reflections can be expressed as (Razavi, 1999)

$$x_{\rm R}(t) = a_1(t)\cos(w_{\rm c}t + \theta_1) + a_2(t)\cos(w_{\rm c}t + \theta_2) + \cdots + a_n \cos(w_{\rm c}t + \theta_n) \quad (7.1)$$

$$= \left[\sum_{j=1}^{n} a_j(t)\cos\,\theta_j\right]\cos\,w_{\rm c}t - \left[\sum_{j=1}^{n} a_j(t)\sin\,\theta_j\right]\sin\,w_{\rm c}t. \quad (7.2)$$

For a large number of multipaths (i.e. large n), each summation has a Gaussian distribution. Denoting the first summation by A and the second by B gives

$$x_{\rm R}(t) = \sqrt{A^2 + B^2}\cos(w_{\rm c}t + \phi), \quad (7.3)$$

where $\phi = \tan^{-1}(B/A)$.

For such a scenario of signals, it can be shown that the envelope of the received signal has a Rayleigh density function given as follows:

$$p(y) = \begin{cases} \frac{y}{\sigma^2} l^{-y^2/2\sigma^2}, & y \geq 0, \\ 0, & y < 0, \end{cases} \quad (7.4)$$

where y is a random variable representing the signal level fluctuation and σ is its standard deviation.

When there is a direct LOS component present, as would be the case when the MS is traveling along a highway or open countryside, together with the Rayleigh or multipath, the received signal becomes Rician. Its probability density function is as follows:

$$p(y) = \begin{cases} \frac{y}{\sigma^2} {\rm e}^{-(y^2+s^2)/2\sigma^2} J_0\left(\frac{ys}{\sigma^2}\right), & y \geq 0, \\ 0, & y < 0, \end{cases} \quad (7.5)$$

where $J_0(.)$ is the modified Bessel function of the zeroth order and s^2 is the mean power of the direct path.

Slow fading related to shadowing is the result of buildings and trees that stand between the BS and MS antennas. It is observed that such slow fading exhibits a log-normal distribution given by

$$p(z) = \begin{cases} \frac{1}{\sqrt{\pi}\sigma z} {\rm e}^{(\log z - u)^2/2\sigma^2}, & z > 0, \\ 0, & \end{cases} \quad (7.6)$$

where μ is the mean of the random variable z and the standard deviation is between 5 and 10 dB.

7.1.2 Doppler Spread

In a wireless channel with multipath Rayleigh fading signals, let the nth reflected wave arrive from an angle θ_n relative to the direction of the motion of the MS antenna. We have shown in Section 2.3 that the frequency of the received signal will go through a shift due to the motion of the antenna. The change in the received signal frequency of this Rayleigh fading signal is known as the Doppler shift. It can be represented by the formula

$$\Delta f_n = \frac{v}{\lambda} \cos \theta_n \tag{7.7}$$

where ν is the speed of the mobile antenna and λ is the wavelength.

Hence, motion of the MS antenna that results in the Doppler shift produces phase shifts of each reflected wave. When the waves arrive at the antenna, since they all have different phase shifts, the amplitude of the resulting composite signal will be modified by the Doppler shift, and hence the velocity vector of the MS.

When the number of multipath waves that arrive at the receiver is large and each wave has its own random angle-of-arrival (AOA; hence, with its own Doppler shift), the baseband power spectrum of the vertical electrical field has the following form:

$$S(f) = \frac{3\sigma}{2\pi f_{\mathrm{m}}}\left[1 - \left(\frac{f - f_{\mathrm{c}}}{f_{\mathrm{m}}}\right)^2\right]^{-1/2}, \quad f_{\mathrm{c}} - f_{\mathrm{m}} < f < f_{\mathrm{c}} + f_{\mathrm{m}}, \tag{7.8}$$

where the AOA is assumed to be uniformly distributed within 0–2π, $f_m = v/\lambda$ is the maximum Doppler shift, f_{c} is the carrier frequency, and σ is the mean signal power. It is seen from Equation (7.8) that the received signal strength will be dependent on the Doppler shift and hence the velocity vector of the MS.

Doppler spread causes time-selective fading since it directly impinges on the phase of the signals. Thus, the instantaneous received signal amplitude varies with time. The signal amplitude in the presence of Doppler fading is characterized by and inversely proportional to the coherence time. We define the coherence time as the time separation over which the channel impulse response at two time instants remains strongly correlated.

7.1.3 Delay Station Spread

Delay spread is a well-known phenomenon at high frequency electromagnetics. In wired systems, we get delay spread due to the high frequency portion of a signal (e.g. a rectangular pulse) traveling much faster than the lower frequency portion of the signal. Hence, at the line termination, the signal looks spread out in the time domain since the lower frequency energy arrives later than the high frequency energy. In wireless systems, time spread occurs due to multipaths, in other words, due to reflection and differences in the distance traveled by each reflected signal. When the distances are different, although the signals travel at the velocity of light, their arrival times will be different and proportional to the distance traveled. Hence, multipath effects create time dispersion and a spreading of the signal. Urban delay spreads of around 3 μs are commonplace. Delay spread results in frequency-selective fading, that is, it is dependent on the frequency. If the variation of the delay is comparable with the symbol period, delay signals from an earlier symbol may interfere with the next symbol, causing intersymbol interference (ISI). It can be characterized by the coherence bandwidth, which is inversely proportional to the delay spread. Coherence bandwidth is defined as the maximum frequency difference for which two frequency-shifted signals are still strongly correlated in terms of either amplitudes or phases.

7.2 Mobile Station Positioning

The importance and usefulness of accurate MS position information has been recognized in recent years. Besides E911 services, it opens up a wide area of possibilities. For example, it can play an important role in improving the system capacity either directly or indirectly, ranging from the development of better beamforming algorithm and enhanced adaptive modulation technique to effective cellular system design.

A hierarchical cell system using the position and velocity of the MS offers a good compromise between an efficient use of available channels while simultaneously keeping the number of handoffs small. In this system, cells of different sizes coexist. Equal-sized cells are grouped into layers, which overlay on top of one another to form a hierarchy. Thus, a layer with large cells can be assigned to fast moving MSs, while a layer with small cells or microcells to the slow moving MSs. This serves two purposes: 1) the cells of small and large radius provide a more economically efficient system for higher and lower traffic densities; 2) subscribers of lower and higher mobility can efficiently be provided with service in the small cells and umbrella cells, respectively.

MS position estimation can be categorized under MS-based and BS-based position-finding techniques. For MS-based techniques, the MS makes use of signals transmitted by the surrounding structures to compute its own position. These structures can be BSs in a cellular network or satellite network in the sky. The latter technique relies solely on the existing infrastructure of cellular BSs.

7.2.1 Global Positioning Satellite

A medium-earth-orbit (MEO) satellite system having about 24 satellites is used in the global positioning satellite (GPS) system (see Figure 7.2). A GPS receiver installed at the MS, as in a taxi for instance, provides the self-position estimate of the MS using a group of satellites in the MEO network. Each satellite transmits a spread spectrum signal to earth on the L-band (centered at 1575.42 MHz). Using an accurate clock for precise timing, the GPS receiver measures the time delay between the signals leaving, say, three satellites in the sight of the MS and arriving at the MS receiver. This allows calculation of the exact distance from the MS to each satellite. Thus, the MS position can be computed using the triangular method, providing coordinates in latitude, longitude, and altitude. In practice, a fourth satellite is used to correct receiver clock errors. GPS is one of the most popular radio navigation aids due to its high accuracy. A commercial GPS receiver costs under US$ 150 with an accuracy of approximately 50 m.

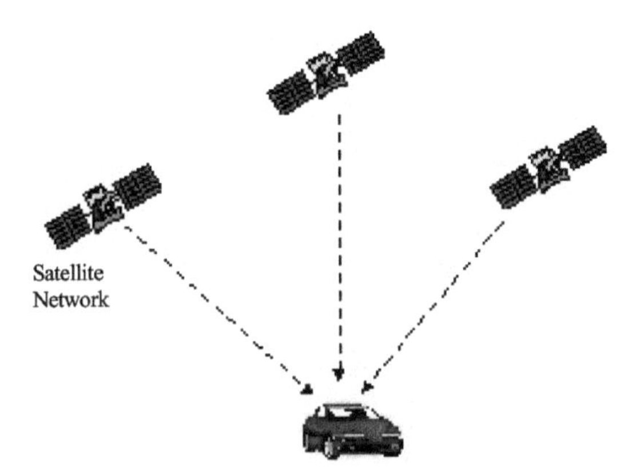

Figure 7.2 Global positioning satellite.

However, GPS has certain disadvantages. The MS has to carry a GPS receiver, which means increased weight, size, and battery drainage. Furthermore, when the L-band electromagnetic signals are blocked or attenuated by buildings, foliage, and heavy cloud cover, the resulting estimate of the MS position may be quite inaccurate due to missed data. One further technical problem is to do with the antenna: the MS antenna is designed to operate in the UHF (0.3–1.0 GHz), L-band (1.0–2.0 GHz), or S-band (2.0–4.0 GHz) wireless communication frequencies, whereas the satellite GPS system operates in the L-band. Where the frequencies differ, antenna performance will also be affected.

7.2.2 MS Positioning in the Cellular Network

7.2.2.1 BS-Based Positioning

In smart antenna technology, BS has played a critical role in making it possible to install large array antennas to achieve adaptive, steerable beamforming where the multiple beams keep track of the MS. With a smart antenna with 10 elements installed at the BS, in the uplink, a signal-to-noise ratio (SNR) improvement of 10 dB, a range increase of 2, a signal to interference noise ratio (SINR) improvement of 20 dB, and a data rate increase of 2 could be achieved. In the downlink, the BS emission can be reduced by 90% and a capacity increase of about 3 is also possible. However, in order to generate these smart beams, the position of each MS should be located. BS-based positioning is implemented mostly using a triangular method based on signal strength, AOA, time-of-arrival (TOA) measurement, or their combination.

Mobile position estimation using signal strength measurement is one of the most well-known methods using the path loss attenuation with distance information. Its primary source of error is multipath fading and shadowing. Variations in the signal strength can be as great as 30–40 dB over a distance of the order of half a wavelength. Signal strength averaging can help, but low-mobility MSs may not be able to average out the effects of multipath fading, and there will still be adverse effects due to shadow fading. The errors due to shadow fading may be handled by using pre-measured signal strength contours centered at the BSs. However, this approach requires that the contours be mapped out for each BS and a data bank be available for use.

This signal strength based approach to position estimation is difficult in digital systems where transmission power is controlled close to 1000 times every second. The code-division multiple access (CDMA) system employs power control in order to combat the near–far effect. The time-division multiple access (TDMA) system uses power control to conserve battery power of the MSs in the upward link. Thus, for such systems, unless the instantaneous transmission power of the MSs is known, it may not be possible to achieve reasonable accuracy.

The AOA methods are sometimes also referred to as direction-of-arrival (DOA) methods. The AOAs of the signal from the MS are calculated at the BS by using adaptive phased antenna arrays at the BS. The main beam of the antenna array is electronically steered until it locks on to the signal arriving at the MS. Two closely spaced antenna arrays are used to scan horizontally to get the exact direction of the peak strength of the incoming MS signal. Since the antenna elements are closely placed, the time delay between the two elements seen by a signal as it propagates across the array may be modeled by a phase shift. This is the basic arrangement for most AOA estimation algorithms. Accurate AOA is estimated when the signal coming in from the MS is a direct LOS signal. If the antenna array beam should lock on to a reflected signal, the AOA estimate will obviously be wrong. Furthermore, the AOA estimator performance degrades as the distance between the MS and the BS increases.

The TOA-based estimation of position is also possible. In free space, the time taken for an electromagnetic wave to travel over a distance is proportional to d since the velocity of the wave is equal to the constant speed of light. Hence, the BS may determine the distance d by first indirectly determining the time that the signal takes to travel from the source to the receiver on the forward or the reverse link. This TOA may be obtained by measuring the time in which the MS responds to an inquiry or an instruction transmitted to the MS from the BS. The total time elapsed from the instant the command is transmitted to the instant the MS responds may be stored, and this time is equal to the sum of the round trip signal delay and any processing and response delay within the MS unit. If the MS microprocessor processing delay for the desired response is known with sufficient accuracy, it can be subtracted from total measured time. The resultant time would give the total round trip delay. Half of the round trip delay would be an estimate of the signal delay in one direction. If we multiply this with the velocity of light, we will get the distance between the BS and the MS. If the MS signal can be detected at a minimum of three BSs, then the MS position can be computed by the triangulation method.

There are certain problems with the TOA method. The estimate of the microprocessor processing time and response delay of MS electronics within the MS may be difficult to determine in practice. Indeed, different manufacturers of MS phones will use different electronics and circuitry, giving rise to quite different microprocessor-discrete electronics response times. Furthermore, LOS between the BS and MS is not available, and severe timing errors could occur. The problem will be difficult to handle when there are many multipath signals arriving at the BS.

7.2.2.2 MS-Based Positioning

This method requires the MS to use signals from several BSs to calculate its position. It is a self-positioning system, where the MS estimates its own position. Again the signal strength (this time from BS to MS) or the TOA information may be used. However, modification to the existing MS software will be necessary since the position computation will be performed at the MS. The control microprocessor must carry the position-estimation software in order to store and process information from four or more BSs (see Figure 7.3).

The MS receiver must be capable of simultaneously processing information from at least four BS frequency channels, either for signal strength or time delay information. Data from three BSs are required for the position estimation, while the fourth is needed to cycle through all available signals to ensure that the best three BS signals are being used for the position estimation. In practical implementation, it is necessary for the network to maintain a BS location database and transfer information to the MS whenever it is requested.

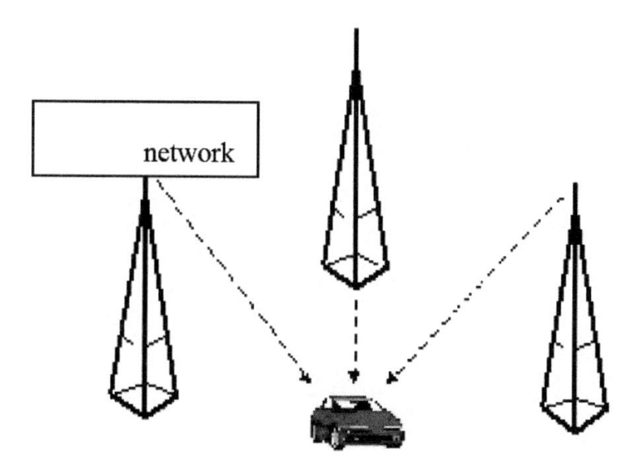

Figure 7.3 MS positioning in the cellular network.

7.3 Position and Velocity Estimation in Cellular Systems

Information about the position and velocity of MSs can be obtained using two available quantities (Stuber, 1996), namely the signal strengths of the MS at different BSs and the corresponding propagation times. However, both quantities are subject to strong fluctuations caused by short-term fading, shadowing, and reflections, which make them useless unless a sophisticated method is used to counter this problem and translate the quantities into the required information.

A new method of estimating the position and velocity of an MS in a cellular network, based on an electric-field strength model, is presented in this chapter. The algorithm developed uses the principle of maximum likelihood estimation and is tested in MATLAB™ simulation experiments for different channel conditions.

7.3.1 Antenna Signal Model

In this section, a model for electromagnetic fields radiated by an infinitesimal element carrying current of any geometrical shape is presented. The geometry is defined in Figure 7.4.

From the radiated electric fields from a finite-sized wire antenna obtained in Chapter 3, we get

$$\begin{aligned}\mathbf{E}(R,t) &= \mathbf{u}_R \frac{h\cos\theta}{4\pi}\left[\left(\frac{\mu_0}{\varepsilon_0}\right)^{1/2}\frac{2[I]}{R^2}+\frac{2[Q]}{\varepsilon_0 R^3}\right]\\ &\quad+\mathbf{u}_\theta\left[\frac{h\sin\theta}{4\pi}\frac{\mu_0}{R}\frac{\mathrm{d}[I]}{\mathrm{d}t}+\left(\frac{\mu_0[I]}{\varepsilon_0 R^2}\right)^{1/2}+\frac{[Q]}{\varepsilon_0 R^3}\right].\end{aligned} \tag{7.9}$$

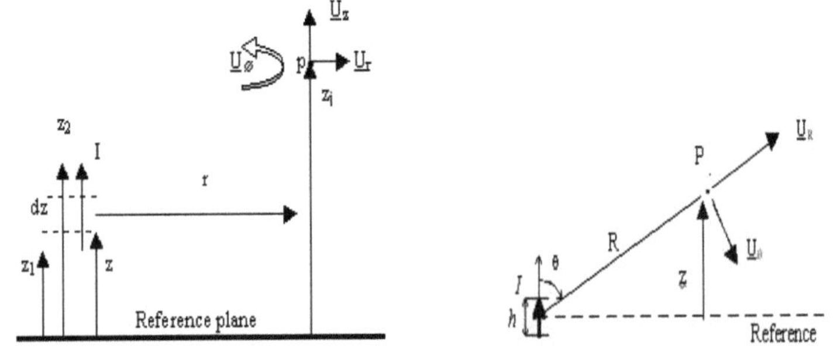

Figure 7.4 An infinitesimal current carrying element and its coordinate system.

In cylindrical coordinates, using cylindrical coordinate unit vectors $\mathbf{u}_r$ and $\mathbf{u}_z$, Equation (7.9) may be written as

$$\begin{aligned}\mathbf{E}(r,z,t) =& \mathbf{u}_r\left[\frac{3h}{4\pi}\left(\frac{\mu_0}{\varepsilon_0}\right)^{1/2}\frac{rz}{(r^2+z^2)^2}[I]+\frac{\mu_0 h}{4\pi}\frac{rz}{(r^2+z^2)^{2.5}}\frac{\mathrm{d}[I]}{\mathrm{d}t}\right.\\ &\left.+\frac{3h}{4\pi\varepsilon_0}\frac{rz}{(r^2+z^2)^{2.5}}[Q]\right]\\ &+\mathbf{u}_z\left[\frac{h}{4\pi}\left(\frac{\mu_0}{\varepsilon_0}\right)^{1/2}\frac{2z^2}{(r^2+z^2)^2}\right.\\ &-\frac{r^2}{(r^2+z^2)^2}[I]-\frac{\mu_0 h}{4\pi}\frac{r^2}{(r^2+z^2)^2}\frac{\mathrm{d}[I]}{\mathrm{d}t}\\ &\left.+\frac{h}{4\pi\varepsilon_0}\left(\frac{z^2}{(r^2+z^2)^{2.5}}-\frac{r^2}{(r^2+z^2)^{2.5}}\right)[Q]\right]. \end{aligned} \tag{7.10}$$

It can be further simplified as the receiving antenna will only pick up the E_z, the vertical polarized signal. Therefore, by ignoring E_r and assuming that $[Q] = 0$,

$$\begin{aligned}\mathbf{E}'(r,z,t) =& \mathbf{u}_z\left[\frac{h}{4\pi}\left(\frac{\mu_0}{\varepsilon_0}\right)^{1/2}\frac{2z^2}{(r^2+z^2)^2}-\frac{r^2}{(r^2+z^2)^2}[I]\right.\\ &\left.-\frac{\mu_0 h}{4\pi}\frac{r^2}{(r^2+z^2)^2}\frac{\mathrm{d}[I]}{\mathrm{d}t}\right]. \end{aligned} \tag{7.11}$$

By setting $h = \mathrm{d}z$, $z = z_j – z$, $\mathrm{d}z = -\mathrm{d}z$ and integrating Equation (7.11), and considering only far-field regions, the electric-field strength radiated by a linear dipole antenna of finite length $(z_2 - z_1)$ is of the form

$$E(r,z,t) = \frac{\mu_0}{4\pi}\left[\frac{z_j - z_1}{\sqrt{r^2+(z_j-z_1)^2}} - \frac{z_j - z_2}{\sqrt{r^2+(z_j-z_2)^2}}\right]\frac{\mathrm{d}[I]}{\mathrm{d}t}, \tag{7.12}$$

$$[I] = Re\left(I_0 \mathrm{e}^{\mathrm{j}w(t-R/c)}\right), \tag{7.13}$$

where I_0 is the amplitude and ω is the frequency of the current flowing through the transmitter antenna. Thus,

$$\frac{\mathrm{d}\,[I]}{\mathrm{d}t} = \mathrm{j}\omega I_0 \mathrm{e}^{\mathrm{j}\omega(t-R/c)}. \tag{7.14}$$

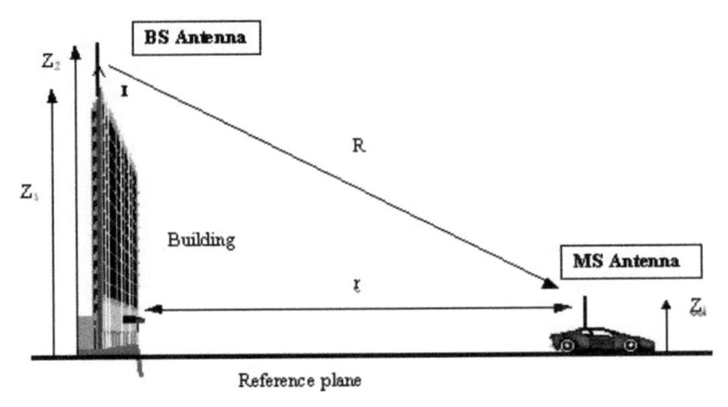

Figure 7.5 *E*-field model parameters.

The signal processor operates only on the magnitude of the measured electric-field strength. Hence, Equation (7.14) is reduced to

$$\left|\frac{\mathrm{d}\,[I]}{\mathrm{d}t}\right| = \omega I_0. \tag{7.15}$$

Therefore, Equation (7.12) can be expressed as

$$E(r,z) = A_0\left[\frac{z_j - z_1}{\sqrt{r^2 + (z_j - z_1)^2}} - \frac{z_j - z_2}{\sqrt{r^2 + (z_j - z_2)^2}}\right], \tag{7.16}$$

where $A_0 = \mu_0 \omega I_0 / 4\pi = 10^{-7}\omega I_0$ and

$$I_0 = \sqrt{\frac{P_\mathrm{r}}{R_\mathrm{r}}}. \tag{7.17}$$

P_r is the power radiated and R_r is the radiation resistance of the transmitter antenna. The parameters of the model are defined in Figure 7.5. The advantage of using this model is that it provides a more realistic picture of a BS–MS communication link as it takes into account the distance r between the BS and the MS as well as their respective heights (z_1, z_2, and z_j) with respect to the ground.

7.3.2 Position and Velocity Estimation Algorithm

The measured instantaneous electric-field strengths E_m in the presence of additive white Gaussian (AWG) noise is modeled by

$$E_\mathrm{m} = E + n, \tag{7.18}$$

where E is defined in Equation (7.16) and n is the AWG noise. Since E is a non-linear function of r, it can be expressed as

$$E(r) = E(r_0) + E'(r_0)\Delta r + 0.5E''(r_0)\Delta r^2 + \cdots. \tag{7.19}$$

Ignoring higher order terms and writing $E_0 = E(r_0)$, let

$$J_0 = E'(r_0) = \frac{\delta E_0}{\delta r_0}, \tag{7.20}$$

$$E(r) = E_0 + J_0 \Delta r. \tag{7.21}$$

Substituting Equation (7.21) into Equation (7.18) gives

$$n = (E_{\mathrm{m}} - E_0) - J_0 \Delta r = \Delta z - J_0 \Delta r. \tag{7.22}$$

We seek to minimize

$$U(\Delta r) = n^{\mathrm{T}} n. \tag{7.23}$$

Substitute for n using Equations (7.21) and (7.22), and let

$$\frac{\partial U(\Delta r)}{\partial r} = 0,$$

and we get

$$\Delta r = (J_0^{\mathrm{T}} J_0)^{-1} J_0^{\mathrm{T}} \Delta z. \tag{7.24}$$

Therefore, r can be estimated using a maximum likelihood estimator (MLE) of the following form:

$$r_{n+1} = r_n + (Jr_n^{\mathrm{T}} Jr_n)^{-1} Jr_n^{\mathrm{T}} (E_{\mathrm{m}} - E_n), \tag{7.25}$$

where E_{m} is a column vector that represents the measured electric-field strengths at the BSs and Jr_n is the Jacobian matrix of the electric-field strengths E_n. The superscript T represents a non-conjugate transpose

$$Jr_n = \frac{\partial E_n}{\partial r_n} = -A_0 \left[\frac{z_j - z_1}{[r_n^2 + (z_j - z_1)^2]^{3/2}} - \frac{z_j - z_2}{[r_n^2 - (z_j - z_2)^2]^{3/2}} \right] r_n. \tag{7.26}$$

Using the estimated r vector, which gives the distances between the BS's positions (PB) and the MS, an estimated position of the MS (PM) can be obtained. However, since the radii of the circles taken from the r estimates are unlikely to arrive at an exact cross intersection, further processing is necessary to reach a closer estimate of the actual position of MS.

Thus, an iteration equation using MLE similar to Equation (7.25) is formed. We have

$$\mathrm{PM}_{k+1} = \mathrm{PM}_k + (Jp_k^{\mathrm{T}}\ Jp_k)^{-1}\ Jp_k^{\mathrm{T}}\ (0 - \mathrm{errp}_k), \tag{7.27}$$

where errp = |PM – PB| – r. Jp_k is defined as

$$Jp_k = \begin{bmatrix} \partial \mathrm{errp}_1 / \partial x & \partial \mathrm{errp}_1 / \partial y \\ \partial \mathrm{errp}_2 / \partial x & \partial \mathrm{errp}_2 / \partial y \\ \partial \mathrm{errp}_3 / \partial x & \partial \mathrm{errp}_3 / \partial y \end{bmatrix}. \tag{7.28}$$

The velocity of the MS (V) can be obtained from sequential PM estimates as follows:

$$V_m = (\mathrm{PM}_m - \mathrm{PM}_{m-1})/T, \tag{7.29}$$

where T is the time interval between two discrete positions of the MS.

7.3.3 Simulation Scenario

The position and velocity estimation (PVE) algorithm is tested by running simulation in MATLAB. The area of interest is a region of 6000 × 6000 m that contains a total of 16 cells, each with a cell radius of 1000 m. A BS, indicated by the alphabet "A" shown in Figure 7.6(a), is located at the center of each cell. An MS travels along a route indicated by the dotted line with its velocities shown in Figure 7.6(b). Twenty sampling points are taken along the route at a time interval of 96 ms. The actual positions of the MS are marked by "x."

As the MS travels along the route, it establishes a wireless communication link with the BS within the cell. This BS is referred to as the control base station (CBS). It is assumed that the desired signal always arrives from the CBS. At least three BSs, including the CBS, are required to compute the position of the MS.

The assumptions made in computing the BS position and velocity are as follows:

1) The network maintains a database of the locations of the BSs and this is sent to the MSs whenever it is requested.
2) LOS is available between the MS and the BSs or a strong direct signal path exists.
3) The CBS used is the BS closest to the MS.
4) No transmission power control in the forward link (downward link) is used for the signal being used for estimation. For example, in CDMA, the pilot tone may be used for PVE.

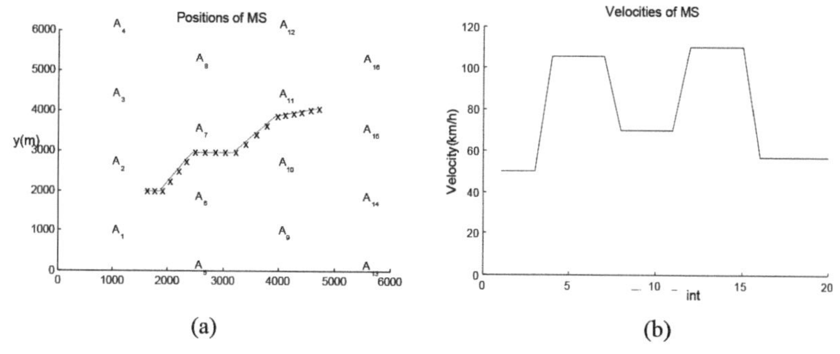

Figure 7.6 Actual positions and velocities of the MS.

Table 7.1 Simulation parameters.

Parameters	Description
$f = 900$ MHz	Signal frequency in GSM system
$c = 3 \times 10^8$ m/s	Speed of light
$z_1 = 60$ m	Height of the lower part of the BS antenna
$z_2 = z_1 + \lambda/2$	Height of the upper part of the BS antenna
$z_j = 1$ m	Height of the MS antenna
$R_r = 73\ \Omega$	Radiation resistance of the BS antenna
$P = 10$ W (M-cells)	Radiation power of the BS antenna
$P = 0.1$ W (P-cells)	

These assumptions are made to simplify the simulation scenario in order to study the performance of the PVE algorithm within reasonable constraints. To further examine the performance of the algorithm, it is also tested in areas with smaller cells (e.g. radius = 200 m). Cell sizes with radii of 1000 and 200 m are referred to as M-cells and P-cells, respectively (see Table 7.1).

7.3.4 Channel Models

The PVE algorithm is tested under different fading channel conditions. The channel models used in the simulation are described in the following subsections.

7.3.4.1 Additive White Gaussian

A Gaussian distributed random generator is used to generate the required random number N_{AWG}, where m is the mean amplitude and σ is the variance of the noise. A_n is a scaling factor used to adjust to the required SNR. All channel models in the simulation assumed zero mean and a variance of 0.161

unless otherwise stated

$$E_{\mathrm{m}} = E + A_n N_{\mathrm{AWG}}(m, \sigma), \tag{7.30}$$

$$\mathrm{SNR} = 10\ \lg \frac{E^2}{N_{\mathrm{AWG}}^2}. \tag{7.31}$$

7.3.4.2 Rayleigh Fading

The Rayleigh fading channel is simulated by (Loo, 1991)

$$N_{\mathrm{Rayleigh}} = \sqrt{N_{\mathrm{AWG}}(m, \sigma)^2 + N_{\mathrm{AWG}}(m, \sigma)^2}, \tag{7.32}$$

$$E_{\mathrm{m}} = N_{\mathrm{Rayleigh}} E + A_n N_{\mathrm{AWG}}(m, \sigma). \tag{7.33}$$

7.3.4.3 Dominant Reflected Path

To study the effects of multipath that results in the $1/r^4$ power decay, a dominant reflected path is assumed, with E_{d} and E_{r} being the direct and reflected components, respectively (see Figure 7.7). E_{d} is the E-field model defined in Equation (7.16). Assuming total reflection on the ground, the reflective coefficient R_{c} is –1. T_{d} is the time delay arrival of E_{r} at the MS with reference to E_{d}. k is a scaling factor calculated by $(1 - \mathrm{SNR}/100)$.

$$E_{\mathrm{m}} = |E_{\mathrm{d}} + kR_{\mathrm{c}}E_{\mathrm{r}}| + A_n N_{\mathrm{AWG}}(m, \sigma), \tag{7.34}$$

$$E_{\mathrm{r}} = A_0 \mathrm{e}^{-\mathrm{j}\omega T_{\mathrm{d}}} \left[\frac{z_j - z_{1r}}{\sqrt{r^2 + (z_j - z_{1r})^2}} - \frac{z_j - z_{2r}}{\sqrt{r^2 + (z_j - z_{2r})^2}} \right], \tag{7.35}$$

$$T_{\mathrm{d}} = (R_{\mathrm{r}} - R_{\mathrm{d}})/c, \tag{7.36}$$

where

$$z_{1r} = z_1 + z_j,$$
$$z_{2r} = z_2 + z_j,$$
$$R_{\mathrm{r}} = \sqrt{(z_2 + z_j)^2 + r^2},$$
$$R_{\mathrm{d}} = \sqrt{(z_2 - z_j)^2 + r^2}.$$

7.3.4.4 Rician Fading

When there is a direct LOS component present, together with the Rayleigh or multipath, it becomes Rician (Loo, 1991). In this case,

$$N_{\mathrm{Rician}} = \sqrt{(E + N_{\mathrm{AWG}}(m, \sigma)^2 + N_{\mathrm{AWG}}(m, \sigma)^2} \tag{7.37}$$

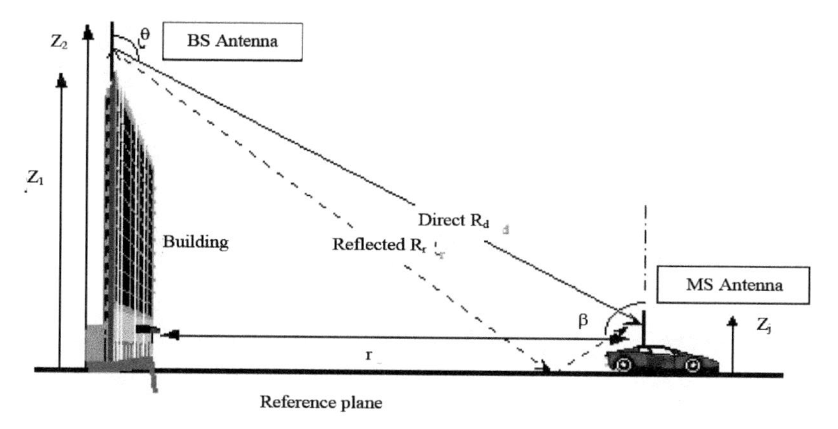

Figure 7.7 Dominant reflected path.

$$E_{\mathrm{m}} = N_{\mathrm{Rician}} + A_n N_{\mathrm{AWG}}(m, \sigma). \tag{7.38}$$

7.3.5 Antenna Radiation Pattern

All antennas for both BS and MS in the simulation are assumed to be omni-directional.

7.3.6 Initial Values

The PVE algorithm requires an initial set of r values (distance from MS to each of the selected BSs) to proceed. These values can be obtained by making use of timing information provided by the networks. It is found that with the help of this timing information to generate the required initial values, the algorithm is able to produce accurate results.

In wireless mobile systems, useful timing information called time advance (TA) measurement is provided which gives a round trip propagation time for the microwave signal to travel between an MS and a particular BS. TA measurement is a technique used in global system for mobile (GSM) to inform an MS how much time in advance of the reference signal it should transmit in order to synchronize correctly at the BS. It reduces the guard period between time slots.

The TA field is coded with eight bits allowing for 63 steps, where one step is one bit period (BP). BP, the fundamental unit of a frame, is equal to 48/13 μs (3.69 μs). The distance corresponding to a duration of BP is $r = cT_{\mathrm{b}}/2 = 554$ m. This is twice the distance between MS and BS as the uplink is timed relative to the frame structure in the downlink. By using the respective TA measurement, the distance between an MS and BSs (r values) can be coarsely estimated. Although these values are not accurate enough to be used for the estimation of the MS's position, they are useful as the initial values of r.

In CDMA systems, similar timing information is also available, known as pilot strength measurement. This information is available and constantly updated at the MS. During each conversation, the MS continuously searches for new pilots, as well as the strengths of the pilots associated with the forward traffics channels. MS reports each pilot's strength and PN phase (or arrival time) to the CBS whenever required. The pilot arrival time is the time of occurrence, as measured at the MS antenna, of the earliest arriving usable multipath component of the pilot. Thus, the reported pilot PN phase can be used to estimate the round trip delay to the BS from which the pilot is transmitted. The time delay information can then be used to compute a set of initial r values.

In the simulation scenario, the only time this information is used to provide the initial r values is when handoff occurred. In such cases, the strongest signal strength received by the MS is assumed to come from the closest BS (control base station). Hence, the handoff is based on some signal strength based handoff algorithms. At any other time (MS travels within a cell), the previous position of the MS is used instead to compute the initial r values.

7.3.7 E-Field Strength Measurement

The GSM network provides signal strength measurement of different BSs measured at a mobile. The corresponding measured E-field strengths can be obtained by the following equation:

$$E_{\mathrm{m}} = \sqrt{\frac{2(120\pi)P_{\mathrm{m}}}{A_{\mathrm{e}}}}, \tag{7.39}$$

where P_{m} is the signal strength and A_{e} is the effective aperture of the antenna. The effective aperture of a half-wave dipole antenna is $0.13\lambda^2$, where λ is the wavelength. The studies are carried out for two typical frequencies used in mobile communications: 900 and 1900 MHz.

Table 7.2 Comparison of estimated r with actual r.

MS position	Estimated r vector (km)	Actual r vector (km)
Position 1	[0.815, 1.044, 1.113]	[0.849, 1.059, 1.097]
Position 8	[0.644, 1.037, 1.438]	[0.623, 1.102, 1.457]
Position 15	[0.849, 0.927, 1.286]	[0.840, 0.955, 1.25]

7.3.8 Simulation Results

The iteration begins with the initial values of r and proceeds until a convergence accuracy of 10^{-4}. It is found that the PVE algorithm can converge very fast, usually in less than eight iteration cycles. The average convergence cycle is about cycle 6. Figure 7.8 shows the convergence curves for the estimated r vector (r_1, r_2, r_3) for MS positions 1, 8, and 15 under AWG noise channel for an SNR of 10 dB. The figure with the tabulated results in Table 7.2 shows that the estimated r is quite accurate as compared to the actual r.

Simulation results for the estimation of MS position and velocity for the Rayleigh channel condition are shown in Figure 7.9. Their respective average position and velocity errors are tabulated and shown in Table 7.3. The estimated MS position is marked by "x," while the actual route taken by the MS is indicated by the dotted line. The overall results shown in Table 7.3 and Figure 7.9 confirmed the good accuracy of the PVE algorithm. Based on the models used, the average position error is largest when a dominant reflected path is considered. This may be caused by a few poor estimates. However, the error may be greatly reduced if a filtering technique such as Kalman filter is employed.

It is observed that the PVE algorithm performs better in an area with smaller cells (cell radius = 200 m). For example, under Rayleigh fading channel for an SNR of 10 dB, the average position error for an area with a cell radius of 1000 m is 68 m, which is about 66% higher than that of an area with smaller cells. However, since the E-field model is derived based on the assumption that the transmitter is in the far-field region, the model could not be used in an area with very small cell structure such as picocell, which has a radius of only a few meters. The E-field model used needs to be redeveloped to take into account the near-field region.

7.3.9 Error Handlers

Like most estimation algorithms, there may be occasions when the PVE algorithm may encounter processing error during computation and fail to

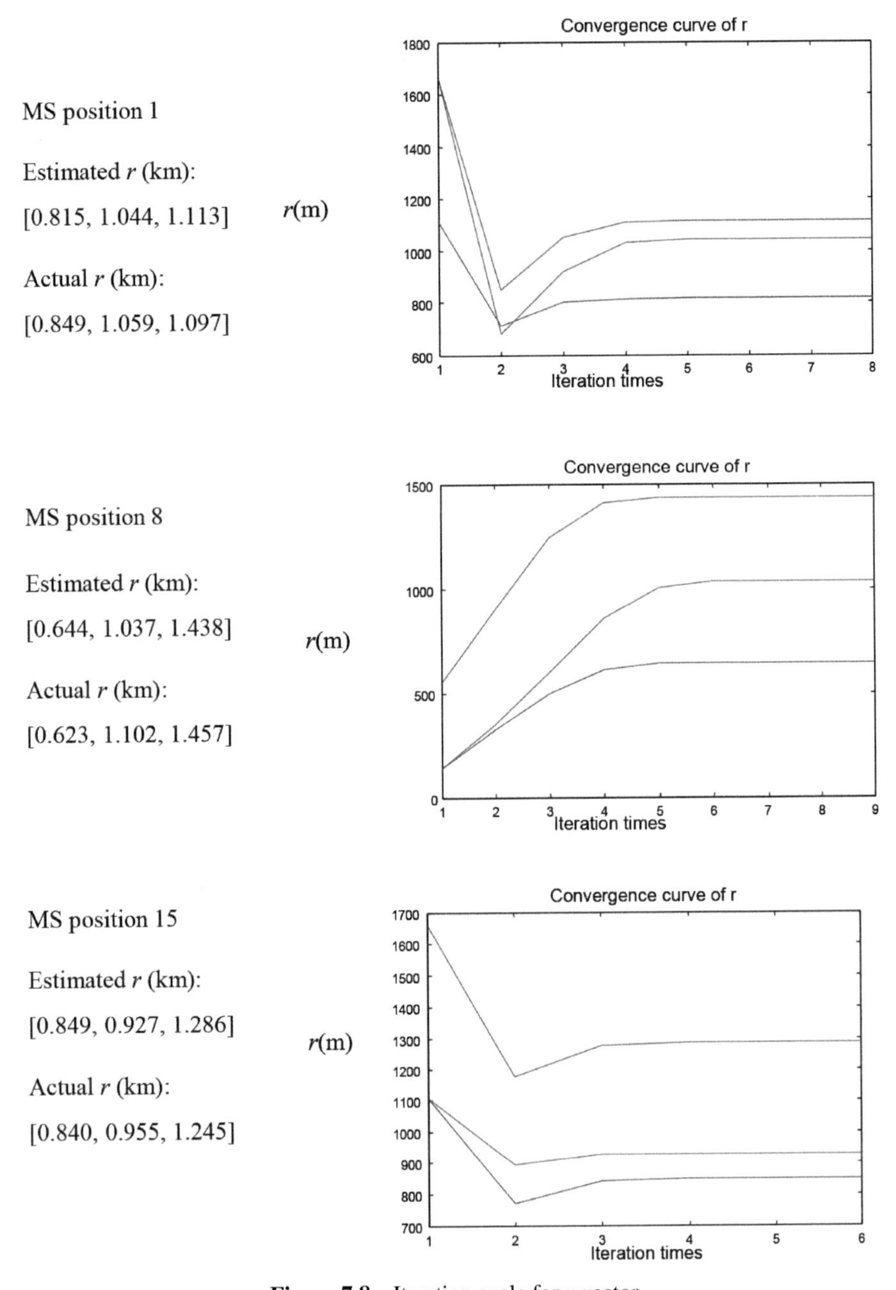

Figure 7.8 Iteration cycle for r vector.

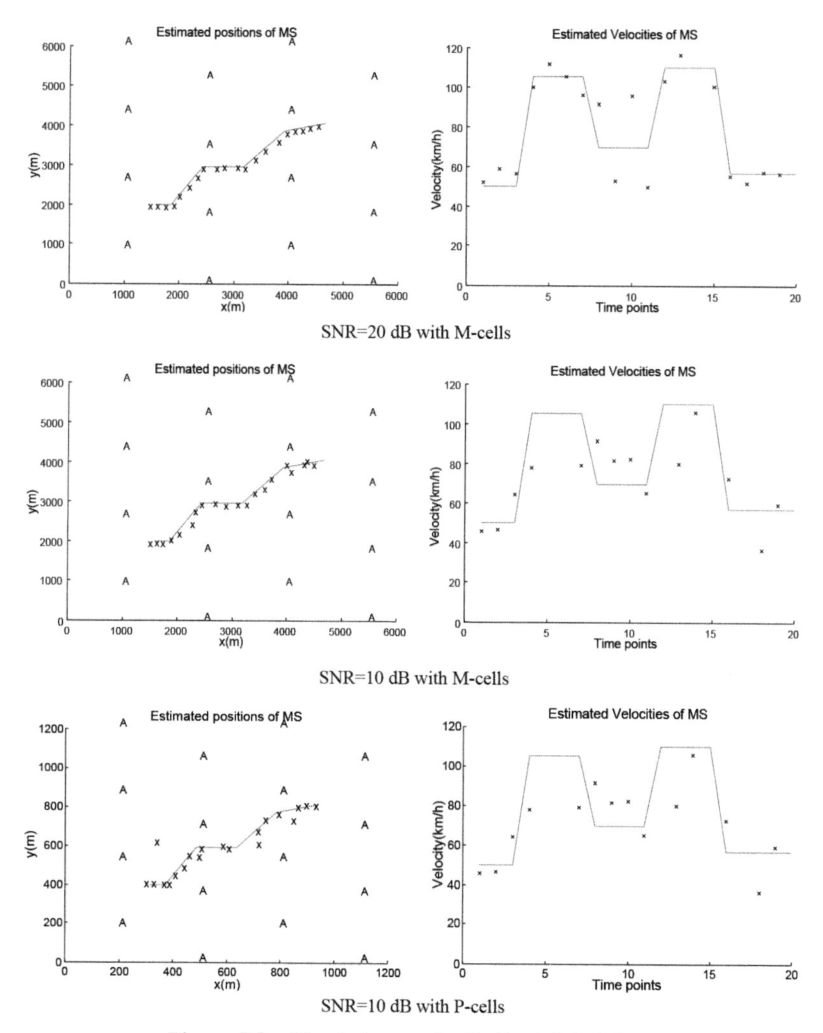

Figure 7.9 Simulation results for Rayleigh fading.

produce accurate estimates. This may happen in the case when low SNR is presented (<10 dB) or when the cell size is small (radius < 200 m). The processing errors encountered are: cell size is small (radius < 200 m). The processing errors encountered are:

1) Jacobian matrix becoming singular during iteration;
2) dividing by zero error when the MS position estimate during iteration is very close to a BS location (<2 m).

Table 7.3 Simulation results.

Channel model	SNR (dB)	Cell type	Average position error (m)	Average velocity error (km/h)
Gaussian	20	M-cell	21	8
	10	M-cell	63	23
	10	P-cell	20	32
Rayleigh	20	M-cell	33	14
	10	M-cell	68	29
	10	P-cell	23	36
Reflected path	20	M-cell	135	42
	10	M-cell	148	47
	10	P-cell	85	68
Rician	20	M-cell	29	13
	10	M-cell	69	29
	10	P-cell	29	54

Several error handlers are introduced in the PVE algorithm to minimize the occurrence of these errors. The first error can be greatly reduced by constantly observing the Jacobian matrix using a reciprocal matrix condition estimator (RCOND) available in MATLAB. When the matrix is estimated as approaching singularity, the iteration is stopped and the latest result is used. The second error can be avoided by shifting the MS position estimate during iteration 5 m away from the BS location whenever the situation occurred and allows the iteration process to continue.

8

Smart Antennas: Mobile Station (MS) and Base Station (BS) Antenna Beamforming

Ng Kim Chong, Stetson Oh Kok Leong, P.R.P. Hoole, and E. Gunawan

Abstract

In this chapter are presented the details of beamforming using model-based electromagnetic signal processor to form a beam from a mobile station (MS) moving through a series of stationary base stations (BSs). The beamforming algorithm is linked to an automatic tracker of the MS position and velocity of the BS with which it needs to maintain connection. The working of the tracker and beamforming algorithms is illustrated for a two-element adaptive array antenna on the MS. The chapter also explains a technique for handing over the control of the MS from one BS to another as the MS moves over the large space or area covered by the wireless system. In the final part of the chapter, the beamforming of a BS antenna that needs to track and stay connected to an MS is presented. In the case of the BS, more antenna elements may be used to form the BS array antenna. Results are presented for two- to nine-element adaptive array antenna.

In this chapter, we shall introduce the concept of smart or intelligent antennas in cellular communications. In a smart antenna, the antenna beam is dynamically changed to enhance the system performance. In particular, by controlling the signal strengths at each element of an array antenna, by changing the weights of an adaptive antenna algorithm, the directivity of the antenna is dynamically controlled. We shall focus on the use of such an antenna on an MS, whilst remembering that the same principles may be applied for a BS smart antenna. Amongst the advantages of using smart antennas, the following are the most important:

1) Increasing the channel capacity through frequency reuse within steerable beams. Since the power required is much less than a fixed antenna,

resulting in a lower carrier-to-interference ratio, the smart antenna can allow channels to reuse frequency channels. Space division multiple access (SDMA) with smart antennas allows for multiple users in a cell to use the same frequency without interfering with each other since the BS smart antenna beams are sliced to keep different users in separate beams at the same frequency.

2) Increasing communication range without increasing battery power. The increase in range is due to a bigger antenna gain with smart antennas. This would also mean that fewer BSs may be used to cover a particular geographical area.
3) Reducing multipath, cochannel interference and jamming signals by forming null points in the direction of unwanted signals. Hence, the link quality can be improved. This could also enable the smart antenna beams to be always focused on the hot spots where the number of subscribers is large in a given area of a cell.
4) Better tracking of the position and velocity of the MSs.

The position–velocity estimator (PVE) algorithm presented in Chapter 7 is further enhanced to include MS antenna beamforming. This is a crucial aspect of smart antennas in cellular communications. The MS estimates its own position and velocity, and simultaneously optimizes its antenna beam for reception and transmission. First, the possibility of combining the PVE algorithm and the least mean square (LMS) beamforming algorithm to perform beamforming and position–velocity estimation is investigated. Next and more importantly, an accurate single module beamforming with position–velocity estimator (BFPVE) algorithm is designed using the principle of maximum likelihood estimation. Based on a two-element antenna array, the proposed algorithms are tested in MATLAB™ for different channel conditions.

8.1 Array Antenna

The adaptive beamforming system makes use of the antenna array to perform signal separation and interference rejection. Hence, to effectively execute these functions, we need to understand the characteristics of the antenna array and how the radiation pattern or beam can be controlled effectively. We shall ignore mutual coupling between elements; the effects of mutual coupling in an array antenna were discussed in Section 3.4. We shall also assume that the bandwidth (30 kHz) of the signal is small compared to the carrier frequency (e.g. 1980 MHz).

An antenna array consists of a set of antenna elements that are spatially distributed at known locations with reference to a common fixed point. By controlling the phase and amplitude of the exciting currents in each of the elements, it is able to electronically steer the main lobe or beam in any particular direction. The antenna elements can be arranged in different geometry, with linear, circular, and planar arrays being very popular. In the case of a linear array, the antenna elements are aligned along a particular axis or straight line. If the spacing of the antenna is equally spaced, it is called a *uniformly spaced linear array*. In a circular array, the elements are arranged around the circumference of a circle with the center as the fixed reference point. In the case of a planar array, the antennas are distributed equally on a single plane.

The radiation pattern of an array is determined by the radiation pattern of its individual elements: their orientation and relative positions in space, and the amplitude and phase of the exciting current. If each element of the array is an isotropic point source, then the radiation pattern of the array will solely depend on the geometry and excitation current. This radiation pattern is called the array factor. If each of the elements of the array is similar but not isotropic, by the principles of pattern multiplication, the radiation pattern can be computed as a product of the array factor and the individual element pattern.

An array antenna formed by using two infinitesimal dipoles separated by distance d is considered and its resultant electromagnetic model is derived. Element 2 of the array antenna includes a weight component that provides the phase shift δ necessary for steering the beam of the antenna in any desired direction (see Figure 8.1). An observation point $P(r,\theta,\phi)$ in the far-field region of the array antenna is defined in the Cartesian coordinate for the purpose of analysis.

The radiation field of an infinitesimal dipole antenna is defined in Equations (8.1) and (8.2) as

$$E(r,z,t) = \frac{\mu_0}{4\pi}\left[\frac{z_j - z_1}{\sqrt{r^2 + (z_j - z_1)^2}} - \frac{z_j - z_2}{\sqrt{r^2 + (z_j - z_2)^2}}\right]\frac{\mathrm{d}[I]}{\mathrm{d}t}, \tag{8.1}$$

$$[I] = \mathrm{Re}\left(I_0 \mathrm{e}^{\mathrm{j}w(t-R/c)}\right). \tag{8.2}$$

The resultant electric field intensity of the two elements is given by

$$E_{\mathrm{T}} = E_1 + wE_2, \tag{8.3}$$

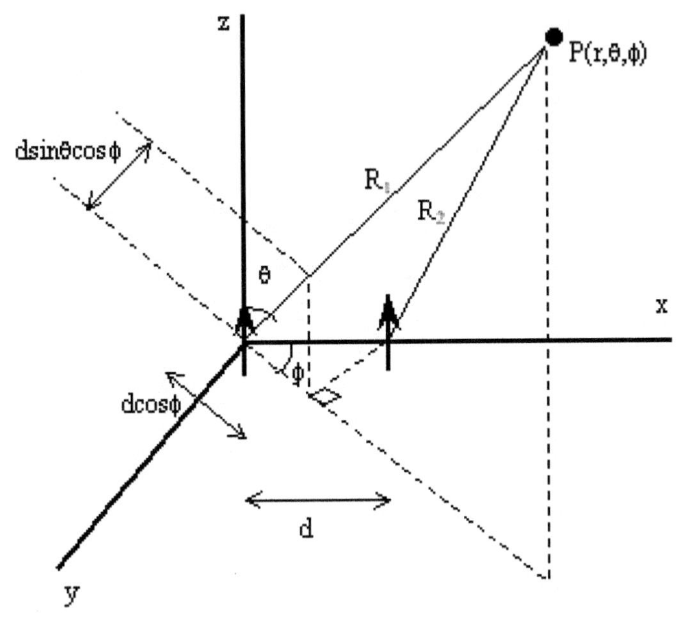

Figure 8.1 Two-element antenna array.

where

$$w = \mathrm{e}^{\mathrm{j}\delta}. \tag{8.4}$$

In the far zone, the magnitudes of E_1 and E_2 are assumed to be equal. This assumption is not valid for the phase of the two signals since a small difference in distance resulted in large changes in phase.

$$E_{\mathrm{T}} = \frac{\mu_0}{4\pi}\left[\frac{z_j - z_1}{\sqrt{r^2 + (z_j - z_1)^2}} - \frac{z_j - z_2}{\sqrt{r^2 + (z_j - z_2)^2}}\right] \times \frac{\mathrm{d}\,[I]}{\mathrm{d}t}\left[1 + \mathrm{e}^{\mathrm{j}(kd\ \sin\theta\ \cos\phi + \delta)}\right]. \tag{8.5}$$

Using Euler's formula and letting $\psi = kd\ \sin\ \theta\ \cos\ \phi + \delta$, the resultant field can be expressed as

$$E_{\mathrm{T}} = \frac{\mu_0}{4\pi}\left[\frac{z_j - z_1}{\sqrt{r^2 + (z_j - z_1)^2}} - \frac{z_j - z_2}{\sqrt{r^2 + (z_j - z_2)^2}}\right] \times \frac{\mathrm{d}\,[I]}{\mathrm{d}t}\left[\mathrm{e}^{\mathrm{j}\psi/2}\ 2\ \cos\left(\frac{\psi}{2}\right)\right]. \tag{8.6}$$

Defining

$$AF = 2\cos\left(\frac{\psi}{2}\right), \tag{8.7}$$

Equation (8.6) becomes

$$E_{\mathrm{T}} = E_1 \, \mathrm{AF} \, \mathrm{e}^{\mathrm{j}\psi/2}. \tag{8.8}$$

In a more general form where each element includes a weight component, the total output of an N-element array with complex weight is given by

$$E_{\mathrm{T}} = \sum_{i=1}^{N} w_i E_i. \tag{8.9}$$

8.2 Adaptive Algorithm

A typical adaptive beamformer is shown in Figure 8.2. In all practical communication channels or links, the incoming signal $x(t)$ usually is a mixture of desired signal, noise, and interference from other users. Based on a set of pre-defined criteria and cost functions, the algorithm optimizes the response of the beamformer or complex weight such that the output signal from the array contains minimum noise and maximum desired signal. The cost function is inversely proportional to the quality of the signal at the output of the array antenna so that when the cost function is minimized, the signal quality is maximized.

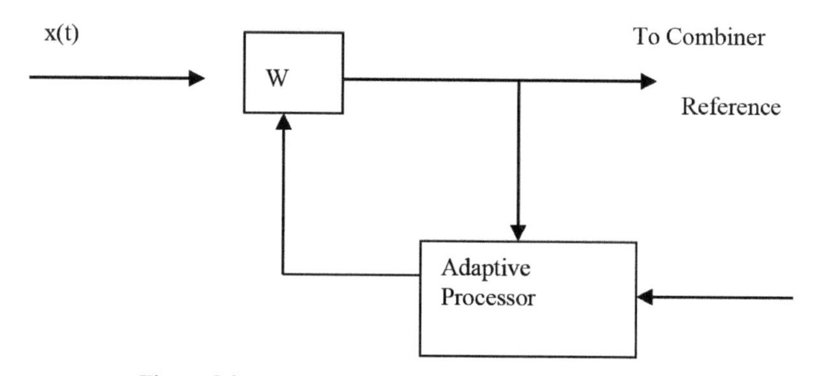

Figure 8.2 Typical adaptive beamformer block diagram.

The criteria can be classified into the following three:

1) minimum mean square error (MMSE) criteria;
2) maximum signal-to-interference ratio (SIR);
3) minimum variance.

As the name implies, least mean square (LMS) error criteria aims to minimize the mean square error between the array output and a reference signal. For the case of maximum SIR, the complex weight is chosen such that the SIR at the output will be maximized. For minimum variance criteria, the desired signal characteristics and its direction are usually unknown. Hence, the output noise variance is suppressed so that there is good reception of the desired signal.

All the above three criteria will resolve to the optimal Wiener–Hopf solution, which yields the same SIR but represented in different forms. The following section briefly describes the LMS criteria, which leads to the LMS algorithm.

8.2.1 Minimum Mean Square Error Criteria

Consider a uniformly spaced linear array as shown in Figure 8.1. Assume that a single desired signal exists and the communication link is polluted only by a single interference source. The desired signal $s(t)$ arrives at the array with a spatial angle θ_{d}, ϕ_{d} and the interference $i(t)$ reaches the array at an angle of θ_i, ϕ_i. The array output $x(t)$ shown in Figure 8.3 is

$$x(t) = s(t)v_{\mathrm{d}} + i(t)v_i, \tag{8.10}$$

where v_{d}, v_i are the array propagation vectors represented by

$$v_{\mathrm{d}}^{\mathrm{T}} = \begin{bmatrix}1 & \mathrm{e}^{-\mathrm{j}kd\ \sin\ \theta_i\ \cos\ \phi_{\mathrm{d}}}\end{bmatrix}, \tag{8.11}$$

$$v_i^{\mathrm{T}} = \begin{bmatrix}1 & \mathrm{e}^{-\mathrm{j}kd\ \sin\ \theta_i\ \cos\ \phi_i}\end{bmatrix}. \tag{8.12}$$

Including the complex weight w_i, the array output will be

$$y(t) = \sum_{i=1}^{N} w_i x_i. \tag{8.13}$$

In matrix form,

$$y(t) = w^{\mathrm{T}} x. \tag{8.14}$$

In LMS criteria, the output is corrected by the complex weight such that MMSE between the output and desired signal is achieved. Therefore, the transmitted signal must be known beforehand, which is an ambiguous statement. However, for many applications, the characteristics of the desired signal may be known with sufficient details. With this knowledge, the receiver can self-generate a reference signal which closely resembles the desired signal or is highly correlated with the original signal. The criteria can be represented as

$$\varepsilon^2(t) = [r(t) - w^{\mathrm{T}} x(t)]^2. \tag{8.15}$$

Taking the expectation on both sides and rearranging the equation,

$$E[\varepsilon^2(t)] = E[r^2(t)] - 2w^{\mathrm{T}} S + w^{\mathrm{T}} Rw, \tag{8.16}$$

$$S = E[x(t) r(t)], \tag{8.17}$$

$$R = E[x(t) x^{\mathrm{T}}(t)], \tag{8.18}$$

where R is usually referred to as the covariance matrix. The MMSE is given by setting the gradient vector of Equation (8.15) to zero,

$$\frac{\mathrm{d}\left\{E\left[\varepsilon^2(t)\right]\right\}}{dw} = 0, \tag{8.19}$$

$$\frac{\mathrm{d}\left\{E\left[\varepsilon^2(t)\right]\right\}}{dw} = -2S + 2Rw. \tag{8.20}$$

Hence, the optimal weight that can achieve MMSE is

$$W_{\mathrm{opt}} = R^{-1} S. \tag{8.21}$$

We note that in the MMSE algorithm, the solution given in Equation (8.21) gives us a single weight vector that is optimal over the collection of measurements over a certain time window. The solution depicted in Equation (8.21) is also referred to as the Wiener–Hopf solution or the optimal Wiener solution. To obtain this solution, the covariance matrix R and correlation matrix S will be required. In practice, both quantities are difficult to compute or may even be impossible to get due to lack of information.

8.2.2 Least Mean Square Algorithm

The mean square error criteria shown in the previous section only served as the cost function in the adaptive algorithm. The various ways of implementing the adaptive algorithms, which select the optimal weight for the beamformer, will have a much greater impact on the system performance. Mainly, speed of convergence, computational power and time required, and hardware complexity are the most critical factors in consideration for the type of algorithm used.

The most common adaptive algorithm being studied is the LMS algorithm. With a reference signal available, the LMS algorithm is simple in nature and requires the least amount of computation time. In the MMSE solution, the error was minimized for a collection of measurements averaged over a period of time. In the LMS approach, the output of the antenna array and a desired response over a finite number of samples is minimized.

To recursively compute and update the weight vectors, the LMS algorithm utilizes the method of steepest descent. According to this method, the updated value of the weight vector at time $n + 1$ is computed using the following relationship:

$$w(n+1) = w(n) + \tfrac{1}{2}\mu[-\nabla(E\{\varepsilon^2(n)\})]. \tag{8.22}$$

The error surface $E\{\varepsilon^2(t)\}$ is always viewed as a bowl or rather as a quadratic function. The definition of such a function is important as it implies that in the error function only a single minimum point exists that will give the optimal solution. Intuitively, one concludes that if the direction of the gradient movement is negative, it will be heading toward the bottom of the bowl, which is the minimum point of the error function.

Recall from Equation (8.20) that, in order to compute the relationship in Equation (8.22), one requires a knowledge of the covariance matrix R and the correlation matrix S, which is not always available.

Thus, the most obvious strategy is to use the instantaneous gradient estimation for the computation in Equation (8.22):

$$\frac{\mathrm{d}[\varepsilon^2(n)]}{\mathrm{d}w} = -2\varepsilon(n)x(n). \tag{8.23}$$

Substitute into Equation (8.22),

$$w(n+1) = w(n) + u\varepsilon(n)x(n). \tag{8.24}$$

This is the LMS updating algorithm presented by Widrow and Stearns (1985). Even under non-stationary signal condition, the algorithm could converge to the optimal Wiener solution, at the expense of slow convergence rate.

The LMS adaptive algorithm is simple and easy to implement. However, it requires a reference signal in order to perform computation. Normally, a pilot signal or training sequence will be transmitted over the channel in order to pre-train the weight. In this way, the performance of the communication system will be degraded. Further, under noisy signal environment like with mobile communication, the algorithm tends to converge of a much slower rate, typically of 200-iteration loop. This is unacceptable for real-time application like voice channel communication. One drawback in the MMSE and LMS methods is that the desired output must either be known or estimated. This knowledge is often obtained by sending a known training sequence on a periodic basis, where the sequence is known to both the transmitter and the receiver. Since this training sequence carries no useful data except for training the adaptive receiver, the frequency spectral it occupies is wasted. Alternatives to MMSE and LMS that do not require such training sequences are the decision-directed algorithms or the blind adaptive algorithm, where an attempt to restore some known property of the received signal is made.

8.3 Electromagnetic Model

As derived in Section 8.1, the E-field of a two-element array is simply the multiplication of a single element pattern with an array factor. From Equation (8.6),

$$\begin{aligned} E_{\mathrm{T}} =& \frac{\mu_0}{4\pi}\left[\frac{z_j - z_1}{\sqrt{r^2 + (z_j - z_1)^2}} - \frac{z_j - z_2}{\sqrt{r^2 + (z_j - z_2)^2}}\right] \\ &\times \frac{\mathrm{d}\,[I]}{\mathrm{d}t}\left[\mathrm{e}^{\mathrm{j}\psi/2}\, 2\, \cos\left(\frac{\psi}{2}\right)\right]. \end{aligned} \tag{8.25}$$

The beamforming signal processor operates only on the magnitude of the measured electric field strength. Hence, Equation (8.25) is reduced to

$$|E_{\mathrm{T}}| = A_0\left[\frac{z_j - z_1}{\sqrt{r^2 + (z_j - z_1)^2}} - \frac{z_j - z_2}{\sqrt{r^2 + (z_j - z_2)^2}}\right]\left|2\, \cos\left(\frac{\psi}{2}\right)\right|, \tag{8.26}$$

where

$$A_0 = \frac{\mu_0 \omega I_0}{4\pi} = 10^{-7}\omega I_0,$$

$$\psi = kd\, \sin\, \theta\, \cos\, \phi + \delta,$$

$$I_0 = \sqrt{\frac{P_{\mathrm{r}}}{R_{\mathrm{r}}}},$$

$|2\cos(\psi/2)|$ is the magnitude of the array factor, P_r is the power radiated, and R_r is the radiation resistance of the transmitter antenna. All parameters are defined in Figures 8.1 and 7.4.

8.4 Tracking and Beamforming with Position and Velocity Estimator (BFPVE)

A single module that performs beamforming and position–velocity estimation simultaneously in each iteration process is being proposed as shown in Figure 8.3.

The measured instantaneous electric field strengths E_m in the presence of additive white Gaussian noise (AWGN) is modeled by

$$E_{\mathrm{m}} = E + n, \tag{8.27}$$

where E is a function of r and w is defined in Equations (8.1) and (8.27).

First, an initial estimated r and w are assumed to be obtainable (described in Section 8.8). From this assumption, E can be made a function of only one of the two parameters (r or w). Thus, an estimator for r and w can be derived similar to the deviation of the position–velocity estimator (PVE) algorithm.

Through series expansion, E can be expressed as

$$E(a) = E(a_0) + E'(a_0)\Delta a + 0.5E''(a_0)\Delta a^2 + \cdots, \tag{8.28}$$

where $a = r$ or w. Ignoring higher order terms and writing $E_0 = E(a_0)$, let

$$J_0 = E'(a_0) = \frac{\delta E_0}{\delta a_0}, \tag{8.29}$$

$$E(a) = E_0 + J_0 \Delta a. \tag{8.30}$$

Therefore, the noise term is given by

$$n = (E_{\mathrm{m}} - E_0) - J_0\Delta a = \Delta z - J_0 \Delta a. \tag{8.31}$$

Figure 8.3 Block diagram of BFPVE.

In order to minimize $U(\Delta a) = n^{\mathrm{T}}\ n$, let $\partial U(\Delta a)/\partial a = 0$, which, with Equations (8.29) and (8.30), yields

$$\Delta a = (J_0^{\mathrm{T}} J_0)^{-1} J_0^{\mathrm{T}} \Delta z. \tag{8.32}$$

Therefore, a can be estimated using a maximum likelihood estimator of the following form:

$$a^{n+1} = a^n + (J_n^{\mathrm{T}} J_n)^{-1} J_n^{\mathrm{T}} (E_{\mathrm{m}} - E_n). \tag{8.33}$$

Replacing a by parameters r and w, Equation (8.33) can be written as

$$r^{n+1} = r^n + (Jr_n^{\mathrm{T}} Jr_n)^{-1} Jr_n^{\mathrm{T}} (E_{\mathrm{m}} - E_n), \tag{8.34}$$

$$w^{n+1} = w^n + (Jw_n^{\mathrm{T}} Jw_n)^{-1} Jw_n^{\mathrm{T}} (E_{\mathrm{m}} - E_n), \tag{8.35}$$

where E_{m} is a column vector that represents the measured electric field strengths at the base stations (BSs). Jr_n and Jw_n are the Jacobian matrix of the electric field strengths E_n with elements given by

$$Jr = \frac{\partial E}{\partial r} = A_0 r \left\{ \frac{Z_2 - Z_j}{(r^2 + (Z_j - Z_2)^2)^{3/2}} - \frac{Z_1 - Z_j}{(r^2 + (Z_j - Z_1)^2)^{3/2}} \right\} \times \left| 2 \cos \frac{\psi}{2} \right|, \tag{8.36}$$

$$Jw = \frac{\partial E}{\partial w} = E_2. \tag{8.37}$$

The algorithm defined by Equations (8.34) and (8.35) is named BFPVE I. Both equations undergo an iteration process simultaneously until the required convergence accuracy is achieved for both r and w.

An alternative approach to generate the required weight is to represent the weight w as a function of the parameter r. Both R and θ shown in Figure 7.2 and Equation (8.26) can be represented as a function of parameter r:

$$R^n = \sqrt{(z_2 - z_j)^2 + (r^n)^2}, \tag{8.38}$$

$$\theta^n = \sin^{-1}\left(\frac{r^n}{R^n}\right), \tag{8.39}$$

where the superscript n represents the value at each iteration cycle number n.

To maximize the desired signal strength, the w required is just the complex conjugate of the element factor of antenna. Hence,

$$w^n = \left(e^{jkd \sin \theta^n \cos \phi^n}\right)^*, \tag{8.40}$$

where * represents complex conjugate. The algorithm defined by Equations (8.34) and (8.35) is named BFPVE I. Hence, BFPVE I uses separated iteration equations for r and w. However, BFPVE II based on Equations (8.34) and (8.40) uses only a single r estimator by setting w as a function of r.

8.5 Simulation Scenario

In order to evaluate the performance of the algorithms, simulations are performed in MATLAB™. A wireless communication scenario is shown in Figure 8.4.

The simulation scenario used here is basically the same as in Chapter 7 used to evaluate the performance of the PVE algorithm, with some additional factors considered for mobile station (MS) adaptive beamforming. As the MS travels along the route, it always beamforms toward the base station

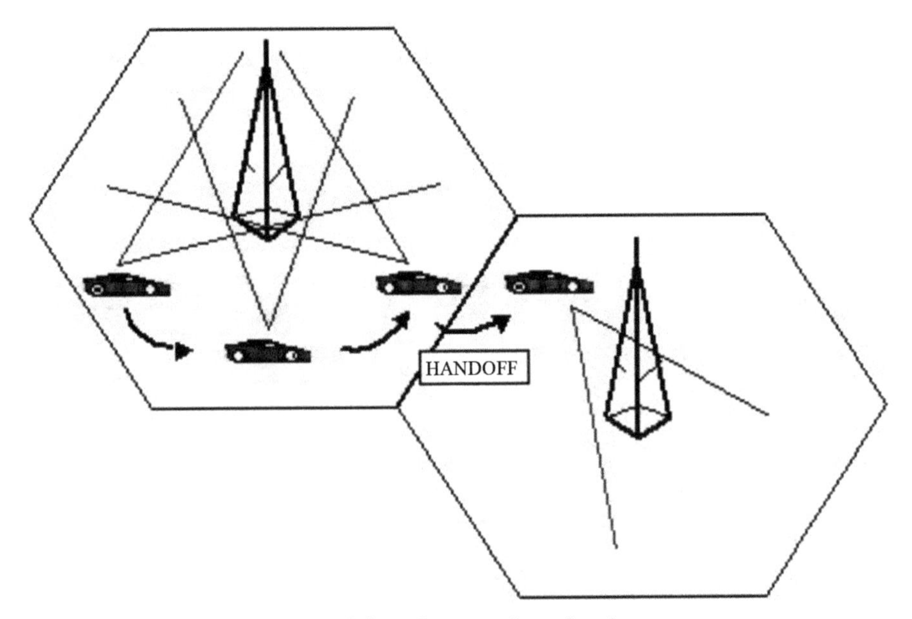

Figure 8.4 MS antenna beamforming.

transceiver (BTS) that handles the wireless communication link between the MS and the radio network. The MS beam pattern constantly steers toward the direction of the BTS so as to maximize the reception of the desired signal while minimizing interference signals from other directions. As it moves into another cell, handoff is performed and the beam is steered toward the new BTS as shown in Figure 8.4.

Reception from three BSs with the strongest field strength is used to compute the parameter r. Thus, r is a vector with three elements $[r_1, r_2, r_3]$, and each value is the horizontal distance between a BS and the MS. As the weight w is derived for maximizing the reception of the desired signal, its value is only dependent on the field strength from that particular BTS where the desired signal is transmitted. Thus, w is a vector with three elements of the same value.

8.6 Channel Models

The channel models used previously in Chapter 7 are based on the assumption of Gaussian distribution noise. However, when beamforming is performed, that assumption is no longer valid. The beam pattern limits the directions of reception of interference signals. To adjust to this condition, the noise may be multiplied by the array factor that forms the beam pattern (see Figure 8.5).

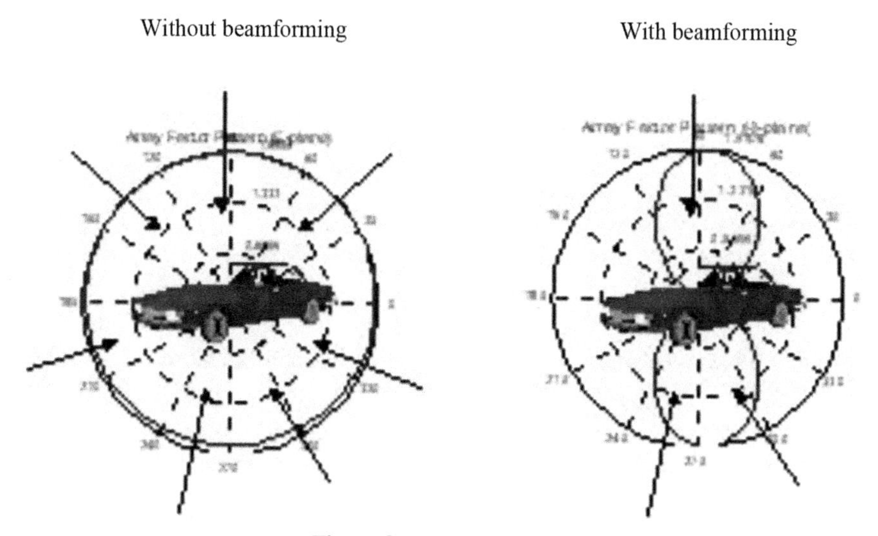

Figure 8.5 Noise channel.

The measured electric field is given by

$$E_{\mathrm{m}} = E + \mathrm{AF}\ N_{\mathrm{AWG}}(m, \sigma), \tag{8.41}$$

where E is the E-field model Equation (8.40), AF is the array factor of the MS antenna, N_{AWG} is the additive Gaussian distributed noise with mean m and standard deviation σ.

8.7 Antenna Radiation Pattern

The formation of the beam pattern depends on the array factor AF that is given by

$$\mathrm{AF} = 2\ \cos\left(\frac{kd\ \sin\ \theta\ \cos\ \phi + \delta}{2}\right), \tag{8.42}$$

where k is the wave number $2\pi/\lambda$, λ is the wavelength, d is the distance spacing between the two elements, θ is the elevation angle of the BS to MS and is a function of parameter r, ϕ is the direction of the desired signal from the BTS, and δ is the phase difference between the two elements contributed by the weight.

Both k and d are fixed parameters, while θ, ϕ, and δ are variables depending on the position of the MS. During iteration, the weight converges to a complex value that is used to compute the phase δ. The angle θ is a function of parameter r as indicated in Equation (8.42). To simplify computation, angle ϕ is assumed to be a known value. However, it can be shown that ϕ can be estimated during each iteration cycle by the iterating r values since the calculated r values represent the position estimation that is converging closer toward the actual MS position. Hence, with the location of the BTS known, the direction of arrival (DOA) of the desired signal ϕ can be estimated.

The overall radiation pattern is just the multiplication of the single element pattern by the array factor pattern. The single element pattern is different in the H-plane and E-plane. The single element radiation pattern in the H-plane is just a circle, whereas it has regions of zero radiation along the axis on which the dipole is placed. As the array factor pattern is identical for both the H-plane and E-plane, the overall radiation pattern in the H-plane is just the array factor pattern. The pattern in the H-plane, which is the *xy*-plane, will show clearly the steering of the beam pattern toward the BTS as the MS travels.

The overall antenna radiation pattern will have different shapes in the horizontal (H-) and vertical (E-) planes. If the maximum gain in the horizontal plane is G_{H} and the maximum gain in the vertical plane is G_{E}, then the maximum gain of the entire three-dimensional beam is given by $G_{\mathrm{B}} = G_{\mathrm{H}}\, G_{\mathrm{E}}$. In general, the carrier to interference and noise ratio (CINR) is proportional to the horizontal beam gain G_{H}. The CINR is typically given by CINR = $(QSG_{\mathrm{H}})/(aN)$, where Q is the reuse factor, S is the spreading factor, a is the vocoder rate, and N is the number of users. The reuse factor Q (<1) is the ratio of received power from all users within a cell to the total interference from all users. The vocoder rate a is associated with the reduction digital signal output rate when a speaker in a voice communication systems is silent. The periods over which the MS does not transmit are given by a (typically 0.45), resulting in a reduction of the multiple access interface level at the BS. For a CINR of 10 dB required for each user, for a smart antenna at the BS with G_{H} = 5 dB, the system can support about 40 users, whereas if an omni-directional BS antenna was used (G_{H} = 0 dB), only about 15 users could have been supported by the BS antenna. If G_{H} of the BS antennas was increased to 10 dB, then more than 100 users could have been supported by the BS receiver. Further discussion of this subject from a BS antenna perspective is found in Liberti and Rappaport (1999).

8.8 Initial Values

The procedure for initial r estimation is the same as that described in Chapter 7. The initial weight (w at n = 0) estimation can be obtained by using the complex conjugate of the element factor which is a function of angles θ and ϕ.

Using antenna element 1 as reference (i.e. element factor = 1), the total E-field measured is

$$E_{\mathrm{T}} = E_{\mathrm{m}} + w\mathrm{e}^{\mathrm{j}kd\ \sin\,\theta\ \cos\,\phi} E_{\mathrm{m}}, \tag{8.43}$$

where $\mathrm{e}^{\mathrm{j}kd\ \sin\,\theta\ \cos\,\phi}$ is the element factor of element 2.

At the start of iteration, initial weight estimation can be obtained by using the complex conjugate of the element factor.

8.9 Simulation Results

The iteration begins with the initial values of r and w and proceeds until a convergence accuracy of 10^{-4} is achieved. BFPVE I needs both r and w

estimators to achieve the required criteria before the iteration loop is stopped, whereas BFPVE II depends only on the r estimator.

Figures 8.6 and 8.7 show convergence characteristics of r and phase angle w for the AWGN channel at a signal-to-noise ratio (SNR) of 10 dB. The average number of iteration loops for convergence is 6. The accuracies of the estimated r vector $[r_1, r_2, r_3]$ are compared with those of the actual values in Table 8.1 for MS positions 1, 8, and 15.

The result shows that the estimated r is reasonably accurate when compared to its actual values. This result is similar for both BFPVE I and

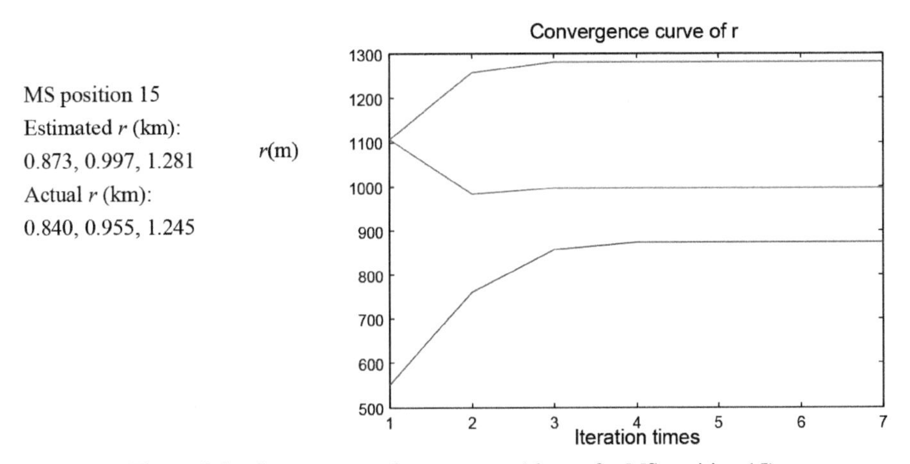

Figure 8.6 Convergence of parameter r (shown for MS position 15).

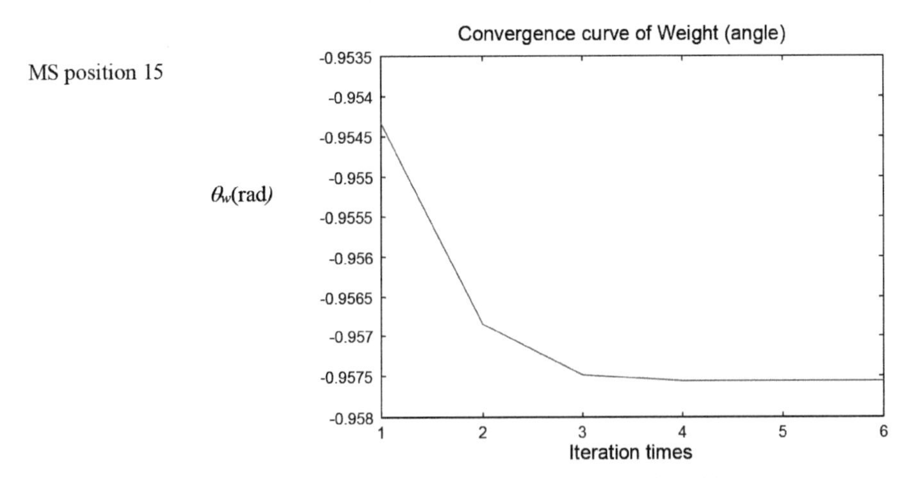

Figure 8.7 Convergence of parameter w (shown for MS position 15).

Table 8.1 Comparison of estimated r with actual r.

MS position	Estimated r vector (km)	Actual r vector (km)
Position 1	[0.808, 1.028, 1.161]	[0.849, 1.059, 1.097]
Position 8	[0.613, 1.139, 1.465]	[0.623, 1.102, 1.457]
Position 15	[0.873, 0.997, 1.281]	[0.840, 0.955, 1.245]

Table 8.2 Simulation results.

Channel model (SNR = 10 dB)		Average position error (m)	Average velocity error (km/h)
AWGN	BFPVE I	55	20
	BFPVE II	56	24
Rayleigh	BFPVE I	70	26
	BFPVE II	74	27
Reflected path	BFPVE I	147	43
	BFPVE II	143	42
Rician	BFPVE I	85	33
	BFPVE II	80	28

BFPVE II. From Table 8.2, the position and velocity estimates obtained using the two estimators can be compared. In Figure 8.8 is shown the estimation using BFPVE II. The results show that the accuracy of BFPVE I is similar as compared to BFPVE II. For example, for AWGN with an SNR of 10 dB, BFPVE I has an average position estimation error of 55 m as compared to approximately 56 m for BFPVE II. From Table 8.2, it can be seen that under a dominant reflected path channel, the estimation is less accurate for all cases. However, using suitable filtering techniques, the accuracy can be significantly improved.

Figure 8.9 shows the three-dimensional radiation pattern for the estimated MS positions. It clearly shows how the beam pattern is steered as the MS travels along the route marked by the dotted line. The beam pattern consists of two lobes, with one of the lobes covering the direction of the desired signal. Besides the additional smaller lobe, the main lobe beam width is quite wide, covering more than the intended direction. This is due to the fact that only two elements are used for the array. For a narrower and better-focused beam, three or more elements are required. With more elements, it is possible that other criteria such as nulling of interference signals from specific directions can also be satisfied. However, for MS beamforming with position estimation, a reasonable wide beam covering at least three BSs is necessary. If the beam is too narrow, the signal from other BSs may be too weak to be used for computation. Furthermore, increasing the number of elements makes the MS antenna cumbersome.

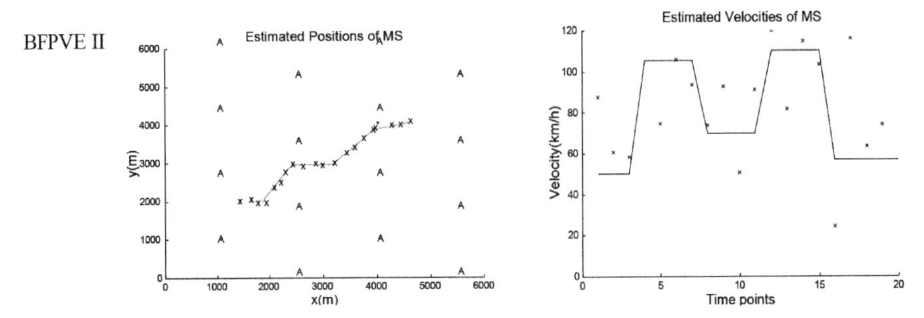

Figure 8.8 Simulation results (Rayleigh channel; SNR = 10 dB).

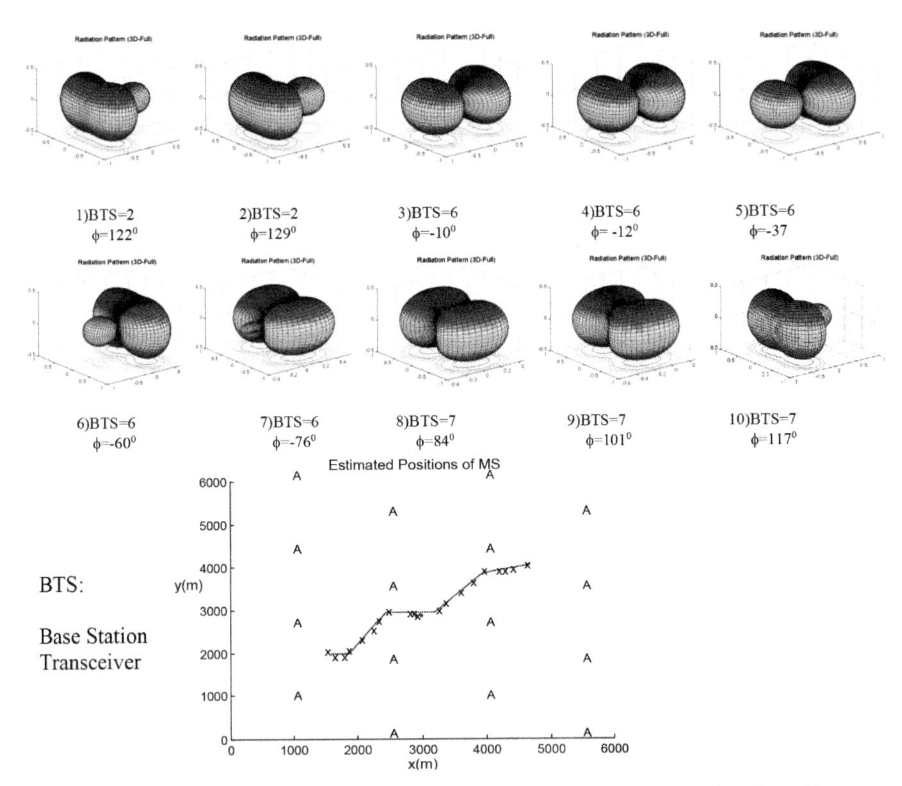

Figure 8.9 Three-dimensional radiation patterns at different estimated MS positions.

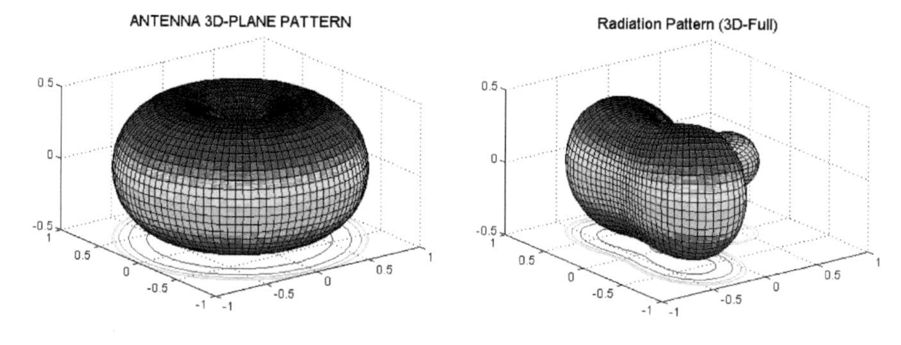

A single dipole antenna Two-element array antenna

Figure 8.10 Advantage of array antenna: power reduction.

Hence, from the results shown, the use of an array antenna at the MS allows its beam pattern to be adjusted such that the gain in the direction of the desired signal is maintained constant while having nulls in other directions. Thus, interference signals or multipath signals with large delay spreads coming from these directions will be rejected. This helps to reduce cochannel interference (CCI) and inter-symbol interference (ISI) and improves the network capacity as the effective signal to interference and noise ratio (SINR) is significantly increased.

Transmission power required at the MS to perform the same basic communication functions is largely reduced. This can be easily seen by looking at the different coverage area of the beam pattern maintained by a single dipole and a two-element array antenna as shown in Figure 8.10. Less power is wasted in covering regions besides the region of interest. The significant reduction in power required helps to extend battery life as well as provides possible reduction in the size and cost of the batteries required at the MS.

8.10 Handover Algorithm in Smart Antenna Systems: The Triangle Method

We noted above that in cellular systems, one crucial consideration in call management is the transfer of a call from one BS to another when a mobile moves into the coverage area of a nearby BS. Conventional handover (or handoff) algorithms use signal strengths to decide if handover is necessary and which BS to hand over. However, the signal strength based algorithms become more complicated in smart antenna systems since the signal strengths received will be a function of the radiation pattern of the BS antenna. Indeed, a handover could be avoided by forming a new beam to the MS so as to increase the uplink and downlink signal strengths. In addition, a non-control base station (NCBS) will receive weak signal strength from a nearby MS because that signal is considered as interference by the NCBS and hence a null will be put in the DOA of the MS. This also complicates the handover algorithm based on signal strength. Therefore, we have proposed a simple distance-based handover algorithm for smart antennas. BS to MS distance is adopted as handover criteria since distance is the most direct parameter in deciding the control BS. It is understood that in areas where there is heavy shadowing, some form of signal strength criteria should be used with the nearest distance criteria.

Received signal strengths at three nearest BSs and the respective beam patterns formed at these BSs are used to find the position of the MS, with the CBS having the strongest received signal strength (see Section 9.2.3). The distance of the mobile from these three BSs is compared from time to time, and if the distance corresponding to the CBS is smaller than the distance between the MS and any of the other two BSs by a threshold of say, 10 m, handover will take place. Such a threshold is necessary to avoid extensive handover due to the fluctuations in the MS position at the edge of the cellular cell. Furthermore, a triangle method is used to select the three nearest BSs from the MS by referring to the previous three nearest BSs.

The global system for mobile (GSM) measures six signal strengths at any one time to decide the control BS. Once the transmission is being handed over to a new control BS, another set of six adjacent BSs will be chosen. Hence, from time to time, measurements from all the BSs within the same base station controller (BSC) coverage area will be compared. This procedure can be resource consuming since the number of BSs controlled by a BSC can be huge.

The estimated position derived from the PVE is used to determine the three nearest BSs. Since any set of the three adjacent BSs forms a triangle, a new triangle will be formed when the MS (the user) moves out of the current triangle. However, only the third BS will be changed when a new triangle is formed. As shown in Figure 8.11, BSs A and B will remain while BS C' will be replaced by BS $C\prime$ when a new triangle is formed.

Hence, our objective is to find the new third adjacent BS when the MS (the user) moves out of the current triangle. In order to do that, we compute and compare three angles formed by the mobile user and the three

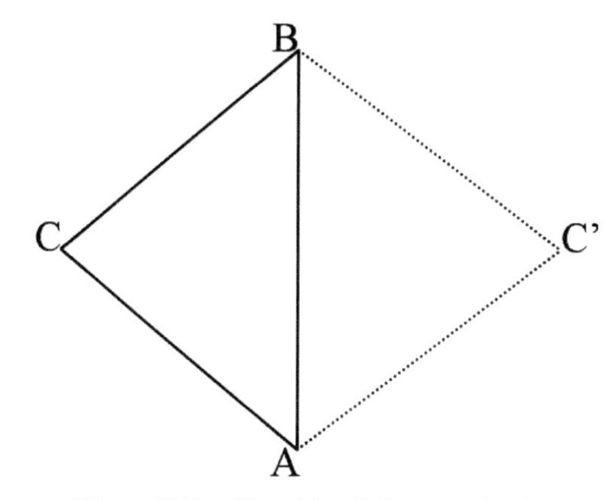

Figure 8.11 The old and the new triangle.

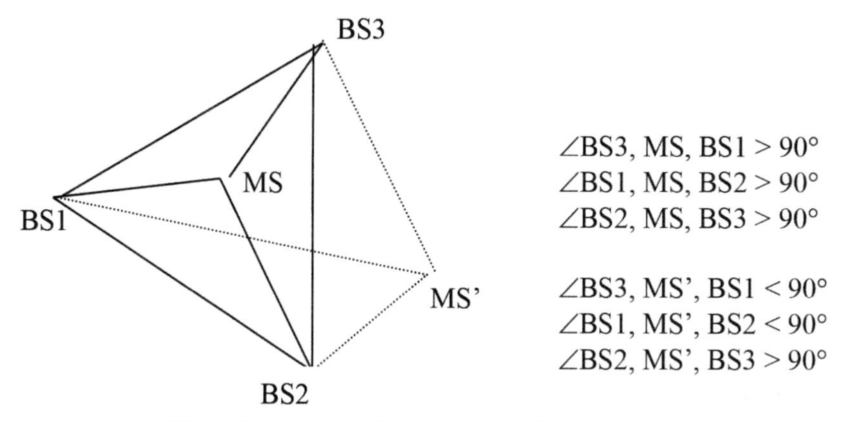

Figure 8.12 Angles involved in the triangle method.

adjacent BSs to check if the MS (the user) is out of the recently occupied triangle.

As shown in Figure 8.12, if the mobile user is inside the control triangle (∠BS1, BS2, BS3), any two of the three angles shown must be larger than 90°. When any two of the angles are smaller than 90°, it is very likely that the MS is out of the control triangle. In addition, the further the MS is out of the triangle, the more likely that any two of the three angles will be smaller than 90°.

Once the MS is out of the "control" triangle, the following formula, with reference to Figure 8.13, is used to calculate the rough position of the new third adjacent BS:

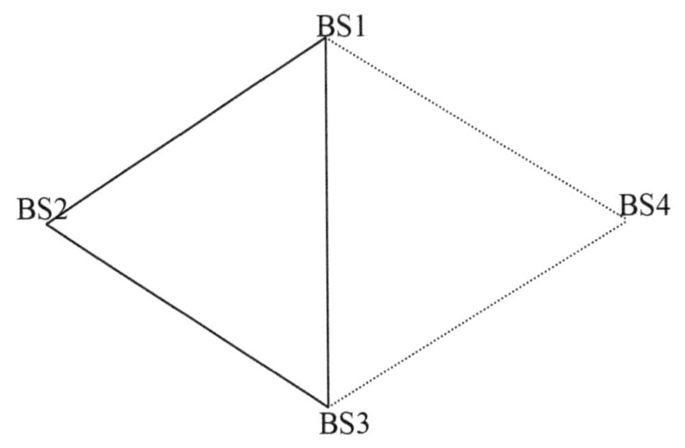

Figure 8.13 New third adjacent base station determination.

$$\begin{aligned} \text{center} &= 0.5\,(\text{BS1} + \text{BS3}), \qquad (8.44)\\ \therefore\ \text{BS4} &= \text{BS2} + 2(\text{center} - \text{BS2}) \\ &= \text{BS2} + (\text{BS1} + \text{BS3} - 2\,\text{BS2}) \\ &= \text{BS1} + \text{BS3} - \text{BS2}. \qquad (8.45) \end{aligned}$$

However, not all the BSs will have exactly the same separation (in our case study, approximately 1732 m). Hence, the calculated position of the new third adjacent BS is compared with all the existing BSs tied to a particular BSC or a mobile switching center (MSC), and the nearest BS will be chosen as the exact new third adjacent BS. Through this method, we only use the pre-determined positions of the existing BSs instead of getting time-varying information from all the BSs within the area of a particular BSC or MSC.

8.11 Base Station Beamforming: Position–Velocity Estimator (Maximum Likelihood Method)

In smart antenna systems, an accurate MS positioning method is crucial for accurate beamforming. Besides, the information of the MS position is essential for the monitoring of the handover between BSs.

For our simulation purpose, we adopted the modified maximum likelihood method. From Section 3.9, we know that the radiated electric field strength from a finite length antenna element may be written in the following form:

$$E_n = E_0 \times \left[\frac{z_2 - z_j}{\sqrt{r_n^2 + (z_2 - z_j)^2}} - \frac{z_1 - z_j}{\sqrt{r_n^2 + (z_1 - z_j)^2}} \right] \times \mathrm{AF}_n \quad (8.46)$$

where E_n = electric field signal strength received at the BTS,

$$E_0 = 10^{-7} \omega \sqrt{\frac{P_\mathrm{r}}{R_\mathrm{r}}},$$

where ω is carrier frequency in rad/s, P_r is power radiated, R_r is radiation resistance of the transmitter antenna, z_2 is height of the BS antenna top, z_1 is height of the BS antenna bottom, z_j is height of the MS antenna center, r_n is distance of MS from BTS, and AF_n is array factor.

Hence, the distance r can be estimated using

$$r_{n+1} = r_n + (Jr_n Jr_n)^{-1} Jr_n' (E_\mathrm{m} - E_n), \quad (8.47)$$

where Jr is Jacobian matrix of the measured electric field strengths, E_m is column vector representing the measured electric field strengths at the BTS, and

$$Jr_n = \frac{\partial E_n}{\partial r_n}. \quad (8.48)$$

The array factor for the smart antennas is given by

$$\mathrm{AF}_n = \left| \mathrm{conj}\left(W\_\mathrm{opt}'\right) \times \begin{bmatrix} 1 \\ \mathrm{e}^{-\mathrm{j}\Phi r} \\ \mathrm{e}^{-\mathrm{j}2\Phi r} \end{bmatrix} \right| \times N, \quad (8.49)$$

$$\Phi r = \frac{2\pi d}{\lambda} \cos\Phi \sin\Theta,$$

$$\Theta = \pi - \tan^{-1} \frac{r}{z_1 + (L/4) - z_j},$$

$$\mathrm{AF}_n = \left| \mathrm{conj}\left(W\_\mathrm{opt1}\right) + \mathrm{conj}\left(W\_\mathrm{opt\,2}\right) \times \mathrm{e}^{-\mathrm{j}\Phi r} + \mathrm{conj}\left(W\_\mathrm{opt\,3}\right) \times \mathrm{e}^{-\mathrm{j}2\Phi r} \right| \times N. \quad (8.50)$$

The optimum weight W_opt is independent of the distance r between the MS and the BS. Thus, for Equation (8.48), we have

$$Jr_n = J_1 + J_2 \quad (8.51)$$

$$J_1 = -E_0 \left[\frac{z_2 - z_j}{\left\{ r_n^2 + (z_2 - z_j)^2 \right\}^{3/2}} - \frac{z_1 - z_j}{\left\{ r_n^2 + (z_1 - z_j)^2 \right\}^{3/2}} \right] r_n \times \mathrm{AF}_n$$

$$J_2 = E_0 \left[\frac{z_2 - z_j}{\sqrt{r_n^2 + (z_2 - z_j)^2}} - \frac{z_1 - z_j}{\sqrt{r_n^2 + (z_1 - z_j)^2}} \right] \times \frac{\partial \mathrm{AF}_n}{\partial r_n},$$

where

$$\frac{\partial \mathrm{AF}_n}{\partial r_n} = \left| -\mathrm{j} \frac{\partial \Phi r}{\partial r} \operatorname{conj}(W\_\mathrm{opt2})\, \mathrm{e}^{-\mathrm{j}\Phi r} - \mathrm{j}2 \frac{\partial \Phi r}{\partial r} \operatorname{conj}(W\_\mathrm{opt3})\, \mathrm{e}^{-\mathrm{j}2\Phi r} \right| \tag{8.52}$$

and

$$\frac{\partial \Theta}{\partial r} = -\frac{z_1 + L/4 - z_j}{r^2 + (z_1 + L/4 - z_j)^2}, \tag{8.53}$$

$$\frac{\partial \Phi r}{\partial r} = \frac{2\pi d}{\lambda} \cos\Phi \cos\Theta \frac{\partial \Theta}{\partial r} \left(\cos\Theta = -\frac{z_1 + L/4 - z_j}{\sqrt{r_n^2 + (z_1 + L/4 - z_j)^2}} \right)$$

$$= \frac{2\pi d}{\lambda} \cos\Phi \frac{(z_1 + L/4 - z_j)^2}{\left[r_n^2 + (z_1 + L/4 - z_j)^2 \right]^{3/2}}, \tag{8.54}$$

$$\therefore \frac{\partial \mathrm{AF}_n}{\partial r} = \frac{2\pi d}{\lambda} \cos\Phi \frac{(z_1 + L/4 - z_j)^2}{\left[r_n^2 + (z_1 + L/4 - z_j)^2 \right]^{3/2}} \cdot |\, \mathrm{j} \operatorname{conj}(W\_\mathrm{opt2})\, \mathrm{e}^{-\mathrm{j}\Phi r} + \mathrm{j}2 \operatorname{conj}(W\_\mathrm{opt3})\, \mathrm{e}^{-\mathrm{j}2\Phi r}. \tag{8.55}$$

After getting the value of r using Equation (8.47), the MS's position can be estimated using the formula given by Equation (7.27) in Chapter 7.

$$\mathrm{PM}_{k+1} = \mathrm{PM}_k + \left(Jp_k' Jp_k \right)^{-1} Jp_k' \left(0 - \mathrm{errp}_k \right), \tag{8.56}$$

where $\mathrm{errp} = |\mathrm{PM} - \mathrm{PB}| - r.Jp_k$ is defined as

$$\begin{bmatrix} \partial \mathrm{errp}_1 / \partial x & \partial \mathrm{errp}_1 / \partial y \\ \partial \mathrm{errp}_2 / \partial x & \partial \mathrm{errp}_2 / \partial y \\ \partial \mathrm{errp}_3 / \partial x & \partial \mathrm{errp}_3 / \partial y \end{bmatrix}.$$

From Equation (7.29) in Chapter 7, the velocity of the MS can be obtained from sequential PM estimates as follows:

$$V_m = (\mathrm{PM}_m - \mathrm{PM}_{m-1})/T, \tag{8.57}$$

where T is the time interval between two discrete positions of the MS.

8.12 Channel Model

Additive White Gaussian Noise: In communication systems, noise is generated in many ways. We concentrate on the white noise or the AWGN. Further details of noise are found in Section 7.3.4. In this section, we are concerned about the correlation of noise and the radiation (beam) pattern. The AWGN can be modeled by a normally distributed random value. For omni-directional transmission, AWGN is taken to be a zero mean value and has a variance of 0.161. In other words, with an SNR of 0 dB and signal power of 1 W, the noise power will be 0.161 W, calculated from the following formula:

$$\text{Noise power} = \text{variance} \times \frac{\text{Signal power}}{10^{\text{SNR (dB)}/10}}. \tag{8.58}$$

However, the use of directive beams in smart antennas will affect the average noise power (or the noise variance) according to the beam pattern formed. This can be seen from Figures 8.14 and 8.15. Contrasting the beams shown in Figures 8.16 and 8.17, it can be seen that after beamforming, noises from certain directions are rejected or attenuated while those that fall within the beam pattern are amplified. The amplification of noise that falls within the narrow beam shown in Figure 8.16 is due to the fact that within that area the array factor will have larger values (e.g. AF = 3.0) than that for omni-directional beams (e.g. AF = 1.0). Therefore, the noise power of the smart antenna system will no longer be the same as that of an omni-directional transmission system.

We propose that the noise power (or noise variance) received by the receiver is directly proportional to the surface area of the beam pattern:

$$\frac{\text{Noise variance}_{\text{ BF}}}{\text{Beam surface area}} = \frac{\text{Noise variance}_{\text{ isotropic}}}{\text{Sphere surface area}} = \frac{0.161}{4\pi}, \tag{8.59}$$

$$\text{Noise variance}_{\text{BF}} = \text{Noise variance}_{\text{isotropic}} \times \frac{\text{Beam surface area}}{4\pi}. \tag{8.60}$$

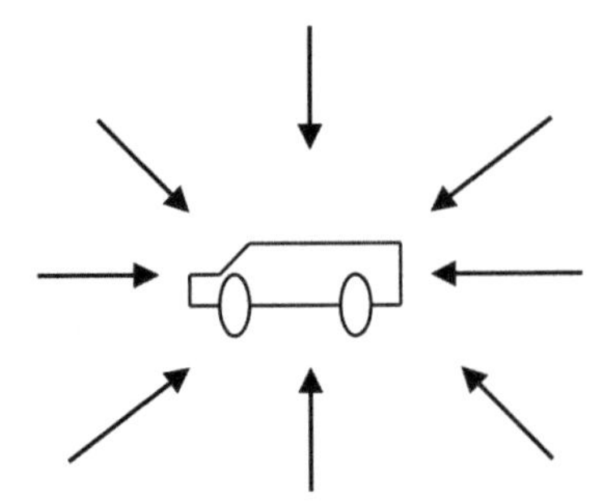

Figure 8.14 Noise from all directions.

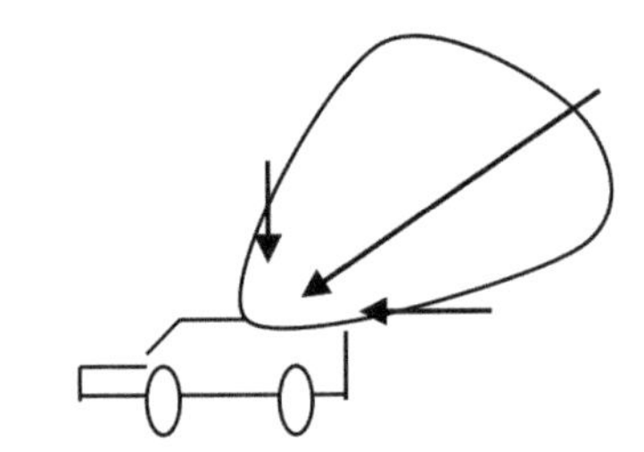

Figure 8.15 Noise after beamforming.

We define noise factor as the ratio of noise power after beamforming (BF) to that before beamforming (i.e. omni-directional):

$$\text{Noise factor} = \frac{\text{Noise variance}_{\text{BF}}}{\text{Noise variance}_{\text{isotropic}}} = \frac{\text{Beam surface area}}{4\pi}. \tag{8.61}$$

We shall illustrate now how the noise factor may be calculated. Consider a three-element (N = 3) array antenna. We want to determine its noise factor by first evaluating its beam surface area using a formula. The array factor is given by

$$\text{AF} = \left|(w_{1\text{r}} + \text{j}w_{2\text{i}}) + (w_{2\text{r}} + \text{j}w_{2\text{i}})\,\text{e}^{\text{j}\Phi r} + (w_{3\text{r}} + \text{j}w_{3\text{i}})\,\text{e}^{\text{j}2\Phi r}\right|, \tag{8.62}$$

where w_r is real part of the respective weight component and w_i is imaginary part of the respective weight component,

$$\Phi r = \cos\left(\frac{2\pi d}{\lambda}\sin\Theta\cos\Phi\right). \tag{8.63}$$

Hence,

$$\text{AF} = \sqrt{A\cos 2\Phi r + B\cos\Phi r + C + D\sin 2\Phi r + E\sin\Phi r}, \tag{8.64}$$

where

$$
\begin{aligned}
A &= 2\left(w_{1\mathrm{r}}w_{3\mathrm{r}} + w_{1\mathrm{i}}w_{3\mathrm{i}}\right), \\
B &= 2\left(w_{1\mathrm{r}}w_{2\mathrm{r}} + w_{1\mathrm{i}}w_{2\mathrm{i}} + w_{2\mathrm{r}}w_{3\mathrm{r}} + w_{2\mathrm{i}}w_{3\mathrm{i}}\right), \\
C &= w_{1\mathrm{r}}^2 + w_{1\mathrm{i}}^2 + w_{2\mathrm{r}}^2 + w_{2\mathrm{i}}^2 + w_{3\mathrm{r}}^2 + w_{3\mathrm{i}}^2, \\
D &= 2\left(w_{1\mathrm{i}}w_{3\mathrm{r}} - w_{1\mathrm{r}}w_{3\mathrm{i}}\right), \\
E &= 2\left(w_{1\mathrm{i}}w_{2\mathrm{r}} - w_{1\mathrm{r}}w_{2\mathrm{i}} + w_{2\mathrm{i}}w_{3\mathrm{r}} - w_{2\mathrm{r}}w_{3\mathrm{i}}\right).
\end{aligned}
$$

Therefore, the surface area of the radiation pattern is given by

$$\text{Beam surface area} = \int_{\Phi=0}^{2\pi}\int_{\Theta=0}^{\pi} |\mathrm{AF}|^2 \sin\Theta \mathrm{d}\Theta \mathrm{d}\Phi. \tag{8.65}$$

An approximate method to get the noise factor is described in the following steps:

Step 1: Generate a number of noise channels (in practice, noise channel number is infinite), say 1000, with random DOA:

$$\mathrm{x}(j) = \times \text{random\_number\_normal}, \tag{8.66}$$

$$\text{rand\_}\Theta(j) = \pi/2 + \text{random\_number} \times \pi, \tag{8.67}$$

$$\text{rand\_}\Phi(j) = \text{random\_number} \times \pi, \tag{8.68}$$

where variance = 0.161, random_number_normal is normally distributed with zero mean value and has a variance of 1, random_number is uniformly distributed within 0 and 1, rand_Θ = angle between the noise channel's direction and the z-axis, rand_Φ = azimuth angle of the noise channel's direction, and j = 1, 2, . . ., 1000.

Step 2: Evaluate the average noise power (without beamforming), *N*_isotropic:

$$N\text{\_isotropic} = \frac{1}{1000}\sum_{j=1}^{1000} x(j)^2. \tag{8.69}$$

It should be approximately 0.161 (variance of the noise channel). This serves as a means to verify whether the noise channel number used is sufficient to simulate the infinite number of noise channels.

Step 3: Use the same noise channels and multiply the noise signal, $x(j)$, with their respective array factor, AF(I), according to the DOA of each noise signal.

Step 4: Evaluate the average noise power received after beamforming, *N_BF*:

$$N\_\mathrm{BF} = \frac{1}{1000}\sum_{j=1}^{1000}[\mathrm{AF}(j)x(j)]^2. \tag{8.70}$$

Step 5: Calculate the ratio of the noise power level after beamforming to the noise power level in the isotropic scenario that is the noise factor:

$$\text{Noise factor} = \frac{N\_\mathrm{BF}}{N\_\mathrm{isotropic}}. \tag{8.71}$$

8.13 Performance Evaluation

In this section, we discuss the analytical tools and computer simulations to evaluate the performance of a code-division multiple access (CDMA) cellular system using smart antennas. The results and simulations used in the performance evaluation are discussed in the later part of this chapter.

8.13.1 System Capacity

We may model a hexagonal cellular system by two concentric circles, with the outer circle being subdivided into six segments (neighboring BSs). The two concentric circle model and the six outer BS segments are shown in Figure 8.16.

Interference of the user in adjacent cells. When an MS is transmitting a signal to its control BS (the CBS), it is also transmitting a signal to all the neighboring BSs as well. The signal will be treated as an interferer by the neighboring BSs. The magnitude of the signal received by the neighboring BSs will be smaller than that received by the CBS. This may be due to path loss and the geometry of the MS beam which may seek to maximize transmission to its own CBS. The identity of the CBS will change as handover occurs with the movement of the MS.

Let us assume that the signal power of the MS received by the CBS is P_0. In order to find the signal power received by a neighboring cell, the distance between the MS and the neighboring BS must be known. The distance can be easily found using cosine rule (see Figure 8.17).

Let the angle-of-arrival of the MS be Θ_a with distance R_0 from the user base station (CBS). Further, the angle subtended by the CBS to the neighboring BS is Θ_b with the distance between them being $2R$.

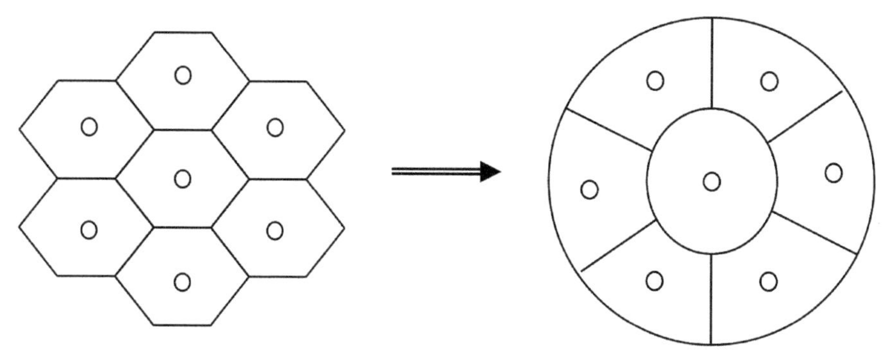

Figure 8.16 Modeling of the cellular system.

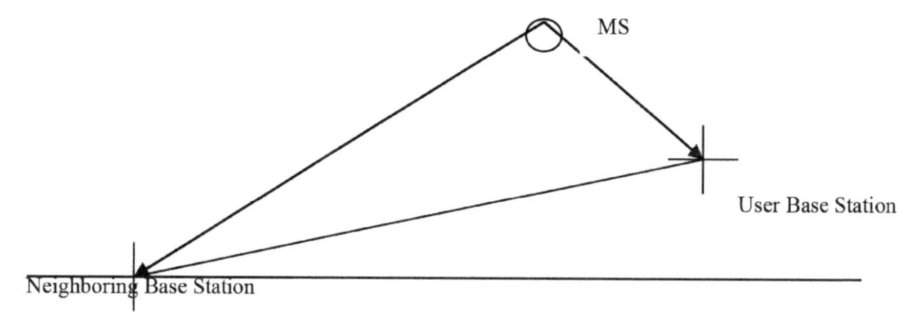

Figure 8.17 MS and adjacent BSs.

Let $P_{\text{user,bs}}$ and $P_{\text{user,ns}}$ be the power received from the CBS and the neighboring BS, respectively, and P_{tx} be the power transmitted by the MS. Using the cosine rule, the distance between the MS and the neighboring BS is obtained from

$$D^2 = R_0^2 + (2R)^2 - 2\,(R_0)\,(2R)\cos\left(\pi - \Theta_{\text{a}} + \Theta_{\text{b}}\right). \tag{8.72}$$

We shall assume 1) perfect power control, 2) the CBS is required to receive a minimum power of P_{req}, and 3) the *log distance path loss model*. Hence, the power received at the CBS is given by

$$P_{\text{user ,bs}} = G_{\text{tx}} \times G_{\text{xx}} \times P_{\text{tx}} \times \left(\frac{\lambda}{4\pi d_{\text{ref}}^2}\right)^2 \left(\frac{d_{\text{ref}}}{R_0}\right)^n \tag{8.73}$$

$$= G_{\text{rx}} \times P_{\text{tr}} \times \left(\frac{\lambda}{4\pi d_{\text{ref}}^2}\right)^2 \left(\frac{d_{\text{ref}}}{R_0}\right)^n \tag{8.74}$$

where G_{tx} = isotropic gain of transmitter antenna = 1, G_{rx} is BS receiver gain, d_{ref} is reference distance for path loss calculation, and n is path loss exponent. We may rewrite Equation (8.74) as

$$P_{\text{user,ns}} G_{\text{rx}} \times P_{\text{tx}} \times \left(\frac{\lambda}{4\pi d_{\text{ref}}^2}\right)^2 \left(\frac{d_{\text{ref}}}{D}\right)^n. \tag{8.75}$$

Therefore, the out-of-cell interference by a single MS user is given by

$$\begin{aligned} P_{\text{outcell-user}} &= P_{\text{pc}} \times \left(\frac{R_0}{D_n}\right)^n \\ &= P_{\text{user ,bs}} \times \left(\frac{R_0}{D}\right)^n, \end{aligned} \tag{8.76}$$

where n is the path loss exponent factor and it is dependent on the specific propagation environment, and P_{pc} is the minimum power required by the CBS in order to achieve perfect power control. Table 8.1 lists the typical path loss exponents for various mobile radio environments.

Since every cell BS smart antenna is being beamformed, the signal to the CBS might be attenuated relative to the maximum array factor amplitude of the cell antenna. Thus, in order to achieve perfect power control, each MS has to transmit more power in order for its CBS to receive the required signal strength. Therefore, Equation (8.46) has to be modified as follows:

$$\begin{aligned} P_{\text{outcell-user}} &= P_{\text{pc}} \times \left(\frac{R_0}{D}\right)^n \times \left(\frac{\text{AF}_{\text{max}}}{\text{AF}_{\text{bs, user}}}\right)^2 \\ &= P_{\text{pc}} \times \frac{(R_0/D)^n}{\text{AF}_{\text{bs, user ,norm}}^2}. \end{aligned} \tag{8.77}$$

Due to the CBS antenna beamforming, the power will be further reduced by multiplication of the array factor and the interference signal power. Therefore, the out-of-cell interference of a single MS is

$$P_{\text{outcell-user}} = P_{\text{pc}} \times \frac{(R_0/D)^n}{\text{AF}_{\text{bs , user , norm}}^2} \times \text{AF}_{\text{ns, user}}^2. \tag{8.78}$$

In our simulation studies, we generated K numbers of MSs with random DOA and distance in each of the six adjacent cells with random distance and angle from its BS. Thus, the total outer cell interference power can be calculated from the summing of the transmission power from all the MSs in the six cells.

Table 8.3 Path loss exponents in wireless communications environment.

(i) Environment exponent	Path loss
Free space	2
Urban area cellular radio	2.7–3.5
Shadowed urban cellular radio	3–5
In-building line-of-sight	1.6–1.8
Obstructed in building	4–6
Obstructed in factories	2–3

Therefore $6 \times K$ users from all the adjacent cells will contribute to the out-of-cell interference. In other words, the total out-of-cell interference will be

$$\begin{aligned} P_{\text{outcell}} &= 6 \times K \times P_{\text{outcell-user}} \\ &= 6 \times K \times P_{\text{pc}} \times \frac{(R_0/D)^n}{\text{AF}^2_{\text{bs,user, norm}}} \times \text{AF}^2_{\text{ns, user}} \\ &= 6 \times K \times P_{\text{pc}} \times \beta, \end{aligned}$$

where

$$\beta = \frac{(R_0/D)^n}{\text{AF}^2_{\text{bs,user,norm}}} \times \text{AF}^2_{\text{ns, user}} \cdot$$

Interference from MSs in the cell under study. Suppose there are *K* MS users in the cell under study. Then the single MS under study will treat the other $K - 1$ MSs as interferers. Since perfect power control is used in the CDMA system, the power of all the MSs received in the cell under study will be the desired power, P_{pc}. Due to BS smart antenna beamforming in the cell under study, those MSs which are not in the direction of the main lobe of the radiation pattern will need to transmit additional power in order to achieve the power of P_{pc}.

Therefore, the total in-cell interference power will be

$$P_{\text{incell}} = (K - 1) \times P_{\text{pc}}. \tag{8.79}$$

From the generation of the interferences for both the in-cell and out-of-cell MSs, we can calculate the reuse factor as follows (Liberti and Rappaport, 1999):

$$\begin{aligned} \text{Reuse factor} &= \frac{P_{\text{incell}}}{P_{\text{incell}} + P_{\text{outcell}}} \\ &\approx \frac{1}{1 + 6\beta} \end{aligned}$$

$$= \frac{(K-1)P_{\text{pe}}}{(K-1)P_{\text{pe}} + 6\text{KP}_{\text{pe}}\beta} \quad (\text{ for K} >> 1). \tag{8.80}$$

CINR is defined after spreading as the ratio of the desired signal to the sum of interference and noise, i.e.

$$\text{CDN} = \frac{P_0}{(1/N)\sum_{i=1}^{R-1} P_k + \sigma_{\text{n}}^2}, \tag{8.81}$$

where P_0 is the power of the desired signal at the input of the despreader, P_k is the power from the other user, ${\sigma_{\text{n}}}^2$ is the noise contribution after despreading, and

$$N = \text{Processing gain} = \frac{\text{Chip rate (chip per second)}}{\text{Information symbol rate(symbol per second)}}, \tag{8.82}$$

After despreading, the noise bandwidth is $1/T_{\text{b}}$. The CINR can be rewritten as

$$\begin{aligned} \text{CINR} &= \frac{P_0 T_{\text{b}}}{(1/N)\sum_{k=1}^{K-1} P_k T_{\text{b}} + \sigma_{\text{n}}^2 T_{\text{b}}} \\ &= \frac{E_{\text{b}}}{N_{\text{i}} + N_{\text{n}}} \\ &= \frac{E_{\text{b}}}{N_0}, \end{aligned}$$

where N_{n} is the power spectral of thermal noise and N_{i} is the power spectral density of the total MAI after despreading.

In our simulation, we can express CINR as

$$\begin{aligned} \text{CINR} &= \frac{NG_{\text{a}}P_{\text{c}}}{v(K-1)P_{\text{c}} + 8vKP_{\text{c}}\beta} \\ &= \frac{NG_{\text{a}}}{v(K-1) + 8vK\beta} \\ &\approx \frac{NG_{\text{a}}}{vK(1+8\beta)} \end{aligned} \tag{8.83}$$

Rearranging, the capacity of the CBS can then be evaluated as shown:

$$\text{Capacity} = \frac{\text{Reuse factor} \times \text{Processing gain }(N) \times \text{Pattern gain }(G_{\text{a}})}{\text{Voice activity }(v) \times E_{\text{b}}/N_0}, \tag{8.84}$$

where

$$\text{Processinggain} = \frac{\text{Chip rate (chipper second)}}{\text{Informationsymbol rate (symbol per second)}}.$$

The voice activity (v) factor arises due to the fact that the MS will not be transmitting continuously, when the speaker pauses in a conversation over the mobile phone (MS). The vocoder will reduce the output rate when the speaker is silent. Pattern gain is the smart antenna horizontal azimuth gain after beamforming. The ratio E_b/N_0 is the one required by the receiving BS, i.e. the CBS.

8.13.2 Loading of Antenna

The fully adaptive array antenna (i.e. the smart antenna) is capable of maximizing its beam array factor in the direction of the signal of interest (SOI) while nulling the beam in the directions of the interferers. However, there is a limitation to the number of interferers that can be nulled due to the antenna array size (i.e. the number of antenna elements). We shall now evaluate the dependence of the number of interferers that the antenna can null on the array antenna size.

8.13.3 Signal to Interference and Noise Ratio

The smart antennas can effectively cancel out the interference in the system if the system is not overloaded. Therefore, in a practical case, not all the interferences can be nulled, and, thus, some interferers will contribute to the SINR. In our simulation studies, we shall evaluate SINR for the whole system and find out how much SINR improvement can be achieved by using smart antennas comparing its performance with a conventional antenna.

SINR is defined as

$$\text{SINR} = \frac{\text{Signal power}}{\text{Total interference} + \text{noise}}. \tag{8.85}$$

8.13.4 Range

A BS smart antenna is expected to provide an increase of range that the BS can cover. The improvement in smart antenna range can be assessed by comparing it with the distance coverage by a conventional antenna system. The antenna gain, which can be numerically determined using Trapezium rule from the surface area of the beam pattern, is used to evaluate the range

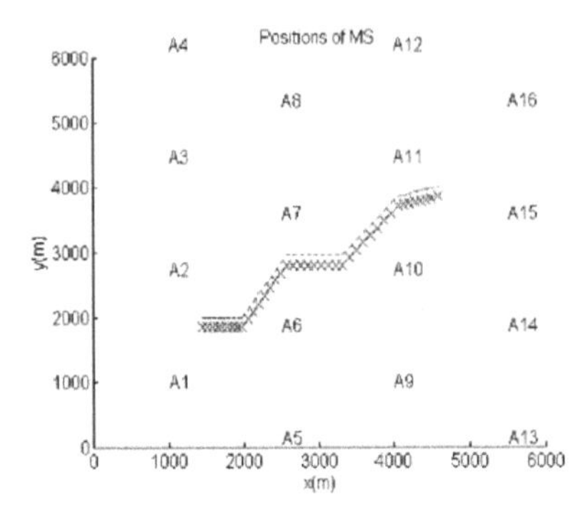

Figure 8.18 MS velocities versus sampling time.

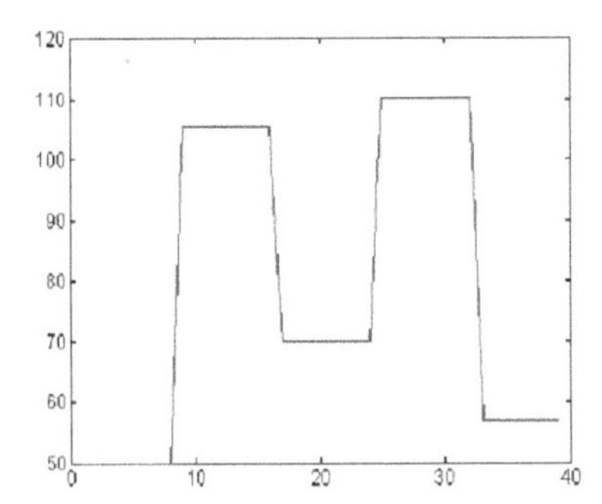

Figure 8.19 MS velocities versus sampling time.

improvement. One of the important factors that affect the range of a system is the loss exponent. In other words, the system will exhibit maximum range in free space and the range will be degraded in a built-up urban area.

8.14 Base Station Beamforming: Simulation Studies

8.14.1 Simulation Scenario

The overall simulation is based on the scenario used in Chapter 7. There are 16 BSs (marked as "A" in Figure 8.18) in the scenario with a cell radius of 1000 m. An MS (marked as "x" in Figure 8.18) travels along the dotted path, with the velocity variation shown in Figure 8.19.

Forty sampling points are used with a time interval of 4.8 s (equivalent to 10,000 sampling intervals adopted in the GSM system time step of 0.48 ms). Some of the parameters used are shown in Table 8.4.

Table 8.4 Simulation parameters.

Parameters	Description
$f = 1250$ MHz	Carrier frequency
$z_1 = 60$ m	Height of the lower part of the BS antenna
$z_2 = z_1 + \lambda/2$	Height of the upper part of the BS antenna ($L = \lambda/2$ for half-dipole antenna)
$z_j = 1$ m	Height of the MS antenna
$R = 73\ \Omega$	Radiation resistance of the MS antenna
$P = 2$ W	Radiation power of the MS antenna

8.14.2 Algorithm

The simulation proceeds as follows.

1) Initialization
2) Choose CBS and BS (1:3)

   (a) get estimation of distances r (initial value for maximum likelihood estimator's iteration) using time advanced (TA) measurement;
   (b) choose CBS by referring to the *estimate_r*. BS (1:3) refers to the three BS with the smallest *estimate_r*;
   (c) generate first set of θ and Φ using exact parameters (r, $z1$, ...);
   (d) generate parameters (*rand_*θ*, rand_interf, rand_*Φ) for random beam patterns for all BSs (*rand_*θ calculation based on θ); *rand_interf* is the random azimuth angle of the interference (assuming only one interferer);
   (e) generate array factors of all the BS with respect to the SOI–MS, using parameters generated in 2(c) and 2(d) (call function *beamform.m*);
   (f) generate E-fields received by all the BS (without noise);
   (g) calculate E_m (with noise);
   (h) choose CBS: in case there are more than one BS (within the three nearest BS) with the smallest *estimate_r* (noting that there is an estimation error of TA of 277 m), CBS is chosen to be the BS with the strongest received E_m;

3) Sampling loop (40 points)

   (a) $kk = 1$;
   (b) if ($kk > 1$),

      (I) generate θ and Φ using exact parameters (*r_used, zj_used, ...*);
      (II) generate array factors of the three nearest BS with respect to the SOI (call function *beamform.m*);

(III) calculate E_m_used (which is the E_m of the three nearest BS);

(c) iteration of *ddr* (iterated value of *r*);
(d) iteration of *pp* (iterated position of MS);
(e) if ($kk > 1$), get velocity;
(f) prepare parameters for the next sampling point;
(g) $kk = kk + 1$;
(h) go back to (b).

3c. Iteration of *ddr*

(I) $k = 1$;
(II) set initial value of *ddr* as the *estimate_r*;
(III) generate $d\theta$ using *ddr*, assuming Φ is known to the system ($d\Phi = \Phi$);
(IV) generate respective AF_used (1:3) and W_opt (1:3) using $d\Phi$ and $d\theta$ (call function *beamform.m*);
(V) generate *EE* (*estimate_E*) and *J* (Jacobian matrix of *EE*) (call function *jascob.m*);
(VI) $ddr = |ddr + (J^T J)^{-1} J^T (Em\_used - EE)|$;
(VII) $k = k + 1$;
(VIII) stop *r* estimation if rcond(*J*) $\leq$ 0.01 (reciprocal condition estimator) (prevent Jacobian matrix from becoming singular);
(IX) go back to (III).

3d. Iteration of *pp*

(I) $k = 1$;
(II) set initial value of *pp* (MS position) as the center of the three nearest BS;
(III) shift *pp* by 10 m in x–y-direction when *pp* is too near (0.1 m) from any of the three adjacent BS;
(IV) evaluate *Errp* (= | *PB* – *pp* | – *ddr*), the position error;
(V) evaluate *Jp*, the Jacobian matrix of *Errp*;
(VI) stop position estimate when cond(Jp) $>$ 500;
(VII) *ppx* (*pp* in *x*-direction) = $ppx + (Jpx^T \times Jpx)^{-1} \times Jpx^T \times (-Errpx)^T$; *ppy* (*pp* in *y*-direction) = $\mathrm{ppy} + (\mathrm{Jpy}^T \times \mathrm{Jpy})^{-1} \times \mathrm{Jpy}^T \times (-\mathrm{Errpy})^T$;
(VIII) $k = k + 1$;
(IX) go back to (III);

3e. Get velocity of the MS
$V_t = (PM_t - PM_{t-1})/T$;

3f. Prepare parameters for the next sampling point

(I) deciding on handover by using *ddr*:
handover occurs when *ddr*(old_CBS) – *ddr*(new_CBS) > 10 m. A tolerance of 10 m is created to avoid excessive handover while the PM fluctuates between two cells;

(II) decide whether or not to change BS smart antenna beam by simulating a beamformed scenario and compare the E-field received by the CBS before and after beamforming: a new beam that is steered toward the MS will be formed when a significant improvement in received E-field is expected;

(III) generate random beam patterns for the two adjacent BSs (exclude CBS), by specifying *rand_Φ* and *rand_interf*;

(IV) decide on a new set of adjacent BSs using triangle method;

(V) update all other parameters (e.g. *PM*, *PB_used*, ...) for the next sampling point;

(VI) set the iterated r value (*ddr*) as the initial value for the next *r*-iteration;

8.15 Results and Discussion

8.15.1 BS Smart Antenna Beams

The array factor (AF) of the smart antenna in the direction of the desired signal is very much higher than that for an isotropic antenna. This is seen from Table 8.5. For a nine-element smart antenna, the AF in the direction of the desired signal is 8.9495, much higher than the AF= 1.0 we would have obtained with an isotropic antenna. In addition to this advantage, the AF is cut down to 0.0064 (from 1 for isotropic antenna) in the direction of the undesired signal. The half-power beam width (HPBW) is 20° and the sidelobe level (SLL) is 1.8889. The beam width for a bigger array antenna will become narrower. If the beam width is too small, the MS may very rapidly fall out of the main lobe if the MS is traveling at a high speed. Thus, there is a need to arrive at a compromise between the beam width and the accuracy of the beam or radiation pattern.

The smart antenna was also tested for an interfering signal that is angularly near to the desired signal. The beam obtained was not as accurate as that for the previous test with the same number of elements. This is due to the fact that the antenna is unable to null the interference and point to the desired signal with the restricted degree of freedom, which is equal to the number of

Table 8.5 Radiation pattern characteristics with smart antenna.

N, number of elements	2	3	5	9
$AF_{desired}$, array factor in the direction of MS	1.9250	2.9496	4.9473	8.9495
$AF_{undesired}$, array factor seen by interferer	0.0443	0.0077	0.0130	0.0064
HPBW, half-power beam width	80°	52°	38°	20°
SLL, sidelobe-level	1.0989	1.8861	2.2085	1.8889

elements. This observed degradation of performance will be important when implementing cell sectorization. Cell sectorization will reduce the number of undesired signal for each smart antenna in any one sector. However, if there is any undesired signal, it means that the undesired signal will be angularly close to the desired signal.

8.15.2 Triangle Method

Table 8.6 shows the location of the three BSs nearest to a moving MS. The SNR was 10 dB. In Table 8.6, *kk* is the sampling number, ranging from 1 to 40; adjacent BS number indicates three nearest BS numbers ranging from 1 to 16 (refer to Figure 8.20); *ddr* is the computed distance of MS from adjacent BS.

Tests were also performed for SNR = 20 dB. The tests performed verified the usefulness of the triangle method. The change of the third nearest BS (shown in bold in Table 8.4) occurs at almost the same sampling number (*kk*) for all the tests performed at different SNR. It can be seen that a new third nearest BS will be "computed" only when more than one angle is less than 90ž (the underlined angle). However, some complex angles (shown in italics) are generated due to the estimation error in *ddr*. This erroneous outcome is because of taking the arc cosine of a value bigger than "1." Even though the occurrence of these complex angles is rare (approximately 5%), there exists a need to further modify the triangle method.

8.15.3 Handover

Figure 8.20 and Table 8.7 show how handover takes place as the MS moves through a region installed with 16 BSs with smart antennas. In Figure 8.20, B is the control base station (CBS), and O is the position of the MS where handover occurred. The dotted track in Figure 8.20 corresponds to the estimated positions of the MS and the dashed track indicates the exact path of the MS.

Table 8.6 Triangle method for locating the three nearest BSs.

kk	Adjacent BS no.	*ddr*	New adjacent BS no.	Computed PM (m)	Angle (*kk*) (°)	Angle (*kk* −1) (°)
12	1, 2, 6	1954, 1227, 518	**7**, 2, 6	(2261, 2407)	61, 165, 57	80, *180 −j27*, 105
18	7, 2, 6	635, 1734, 999	7, **10**, 6	(2692, 2920)	79, 73, *180 −j40*	96, 80, *180 −j25*
27	7, 10, 6	1228, 634, 1668	7, 10, **11**	(3592, 3153)	134, 85, 71	149, 94, 80
34	7, 10, 11	1686, 1294, 616	**15**, 10, 11	(4183, 3944)	70, 126, 84	75, 144, 98

SNR = 10 dB; noise type = AWGN.
Seven complex angles out of 117 computed angles.

Table 8.7 Handover of MS control. The CBS is highlighted.

Sampling no.	*ddr*	*(ii) ()ddr* (Sampling no. – 1)
(iii) 3	(iv) 1251, 983, **940**	(v) 1109, **893**, 995
16	**772**, 1475, 1046	864, 1376, **773**
23	1007, **962**, 1141	**802**, 986, 1235
31	1472, 900, **875**	1356, **806**, 999

SNR = 10 dB.

Further tests at different SNRs indicated that all the handovers take place almost at the same sampling number irrespective of the SNR of the entire region. No experiments were carried out for SNR variations within the area covered by the 16 BSs. The handover algorithm can be improved by taking into account the electric field or power strength received by the three nearest BSs.

8.15.4 BS-Based Position–Velocity Estimator

In Chapter 7, we discussed MS-based position–velocity estimation of the MS using electric field strengths of signals from the BS picked up at the MS. Figure 8.21 shows the BS-based position–velocity estimation results for SNR = 20 dB using electric field strengths of signals from the MS picked up at the BS. With an SNR of 20 dB, the average position estimation error is approximately 23 m. Besides, it is also shown that all estimated positions

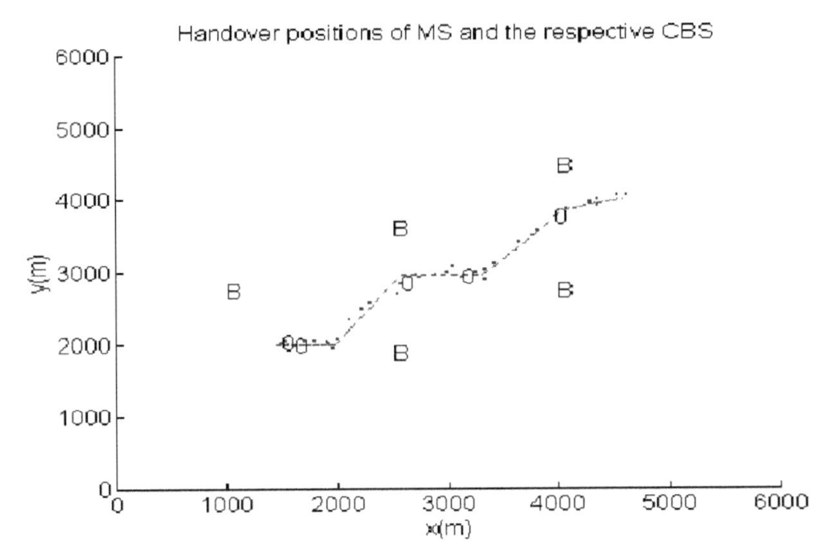

Figure 8.20 Handover of MS control.

are accurate to 70 m. The velocity estimator achieved an average velocity estimation error of 24 km/h.

For SNR = 10 dB, it was found that the position estimator achieved quite satisfactory results, with an average estimation error of 78 m, which is much better compared to the estimation error of 277 m with the time delay method. Besides, it is also shown that all the position estimation errors using BS base position estimation using electric field measurements are below 200 m. In other words, all positions are successfully estimated.

However, the velocity estimator for SNR = 10 dB was quite poor. The velocity estimation error of almost 100% was observed at some point, and this is simply intolerable. This error is accumulated from the position estimation error on which the velocity estimator depends.

8.15.5 AWGN Model for Smart Antenna Systems

Table 8.8 shows the effect of interference on the array factor of the smart antenna. From the data shown in Table 8.8, it is seen that the noise factor of an AWGN channel after beamforming is bigger than 1. In other words, the channel noise power (or variance) is increased after a beam is formed – an unexpected result. This can be explained by observing the beam shown in Figure 8.22. The beam, or radiation pattern, shown in Figure 8.22 is for

SNR = 20 dB.

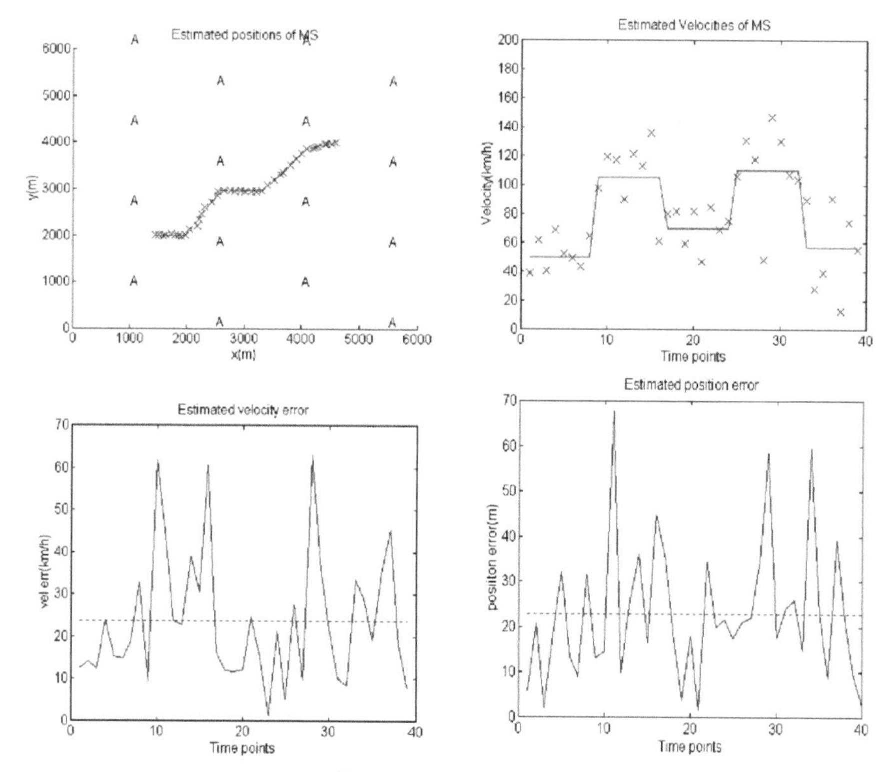

Figure 8.21 PositionŰvelocity estimation with BS smart antennas.

the case when the desired signal is at 131.5° and interference comes from 108.0°. Observe that the beam pattern formed is bigger than that for the single isotropic element case where we get a sphere with unity array factor. Therefore, the noise factor tends to increase with beamforming due to the increase in beam surface area that could receive more noise power.

However, the increase in noise factor does not mean that the beamforming will impair the signal transmission. As the signal power is proportional to the square of the array factor, it can be seen that the increase in signal power is much bigger than the noise factor. Therefore, the overall effect of beamforming will be to increase the SNR.

Besides, it is verified that the noise factor is also equal to the area ratio beam area/(4π). In other words, the bigger (in surface area) the beam that is formed, the larger the value of the noise factor. The small discrepancy

Table 8.8 Effect of radiation pattern area on noise at the receiver.

Signal (θ, ϕ)	Interference (θ, ϕ)	Noise factor	Area/(4π) of beam	Array factor^2
91.7, 43.4	91.7, 143.8	2.85, 2.99, 2.69, 2.89, 3.22	3.42	8.65
91.7, 24.9	91.7, 139.4	4.27, 4.27, 4.03, 4.32, 4.31	4.93	8.49
91.7, 131.5	91.7, 108.0	4.17, 4.13, 3.79, 4.04, 4.20	4.62	8.53
91.7, 154.0	91.7, 11.4	16.31, 15.87, 16.38, 15.86, 15.46	17.90	6.77
91.7, 54.1	91.7, 86.8	2.78, 2.70, 2.48, 2.50, 2.41	3.11	8.69
91.7, 170.0	91.7, 80.9	2.27, 2.21, 2.08, 2.13, 2.24	3.11	6.61

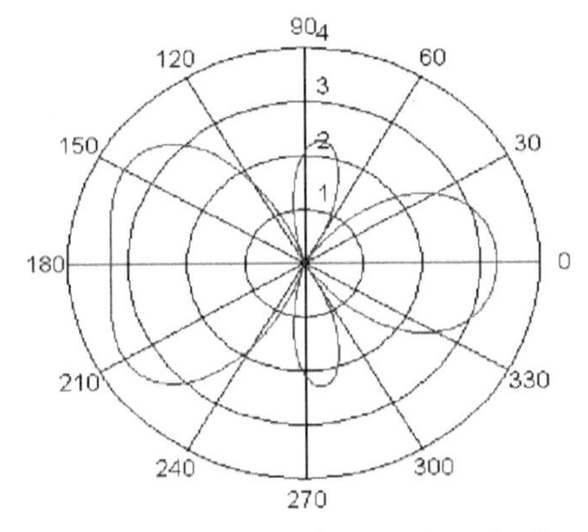

Figure 8.22 Illustrating increase in noise factor.

between the noise factor and the area ratio is due to the approximation error introduced while computing the beam area numerically and the simulation error created while generating noise factor.

For the case when the signal is at (91.7°, 154°) and interference comes from (91.7°, 11.4°), it is seen from Table 8.6 that the noise factor is larger than the (array factor)^2 of the signal. This is due to the fact that we did not consider the smart antenna being used in a sectored form. Even though the beam is not exactly symmetrical at the 90°–270° axis (it is exactly symmetrical along the 0°–180° axis), the algorithm will not be able to create a large array factor at ϕ when a null is expected at (180°–ϕ). In these unfavorable circumstances, a large beam will be formed resulting in a large noise factor. But the value of the array factor for the desired signal is still relatively small. This situation can be avoided if sectioning is used. Figure 8.23 shows a sectored arrangement of smart antennas. Figure 8.23(a) shows a BS arrangement with three sectors. Each sector has a smart antenna with five elements that will cover 120° each. Any sidelobe or backlobe radiation in the backward direction will be absorbed

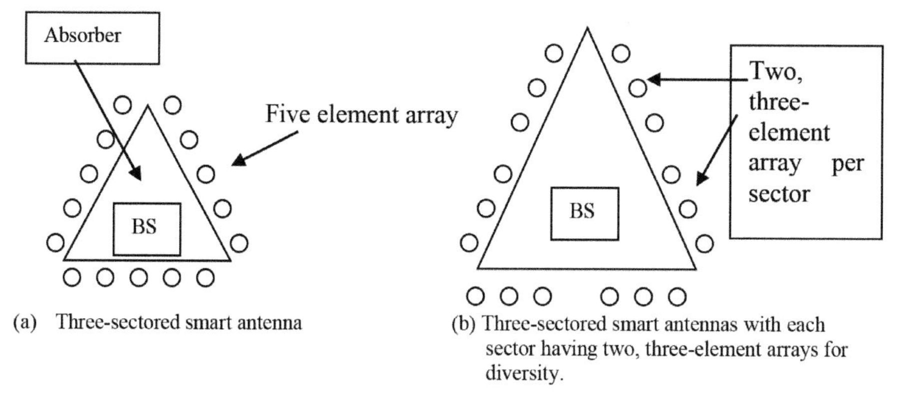

Figure 8.23 Sectored smart antenna arrangements. (a) Three-sectored smart antenna. (b) Three-sectored smart antennas with each sector having two, three-element arrays for diversity.

to prevent interference with the antennas of the other sectors. Figure 8.23(b) shows a three-sectored BS arrangement, where each sector is divided into two separate arrays so as to allow for better diversity and hence better noise or interference cancelation.

8.15.6 Performance Evaluation

8.15.6.1 Capacity, SIR, and Range

The simulation is tested for the route shown in Figure 8.20 having 40 discrete positions. Each BS is assumed to have three-element arrays. Figure 8.24 shows the range, capacity, and SIR performance of the three-element BS antennas compared to isotropic BS antennas. Sectoring, which will improve the performance of the smart antenna further, was not used. With BS beamforming performed by a simple three-element smart antenna, it can be seen from Figure 8.24(a) that the BS can cover a longer range.

It can also be seen that with beamforming of the BSs, the range (Figure 8.24(b)) and SIR (Figure 8.24(c)) are improved when compared to an isotropic antenna. These observations make the use of BS smart antennas attractive. The simulation results reveal that for 10 users in each cell in an urban environment ($n = 4$), a mean range improvement of 1.2799 dB, and a mean capacity of 41 MS users and an SIR improvement of 2.9394 could be obtained. However, the simulation results shown are based on the assumption that perfect power control is achievable.

The limitation of BS beamforming in multiple directions may demand an increase in the MS transmitted power if an MS is not positioned at the peak of the BS antenna beam. This may result in a shortened battery life. To overcome this, the BS may be sectored into three sectors, each covering 120° as shown in Figure 8.23(a). Depending on the position distribution of the users in the cell, the BS smart antenna must maximize its beam pattern for all the MS users in the cell in order to minimize the transmission power required of the MS user.

It was shown in Section 9.4.1 that HPBW will decrease as the number of antenna elements increases. Thus, the BS antennas can be designed as follows:

1) Each cell is sectored to 120° each.
2) Each sector will have a reasonably large array so that a narrower beam can be formed.
3) Each sector is again sectored into eight subsectors (for simplicity), each covering 15°.

The advantage of sectoring is that the array factor will be large for all MS users instead of a few that might be clustered around a single BS antenna beam peak. This means the required MS user transmission power will be less. However, the disadvantage is that such sectoring is only possible where MS users are usually clustered within a few subsectors not evenly distributed in the sectors. Handover will become more complicated with sectoring.

8.15.6.2 Loading of Antenna

In this section, we study the effect of interferers on the loading of the smart antenna and its accuracy in generating the desired radiation pattern. When the number of interferers increases, some of the interferers will not be nulled by the adaptive smart antenna system. This is because an array antenna system of N number of elements can only effectively null $N - 1$ number of interferers. In addition, the direction of the main beam will be slightly shifted if the direction of any interferers is too close to the direction of the desired signal.

Table 8.9 Relative interference power level at the receiver as interferers increase in number using a four-element antenna.

Angle of interference (°)	Total number of interferers			
	2 (dB)	3 (dB)	4 (dB)	5 (dB)
30	−46.03	−42.74	−20.11	−18.86
90	−60.55	−50.39	−32.60	−22.94
135		−44.08	−21.04	−14.96
150			−16.43	−13.25
45				−8.55

This will tend to cause the gain in the direction of the desired signal to degrade.

Table 8.9 shows the interference power level with reference to the desired signal power level at the receiver electronics using a four-element array antenna. The desired signal is at 60°. It is obvious that the performance of the smart antenna will depend on the number and position of the interferers. As the number of interferers increases, the interference power level approaches that of the desired signal power.

9

Real- and Complex-Valued Artificial Intelligence Weight Optimization Algorithms for Smart Antennas in 5/6G Wireless Systems: Linear and Nonlinear Arrays

K.S. Senthilkumar[1], K. Pirapaharan[2], H. Kunsei[3], S.R.H. Hoole[4], and P.R.P Hoole[5]

[1]Department of Computers and Technology, St. George's University, Grenada, West Indies
[2]Faculty of Engineering, University of Jaffna, Jaffna, Sri Lanka
[3]Papua New Guinea University of Technology, Lae, Papua New Guinea
[4]FIEEE as Professor of Electrical Engineering (Retired), Michigan State University, USA
[5]Wessex Institute of Technology, Southampton, UK

Abstract

Artificial neural network (ANN) has been applied in many fields including wireless systems. ANN has been used in the design and optimization of different antenna types. The dynamic ability of the ANN to adapt a system variable in any given application is beneficial in smart antenna design for 5/6G wireless systems. This chapter presents a novel ANN algorithm, the single neuron weight optimization model (SNWOM), that optimizes the radiation patterns of uniform linear and nonlinear array (ULA) smart antenna in a desired direction. The robustness of the algorithm was compared against the least mean square (LMS) algorithm for three different functions, namely, the hyperbolic tangent, bipolar, and squash or Elliot functions, for varying number of antenna array elements. SNWOM showed excellent performance as a smart antenna beamformer. The benefit of SNWOM includes fast convergences and demands less hardware resources to offer the favorable performance.

Keywords: 5G, smart antenna, artificial neural network, activation function, beamforming, least mean square (LMS), uniform linear array, single neuron weight optimization model (SNWOM).

9.1 Introduction

Neural computing began with the development of the McCulloch–Pitts network in the 1940s. Many problems in artificial neural network (ANN) come out of the paradigms which serve as the basis for the complete machine learning discipline. Some of these basics are based on mathematics and others are based on observation and experimentation. Several constraints arise out of the paradigms of neuro-computing. It is a fact that at some point, each neural network should be implemented in some real-world application. On that line, we worked on a single-layer perceptron (SLP) model to optimize the weights of a dipole array antenna to steer the beam to desired directions. The objective of this work is to reduce the complexity by using a single neuron neural network and utilize it for adaptive beamforming in array antennas.

Adaptive array antenna is most popular in the present world [1]–[3]. The array antenna is made up of a collection of individual antennas discussed in Chapters 2 and 3, including the dipole antenna, the patch antenna, the horn antenna, the Cassegrain reflector antenna, and the helical antenna. One example of the elements making up the array antenna is the simple electric dipole antenna presented in Chapter 2. A four-element array antenna, for instance, could be made up of four electric dipole antennas. Present world applications require much faster beam steering that cannot be achieved using a mechanical systems. Hence, it is required to use more consistent and much faster electronic beam steering techniques such as adaptive arrays. The requirements for almost identical elements result lack of flexibility. On the other hand, adaptive beamforming methods by means of weight optimization are capable of managing the complexities of distinct elements. The adaptive array can detect, track, and allocate narrow beams in the direction of the desired users while nulling unwanted sources of interferences. There are well-known traditional techniques for adaptive beamforming in array antennas. The long-term evolutionary (LTE) and fifth generation (5G) communication system is the latest wireless communication technology in use which provides high speed and high capacity wireless communication when compared to the 3G wireless systems. What is referred to as the LTE 4G systems is still limited in many areas. Smart antennas and the multiple-input and multiple-output (MIMO) system are one of the available ways to increase the rapidly increasing demand on capacity in 5G systems. The peak data rate is proportional to the number of antennas at the sending and receiving ends.

Soft computing techniques, namely ANN, fuzzy logics, genetic algorithms (GAs), etc., provide low cost solutions and robustness to different complex real-world problems. ANN is a powerful information processing paradigm that tries to simulate the structure and functionalities of the biological brain systems. The ANN is used to deal with many applications, and they have proved their effectiveness in several research areas such as image recognition, speech recognition, signal analysis, process control, robotics, and many more. The true power of neural networks lies in their ability to represent both linear and nonlinear relationships. ANNs, like humans, learn by example. Training a neural network is, in most cases, an exercise in numerical optimization of usually a nonlinear function. The basic building block of every ANN is an artificial neuron or perceptron that is a simple mathematical model. ANN is a powerful technique to be used because of its general purpose nature, fast convergent rate, and large-scale integration implementations where the mathematical relationship between input and output can be reliably established. The ANN is able to approximately model the input–output relationship by optimizing the weights through using known input–output training pairs. Once the training is done, it is able to obtain the needed antenna radiation beam for a given set of inputs by adaptive signal processing [4]–[7]. The use of both real-valued and complex-valued activation functions are explored in this chapter [8]–[23].

Here we present a simple single neuron model to optimize the weights of an adaptive array antenna which leads to adaptive, smart beamforming toward the desired users. We focus on beam steering using a fast neural network adaptation to direct the beam toward particular users and/or to steer nulls to reduce interference. This is a crucial role of a smart antenna which is able to provide electrical tilt, beam width and azimuth control suitable for handling moving traffic patterns [24]–[30]. The smart antenna solution is far more versatile, and cheaper required low memory, fast beam steering techniques such as that reported in this chapter. They are used at the base stations and increasingly in the mobile stations as well. In parallel, the development of cost-effective fast cell site addition, increasing the number of cell sectors and bandwidth, and better air interface capabilities will be critical in the 5G wireless systems in the future.

9.2 Processing Element

In a simple neural network, the user collects descriptive data and then uses algorithms to automatically train the pattern of the data. The processing element (PE), the smallest unit in the network, receives a number of inputs. Each input comes through a connection that has a weight and this weight represents the biological neuron. The system is inherently parallel in the sense that many PEs can carry out their computations at the same time. In each PE, the weighted sum of the inputs is formed, and the threshold is subtracted, to compose the activation of the PE and the resulting signal is passed through an activation function to produce the output. The weight is adjusted using perceptron learning rule to train the network. The adjustment can be put into the following equation:

$$w(i+1) = w\,(i) + \eta * (d\,(i) - y\,(i)) * x(i) \tag{9.1}$$

where η is the step size or learning rate, $y(i)$ is the PE output, and $d(i)$ is the desired response. In the perceptron learning algorithm given in Equation (9.1), $y(i)$ is the output of the nonlinear system and the algorithm directly minimizes the difference between the output of the PE and the desired outcome. The common practice is to have the network learn the appropriate weights from a set of training data. The PE learns only when its output is wrong, and the weights remain the same if $y(n) = d(n)$. There are many forms of neural networks. The most common architectures are as follows:

1) Single-layer feed-forward ANNs also known as single-layer perceptron: one input layer and one output layer of PEs.
2) Multi-layer feed-forward ANNs also known as multi-layer perceptron (MLP): one input layer, next, one or more hidden layers, and finally one output layer of PEs.

The neural network can be a SLP, or MLP, but we need to develop a systematic procedure for determining appropriate connection weights. Initially, a random distribution of weights can help to minimize the chances of the network becoming stuck in local minima and to break the symmetric problem. Moreover, small initial weights avoid immediate saturation of the activation function.

9.2.1 Single-Layer Perceptron

Rosenblatt's perceptron is a pattern recognition machine that was first designed in late 1950s for optical character recognition problem. It consists

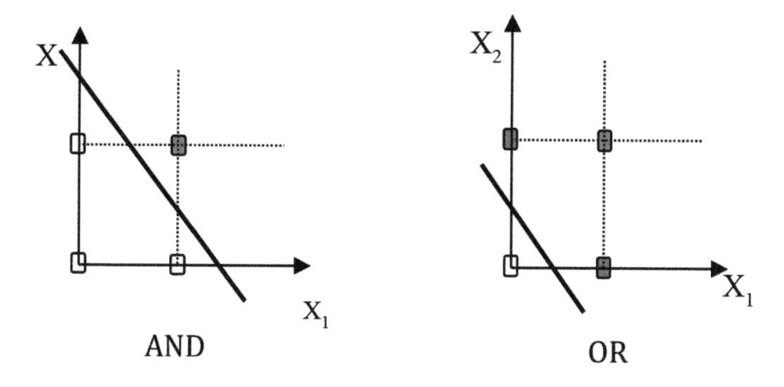

Linearly separable

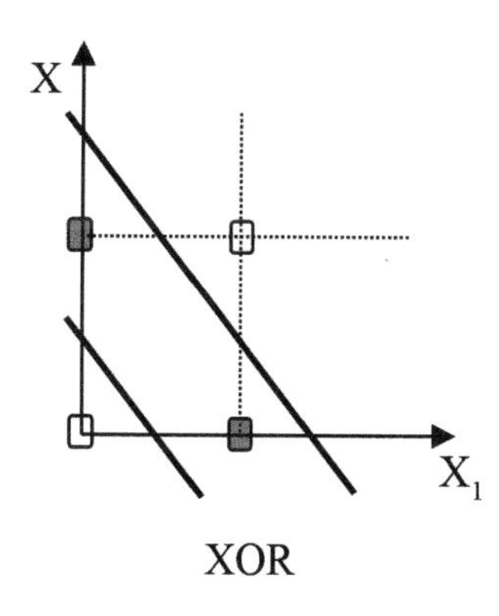

Nonlinearly separable

Figure 9.1 AND, OR, and XOR problem diagram.

of binary activations and trained to recognize linearly separable patterns in a finite number of steps. The problems with input patterns which can be classified using straight lines or a single hyper plane are called *linearly separable problems*. The problems which can be classified, but not by a straight line, are called *nonlinearly separable problems*. Simple problems, such as AND and OR, are linearly separable and a simple problem such as XOR is linearly separable as depicted in Figure 9.1.

A simple SLP is depicted in Figure 9.2. We can write the equation for activation of the PE as

$$y_1 = f[w_1x_1 + w_2x_2 + b_1] \tag{9.2}$$

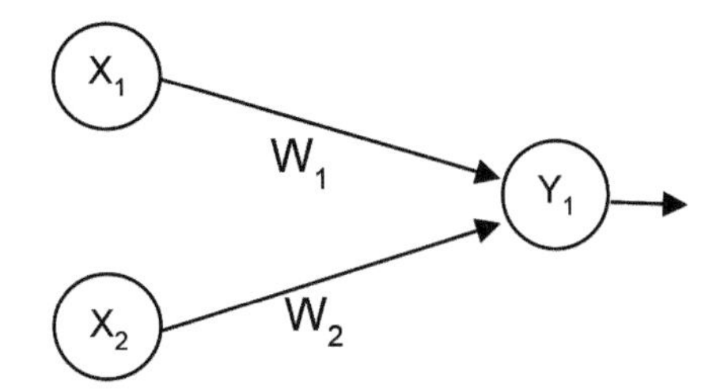

Figure 9.2 Single-layer perceptron.

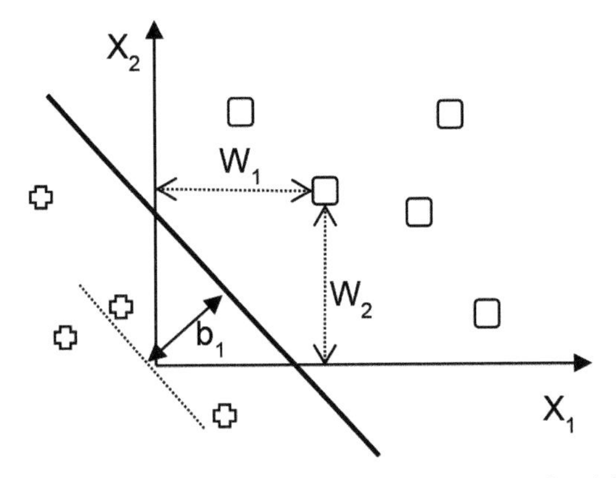

Figure 9.3 Geometric representation of patterns and weights.

We can see from Figure 9.3 that the weights (w_1, w_2) determine the slope of the line and the bias (b_1) determines the offset.

In a two-dimensional plane, the surface is

$$w_1.x_1 + w_2.x_2 - b_1 = 0. \tag{9.3}$$

Hence,

$$x_1 = \frac{b_1}{w_1} - \frac{w_2}{w_1}x_2. \tag{9.4}$$

Equation (9.4) defines a straight line.

The final result of the SLP depends on the activation function in a way that explicitly minimizes the output error and it is not controlled linearly by all the input samples. The learning rule takes a finite number of steps in SLP to reach an optimal solution for linearly separable problems. The algorithm is very simple and adds a quantity proportional to the product of the error to the current weight and activation available at the PE. The algorithm used in a SLP is as follows.

1) Initialize weights with small random values for each connection
2) For each training pair $(x, d(i))$
   (a) Calculate actual output $y(i)$
   (b) Calculate error, $\delta = (d(i) - y(i))$
   (c) Use the error to update weights using Equation (9.1)
3) Repeat step 2 until convergence or error δ is closed to zero

When we use SLP, if the problem is not linearly separable, then the training will never stop and rapid changes in the weights close to the end of the training affect the classification performance too. In this SLP type, the perceptron learning algorithm searches only for a satisfactory answer; hence, the network may not perform well on data that is not included in the training data. The linear activation function is much more sensitive to the learning rates than a nonlinear function in SLP. The nonlinearity keeps the small output values between -1 and 1 no matter what the value of the weights are. The number of outputs in a SLP is normally determined by the number of classes in the data set. An SLP is found to be useful in classifying a continuous-valued set of inputs into one of two classes only. The problem that is not linearly separable cannot be solved by SLP; hence, we could say it is a physical implementation of the linear pattern recognition machine.

We conclude that the SLP is very efficient and can only classify linearly separable problems but not very effective because the convergence of the activation function stabilizes as soon as the last sample is classified without error. Because of the linearity of the system, the training algorithm will converge to the optimal solution. Obviously, this will solve the problem in the training data, but it may not produce the best possible classification for data not in the test set. Therefore, the solution to these issues is by introducing hidden layers in the network.

9.2.2 Multi-Layer Perceptron

MLP extends the network with one or more hidden layers. In other words, there are layers of PEs that are not directly connected to the external environment. The linear nonseparable problem in SLP can be overcome by introducing the MLP. The advantage of having the MLP instead of the SLP is the ability to tune each PE to respond at least to a region in the input space. For binary units, one can prove that this architecture is able to perform any transformation, given the correct connections and weights. Conceptually a single hidden layer MLP as depicted in Figure 9.4 is a cascade of SLP. Let us

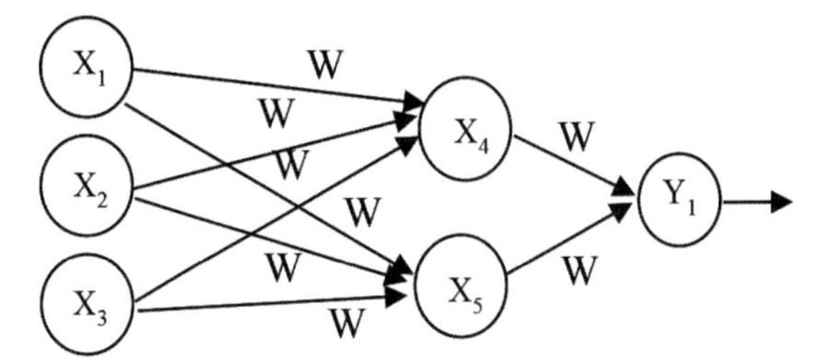

Figure 9.4 Multi-layer perceptron.

label the input layer PEs as X_1, X_2, and X_3 and the output of each hidden PEs as X_4 and X_5 (there are two PEs in the hidden layer). Y_1 is the final output of the PE in the output layer. The overall input–output map is driven as

$$y_1 = f[w_7 x_4 + w_8 x_5 + b_3] \tag{9.5}$$

$$\begin{aligned} y_1 &= f[w_7 * (f(w_1 x_1 + w_2 x_2 + w_3 x_3 + b_1)) \\ &\quad + w_8 * (f(w_4 x_1 + w_5 x_2 + w_6 x_3 + b_2)) + b_3]. \end{aligned} \tag{9.6}$$

The output layer can consist of one or more PEs, depending on the problem addressed. In most classification applications, there will be either a single PE output or the same number of PEs in the output layer as the classes. There is no explicit error in the hidden layer PEs of the MLP. This is known as the credit assignment problem. The important issues in MLP are to appropriately set the number of PEs in the hidden layer and the number of hidden layers. If the network has too many hidden PEs, or it has too few, then the problem becomes challenging. Each neuron in one layer has direct connections to the neurons of the subsequent layer. It is possible to define networks that are partially connected to only some PEs in the preceding layer. However, for most applications, fully connected networks do better. The use of a smooth, nonlinear activation function is essential for a multi-layer network employing gradient-descent learning. In most applications, a sigmoid function is used as the activation function.

9.3 Adaptive Array Model

A simple array of dipoles placed in a straight line is shown in Figure 9.5 and is considered as a linear array. We have tested array beamformer with five and seven elements placed in the straight line. The array radiation equation with respective coefficients for five-element array could be given as in [7]:

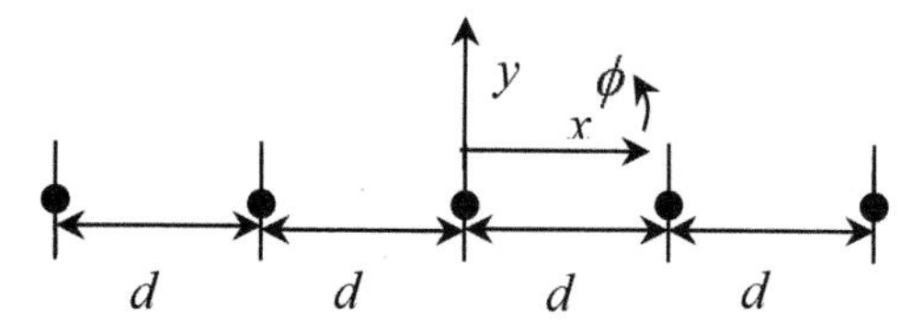

Figure 9.5 Schematic diagram of the five-element array model.

$$\begin{aligned} &w_1 e^{2j\beta d\cos\varphi} + w_2 e^{j\beta d\cos\varphi} + w_3 + w_4 e^{-j\beta d\cos\varphi} \\ &\quad + w_5 e^{-2j\beta d\cos\varphi} = f(\varphi). \end{aligned} \tag{9.7}$$

Similarly, we can write a seven-element array antenna radiation field as follows:

$$\begin{aligned} &w_1 e^{3j\beta d\cos\varphi} + w_2 e^{2j\beta d\cos\varphi} + w_3 e^{j\beta d\cos\varphi} + w_4 \\ &\quad + w_5 e^{-j\beta d\cos\varphi} + w_6 e^{-2j\beta d\cos\varphi} + w_7 e^{-3j\beta d\cos\varphi} = f(\varphi). \end{aligned} \tag{9.8}$$

In Equations (9.7) and (9.8), the function $f(\phi)$ is the desired smart antenna beam pattern.

In adaptive array antenna design, the placement of antenna elements (e.g. electric diploes) can be in any manner since the current amplitude and the phase could be adjusted to get the desired radiation patterns. However, analytically, we can show that any arbitrary set of dipoles arranged in a straight line will produce a radiation pattern that is symmetrical on both sides of the plane where the dipoles are placed. As a result, the placement of dipoles must be chosen based on the desired radiation patterns. If the set of desired radiation patterns are symmetrical over a common axis, then the dipoles can be placed along that common axis (linear array) where all the current components will be in phase with different sets of amplitudes for respective radiation patterns. Alternatively, when the set of desired radiation patterns are unsymmetrical over a common axis, the placement of diploes should not be in a common axis (unless a reflector is used) while the current components will be at different phases and amplitudes. Consequently, the in-phase and the different-phase current components will result in real and complex optimized weight values, respectively. Therefore, we have proposed two types of activation functions for optimizing real and complex weights. Accordingly, we model a general setup as shown in Figure 9.6 where n numbers of antenna elements (e.g. electric diploes) are placed arbitrarily.

For the arbitrary placement of dipoles in Figure 9.6, the currents will be different in phase amplitude and it can be represented by complex current

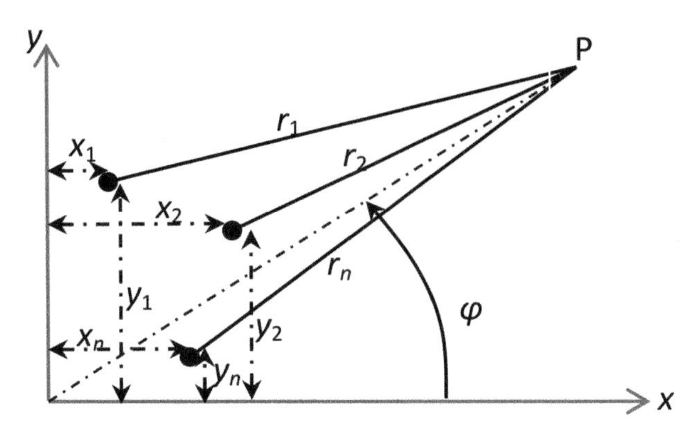

Figure 9.6 Schematic diagram of dipole placement.

phasors. The respective complex current phasors of the dipoles are taken as $I_1, I_2, \ldots, I_n$. Hence, the electric field (far-field) at the observation point P would be given by

$$E = A_0 I_1 e^{-j\beta r_1} + A_0 I_2 e^{-j\beta r_2} + \cdots + A_0 I_n e^{-j\beta r_n} \tag{9.9}$$

where A_0 and β are a constant and the phase constant, respectively.

Substituting for $r_1, r_2, \ldots, r_n$ in terms of the distance from origin, Equation (9.9) can be simplified to

$$\begin{aligned} E = {} & w_1 e^{j\beta(x_1 \cos\phi + y_1 \sin\phi)} + w_2 e^{j\beta(x_2 \cos\phi + y_2 \sin\phi)} + \cdots \\ & + w_n e^{j\beta(x_n \cos\phi + y_n \sin\phi)} \end{aligned} \tag{9.10}$$

where w_1, w_2, and w_n are the complex weights and proportional to the complex current phasors $I_1, I_2, \ldots, I_n$, respectively. To achieve the objective of forming a resultant single beam, the values of the complex weights w_1, $w_2, \ldots, w_n$ need to be optimized such that the resultant field must match a desired, specified single beam function$f\,(\varphi)$. Thus, Equation (9.10) can be written as

$$\begin{aligned} & w_1 e^{j\beta(x_1 \cos\phi + y_1 \sin\phi)} + w_2 e^{j\beta(x_2 \cos\phi + y_2 \sin\phi)} + \cdots \\ & + w_n e^{j\beta(x_n \cos\phi + y_n \sin\phi)} = f\,(\phi)\,. \end{aligned} \tag{9.11}$$

We shall show why the linear array tends to produce a symmetrical beam, thus making it problematic to get a single beam antenna with a linear array perceptron neural network. When all the antenna elements (e.g. dipoles) are in a straight line, the radiation pattern is symmetrical over the axis of the dipole placement. Hence, a rotatable single beam cannot be obtained.

Consider all the dipoles to be in a straight line along the x-axis. Thus, $y_1 = y_2 = \cdots = y_n = 0$. Hence, Equation (9.10) is simplified to

$$E(\phi) = w_1 e^{j\beta\, x_1 \cos\phi} + w_2 e^{j\beta\, x_2 \cos\phi} + \cdots + w_n e^{j\beta\, x_n \cos\phi}. \tag{9.12}$$

The electric field expression, when the angle φ is in the opposite direction, could be obtained by replacing φ with –φ in Equation (9.12), yielding

$$E(-\phi) = w_1 e^{j\beta\, x_1 \cos(-\phi)} + w_2 e^{j\beta\, x_2 \cos(-\phi)} + \cdots + w_n e^{j\beta\, x_n \cos(-\phi)}. \tag{9.13}$$

Since $\cos(-\varphi) = \cos\varphi$, replacing $\cos(-\varphi)$ in Equation (9.13) with $\cos\varphi$, we obtain

$$E(-\phi) = w_1 e^{j\beta\, x_1 \cos\phi} + w_2 e^{j\beta\, x_2 \cos\phi} + \cdots + w_n e^{j\beta\, x_n \cos\phi} = E(\phi). \tag{9.14}$$

Since $E(-\varphi) = E(\varphi)$, the electrical field is symmetrical over the x-axis whatever the weight values may be. Thus, the smart antenna beams obtained in Chapter 8 are seen to be symmetrical about the single axis along which the antenna elements are placed since the linear arrays were used. Hence, it is impossible to get a single beam on one side of the x-axis. The only possible solution for the weights of a linear array will produce beams in both the positive and negative directions of the x-axis along which the linear array antenna elements are placed. In that case, the beam cannot be rotated to other directions, without always having a mirror beam about the x-axis. The only way to get rid of the mirror, symmetrical beam is to place a reflector along the plane of the linear array elements, in which case the beam will be folded to double the radiation and reception strengths above the plane. Therefore, a rotatable single beam cannot be obtained by means of weight optimization when all the antenna elements (e.g. dipoles) are placed in a straight line.

When the desired radiation patterns are unsymmetrical on common axis, then the placement of the dipoles cannot be along a single, common axis as discussed earlier. Bodhe *et al.* have proposed a rectangular array structure to provide a solution for such condition. However, it is possible to construct an array with minimum three elements that need not be placed in single axis, resulting in complex weight values. Therefore, we have considered a minimum of three-element array smart antenna along with four- and six-element array smart antenna to compare the accuracy of beamforming as

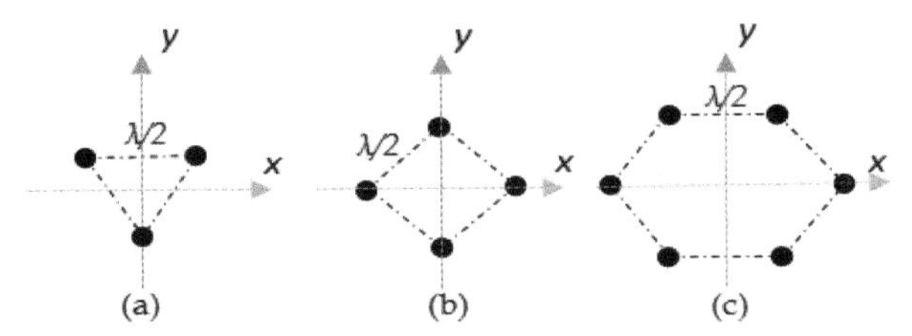

Figure 9.7 Schematic diagram of array models. (a) Equilateral triangular model. (b) Square model. (c) Regular hexagonal model.

shown in Figure 9.7. The desired smart antenna radiation beam function is selected to be $f(\varphi) = \mathrm{sinc}(\varphi - \varphi_0)$ to form a single beam, where φ_0 is the desired angle.

9.4 Single Neuron Weight Optimization Model

One of the important issues in ANN is that the activation functions of the neurons of the neural networks must be bounded. Otherwise any computation performed will become unstable. Hence, only a limited class of functions can be used in ANN. The use of GAs for ANN training is a promising development. But GA requires much time for training. Architecture selection is a very important step when working with neural networks. If the activation function is unknown, the network should be able to approximate any arbitrary nonlinear function.

In the perceptron model as shown in Figure 9.8, a single neuron with a linear weighted net function and a threshold activation function, also known as transfer function, is employed. The model has three parts. The first part, consisting of inputs ($x_1, x_2, \ldots, x_n$), are multiplied with individual weights ($w_1, w_2, \ldots, w_n$). The second part of the simple perceptron consists of the net function that sums all weighted inputs and bias as follows:

$$z = b + \sum_{k=1}^{n} w_k x_k. \tag{9.15}$$

In the final part of the perceptron ANN, the sum of previously weighted inputs and bias is passed through a transfer, or activation, function to get the output. In case of linear activation function, the artificial neuron is doing a simple linear transformation over the sum of weighted inputs and bias b. There is no single best method for nonlinear optimization and optimization is based on the characteristics of the problem to be solved.

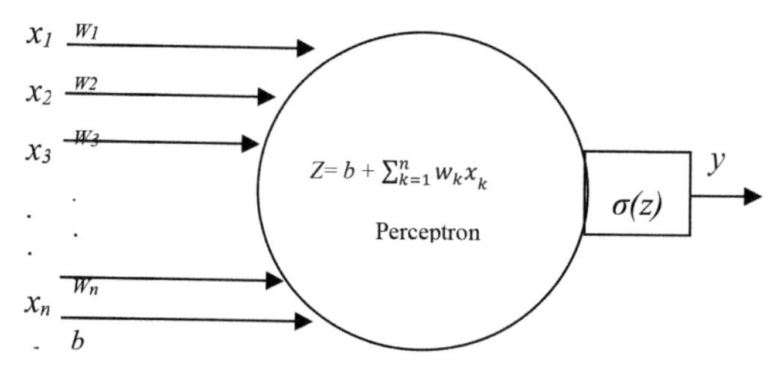

Figure 9.8 Perceptron model for weight optimization.

We simplify the calculation complexity to reduce the processing delay. Hence, we have used a single neuron for this problem and a nonlinear sigmoid activation function σ to get the output y as follows:

$$y = \sigma(z) \quad = \frac{1}{1 + e^{-z}}. \tag{9.16}$$

In order to train the weights to match the desired output y_0, the deviation Δ is obtained and the weights are iterated until the deviation reaches the trained means error (TMR) which is set below the predefined value. The deviation and TMR are defined in the following equations, respectively:

$$\Delta = y_0 - y \tag{9.17}$$

$$\text{TMR} = \frac{\Delta}{y_0} * 100. \tag{9.18}$$

The weights are adjusted in every iteration using the deviation and the selected learning rate, also known as coefficient (η). The adjusted new weights are

$$w_i = w_i + (\eta \times \Delta \times x_i). \tag{9.19}$$

The iteration continues until either the deviation TMR is below the predefined TMR_m value or the allowed maximum number (N) of iterations is reached. The single neuron weight optimization model (SNWOM) flowchart is given in Figure 9.9. For training, we have used a set of angles θ in the range of 0°–360° toward which the peak of the smart antenna beam is required to point. During testing to determine whether the trained ANN gives us the required single beam pointing in any required direction, we have used another set of angles θ in the range of 0°–360°.

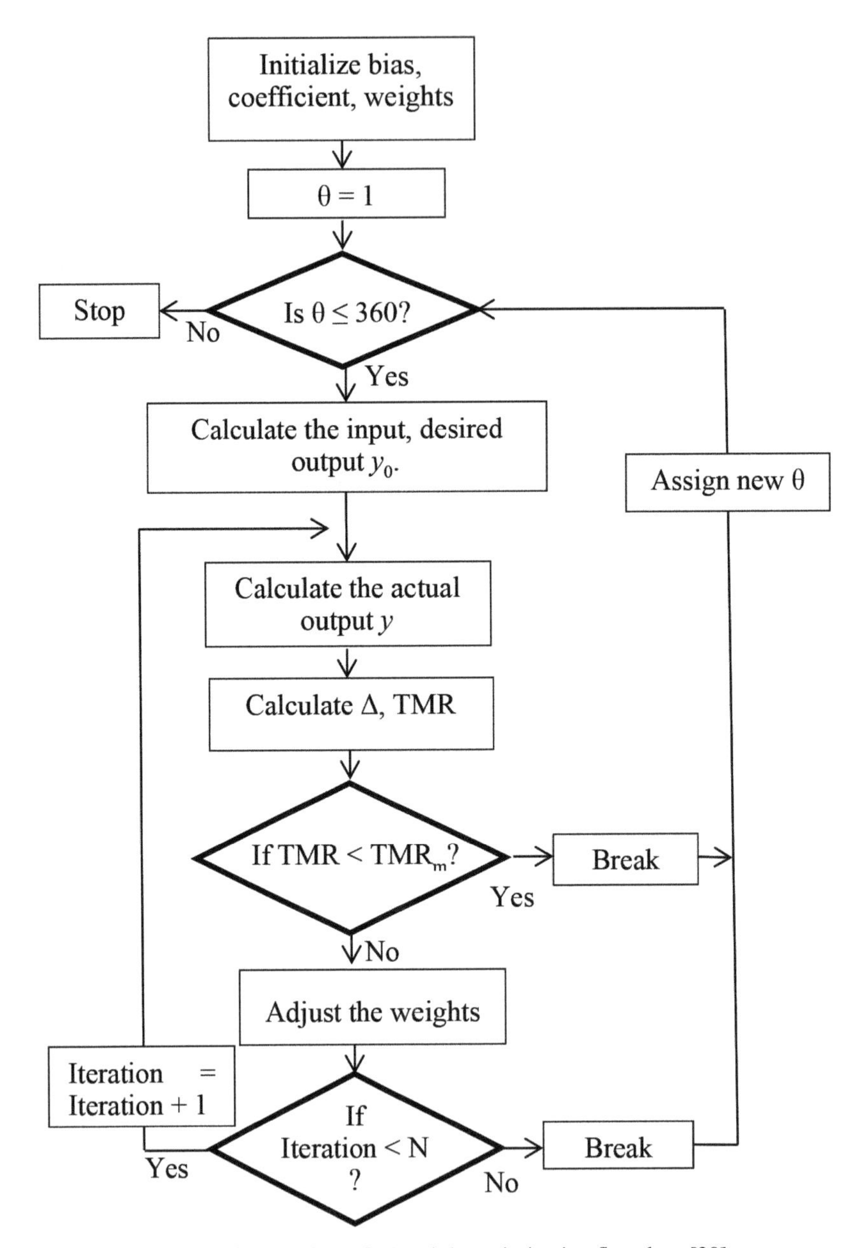

Figure 9.9 The SNWOM weight optimization flowchart [30].

9.4.1 Real-Valued Neural Network

There are dozens of ANN architectures that have been tried out and it is proposed that real-valued neural network (RVNN) is capable of solving the real-time problems posed to artificial intelligence (AI). Activation function plays an important role in neural network theory. The search continues for workable activation functions that satisfy requirements. The main requirements with real-valued functions are that the functions must be continuous and differentiable and, in an ideal case, that it is also bounded.

In our experiment, the smart antenna array elements are placed on a straight line by fixing the desired antenna beam function $f(\phi)$ as $\cos 2\phi$ and taking the distance between two adjacent elements as half wavelength. In order to have the comparison between accuracy of weights optimized using the SNWOM method with the weights optimized using the traditional least mean square (LMS) method, the weights are calculated for five-element and seven-element array smart antenna using LMS optimization. Having obtained the optimized weights after convergence, we have drawn the radiation patterns using the optimized weights and compared them with the radiation patterns of the desired beam for the five-element array as shown in Figure 9.10(a). The perceptron ANN generated antenna beam is seen to closely match the desired beam to which it is optimized. The beam is seen to give maximum radiation in the desired direction and null points that match the desired beam's null points. The beam width is slightly wider, thus leading to some interference when receiving as well as indicating a measure of power wastage when transmitting. It is seen that there is maximum radiation over a wider area than what is required by the desired beam. However, as shown in Figure 9.10(b), this is rectified by increasing the number of elements from five to seven, thus getting greater accuracy at the cost of computational time and the cost of adding extra elements. As we have expected, with increased number of elements, the adaptive array beamforming is much closer to the desired beam. However, the amplitudes in the 0° and 180° are better for the five-element array antenna than for the seven-element array. This is due to the characteristics of the desired beam selected. In order to have the comparison of accuracy of weights optimized using the SNWOM method with the weights optimized using the traditional LMS method, the weights are calculated for five-element and seven-element array smart antenna using LMS optimization. The radiation patterns for five and seven elements optimized from LMS methods are shown in Figure 9.11(a) and (b), respectively.

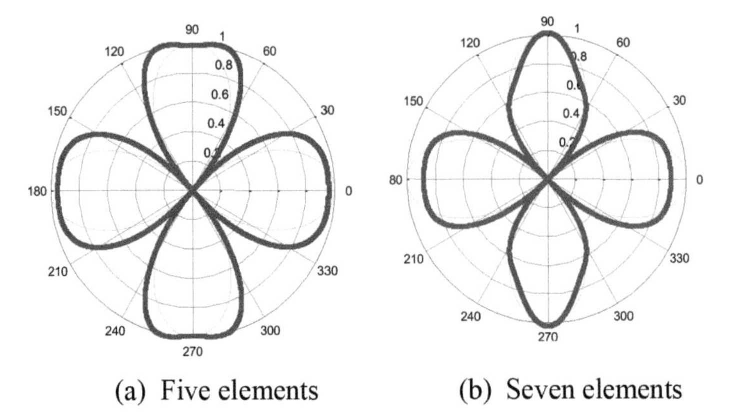

(a) Five elements (b) Seven elements

Figure 9.10 Comparison of radiation pattern between optimized beam and desired beam obtained by SNWOM [28].

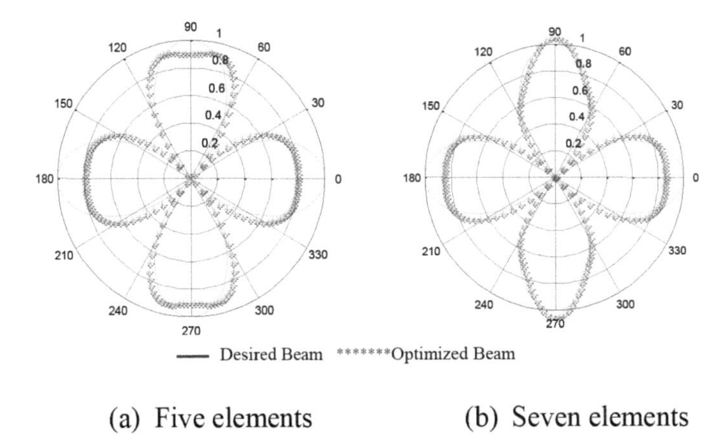

(a) Five elements (b) Seven elements

Figure 9.11 Comparison of radiation pattern between optimized beam and desired beam obtained by LMS [28].

Comparing Figures 9.10(a) and 9.11(a) shows that the results obtained from SNWOM has better match than the LMS method, although significant differences cannot be observed between Figures 9.10(b) and 9.11(b). It should be remembered that most multi-layer neural network beamformers, as well as the optimization procedures that use algorithms such as the LMS methods, require heavy computational time and memory to store multi-layer weights, for instance. But the perceptron ANN driven smart antenna requires little memory and gives fast convergence when training and rapid generation of the beam while operating in real time.

In order to make the comparison of accuracy between five- and seven-element array smart antennas, the error between the desired and optimized values for corresponding angles is shown in Figure 9.12(a) and (b), respectively. The error analysis clearly displays that when the number of elements increases, the error reduces, though with oscillations. With the five-element array, both end-fire beams and the broadside beams have larger beam widths than the desired beam width. With regard to peak radiation, the broadside antenna beams match the desired peak radiation better, whereas with the end-fire antenna beams, the maximum field strength is less than the desired maximum field strength (see Figures 9.10(a) and 9.12(a)). With seven elements, there is almost a perfect match in the case of the broadside antenna beams, whereas for the end-fire antenna beams, the widths of the perceptron ANN generated and desired antenna beams match; the peak radiation drops to lower values than for the desired maximum (see Figures 9.10(b) and 9.12(b)).

Further, to test the precision of the SNWOM method with a variety of desired functions, we selected the desired function as [24]

$$f(\varphi) = \frac{1}{9}\left|3 + 4\cos(\pi\cos\varphi) + 2\cos(2\pi\cos\varphi)\right|. \tag{9.20}$$

To compare the accuracy of the weights obtained from the SNWOM method by using the above desired radiation beam function, the results were compared with the results obtained using the LMS method as depicted in Figure 9.13. We observe the close match between desired and optimized radiation patterns. It is evident from the results that the desired narrow beam could not be achieved using the five-element array, while it was feasible with the seven-element array. Therefore, it can be seen that the narrow desired beam requires more number of antenna (e.g. dipole) elements. The sidelobes of this broadside antenna are relatively small, as seen in Figure 9.13(b). The SNWOM was found to be slightly inferior to the LMS beam, in that it gave rise to slightly larger sidelobes in the 0° and 180° directions as seen in Figure 9.12(b).

When linear array antenna needs to be used as a single beam antenna, a reflector placed along the array plane may be used to flip over the unwanted of the two main beams.

The errors between desired and optimized values in relation to the angle for five-element and seven-element array antennas are shown in Figure 9.14(a) and (b), respectively. Similar to previous results, the range of error reduces while the frequency of oscillations increases with increasing number of elements. In this broadside array antenna, using the five-element array,

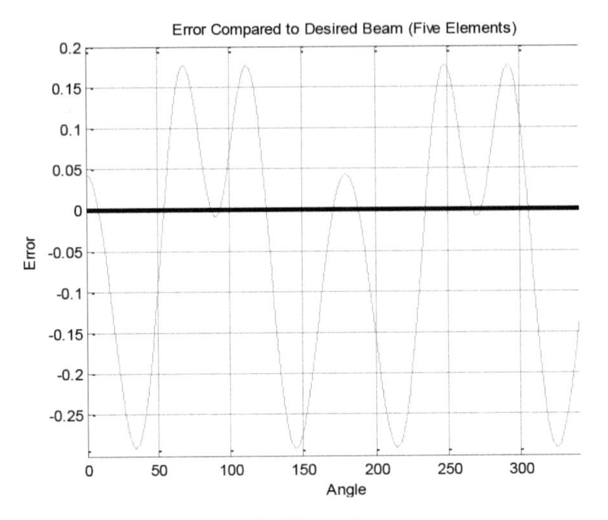

(a) Five elements

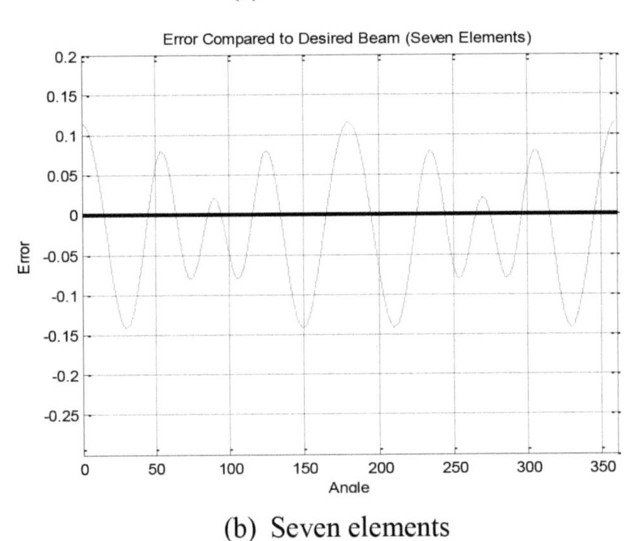

(b) Seven elements

Figure 9.12 The error between desired and optimized beam with the corresponding angle [28].

the beam width of the perceptron-generated array factor is sharper than the desired beam width (see Figures 9.13(a) and 9.14(a)). But the end-fire sidelobes are very small. Hence, the power degradation takes place along the direction of the main beams.

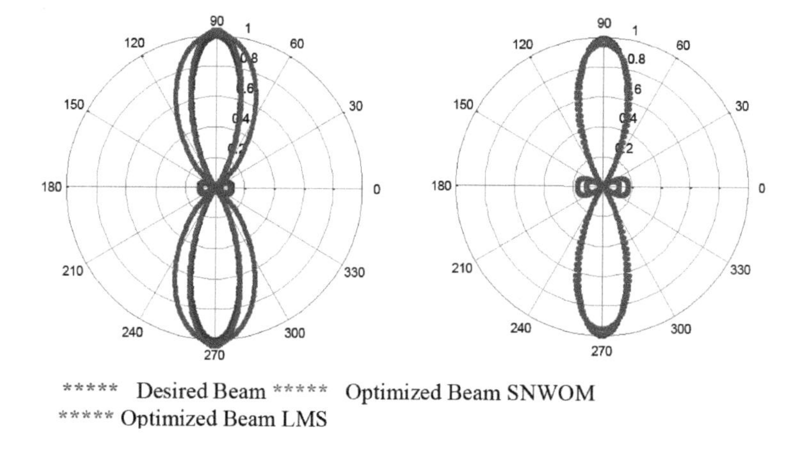

(a) Five elements (b) Seven elements

Figure 9.13 Comparison of radiation patterns obtained from SNWOM and LMS methods.

For the seven-element broadside array smart antenna (see Figures 9.13(b) and 9.14(b)), while the perceptron-generated broadside antenna beams almost perfectly match the two broadside beams, the power degradation along the end-fire directions becomes significant due to larger sidelobes in the end-fire directions.

We have optimized the smart antenna weights for three, four, and six elements to investigate the actual output using the above SNWOM model with initial weights, bias, and learning rate with the appropriate activation function. The radiation pattern in Figure 9.15 shows the results obtained when conducting SNWOM optimization. The radiation pattern illustrates significant match between the desired pattern and the optimized pattern.

It can be observed from Figure 9.15 that as the number of elements increases, the optimized beam patterns are better matched to the desired beam for three- to four-element array antennas. In addition to beam pattern matching, the beam width is also reduced. A narrow beam will have a greater distance coverage while utilizing less power when compared to an omni-directional antenna. However, the complex weights optimized using the LMS method for the same cases give superior match when the number of elements is increased as shown in Figure 9.16 (which is Figure 5.9, in Chapter 5, repeated here for completeness).

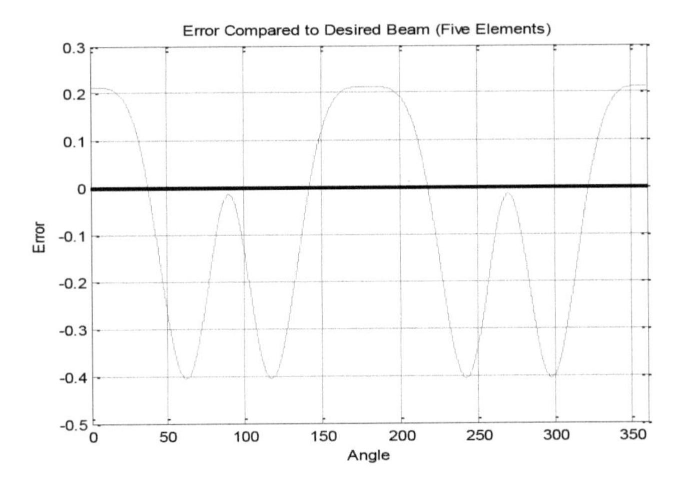

(a) Five elements

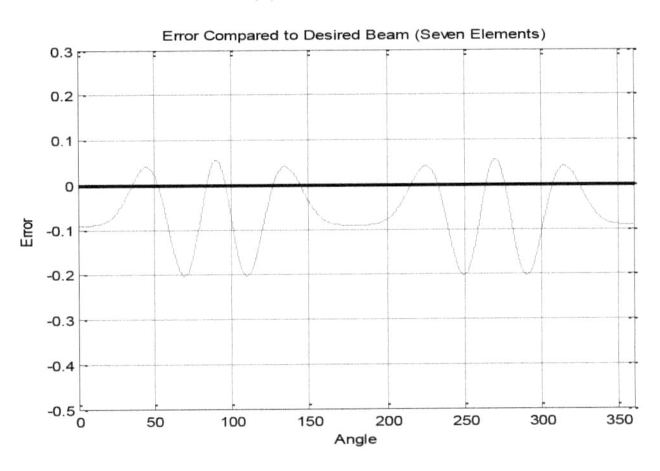

(b) Seven elements

Figure 9.14 The error between desired and optimized beams with the corresponding angle [29].

9.4.2 Complex-Valued Neural Network

Many ANN architectures generally operate on real values. But there are applications where consideration of complex-valued inputs is quite desirable. The main motivation behind these attempts was the observation that real-valued data is often best understood when embedded in the

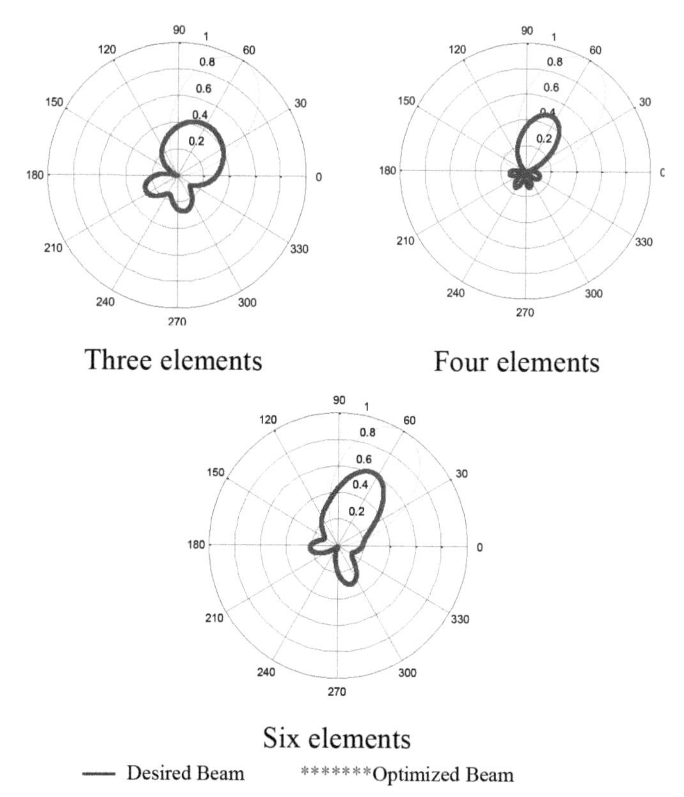

Figure 9.15 Comparison of radiation patterns obtained by SNWOM optimization [27].

complex domain. For example, waves are meaningfully represented by their Fourier coefficients. Training the neural networks using complex-valued inputs was performed using techniques like the back-propagation, Hopfield model, and perceptron learning rules. Their performances were tested using pattern classification, signal processing, and time series experiments, and its generalization capability was found to be satisfactory. In complex-valued neural networks (CVNNs), input, output, threshold, and weights are complex values and selecting activation function is a challenging part. Because of the neural network's outstanding capability of fitting on nonlinear models, the use of ANN has increased in the recent past. New types of complex-valued sigmoid activation function for multi-layered neural network have been explored. Their simulation results proved that their proposed network reduced 54% of testing time compared to neural network using normal sigmoid activation function given in Equation (9.15).

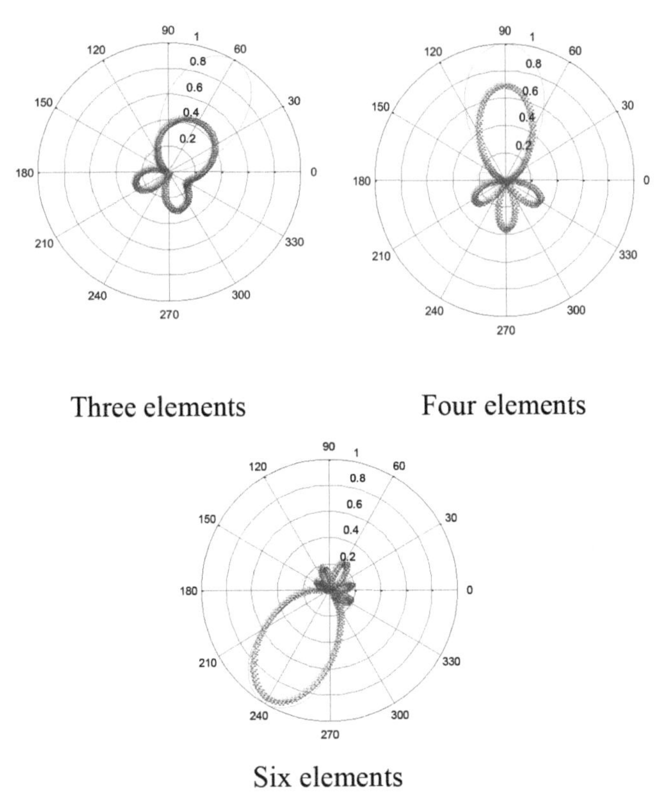

Figure 9.16 Comparison of radiation patterns obtained by LMS optimization [26].

CVNN algorithm and complex LMS algorithm were first discussed in 1975. Following their introduction, a complex-valued learning algorithm for signal processing application was developed. When a neural network algorithm incorporating different activation functions was explored, it was shown that if real-valued algorithms are simply done as complex-valued algorithms, then singularities and other such unpleasant phenomena may arise. Activation functions for digital very large-scale integration (VLSI) neural networks have been explored, requiring less hardware than the conventional RVNN. The relative performance of CVNN using different error functions was studied using 16 different error functions. The function's form has been retained to that of real error functions while extending to the complex domain. This was done to make sure that the error computed kept the same formula even while operating in the complex domain. Simulations show that the error functions can indeed be treated as a parameter for training

CVNN. Further, it was observed that the performance of the CVNN depends on the architecture of the network for some error functions. The input data should be scaled to some region in complex domain, and to overcome the implementation problem, a split sigmoid activation function could be used for training the network.

A CVNN is more powerful than a real-valued neuron since it is capable of solving linearly nonseparable problems too. Mathematical theory proves that the decision boundary of a single complex-valued neuron consists of two hyper surfaces that intersect orthogonally and divide a decision region into four equal regions. But a real-valued neuron can solve only linearly separable problems since it gives a straight line as a decision boundary. Hence, an activation function of such a neuron saturates in two regions only. Designing a neural network to process complex-valued signals is a challenging task since a complex nonlinear activation function (AF) cannot be both differentiable and bounded everywhere in the complex plane. In real case, the activation function is usually chosen to be a continuous, bounded, and nonconstant function. The traditional approach to overcome this problem is to use a pair of functions for the real and imaginary components. In the complex case, any regular analytic function cannot be bounded unless it reduces to a constant. The only bounded and continuously differentiable function in the complex plane is the constant function. The average learning speed of CVNN algorithm is faster than that of RVNN algorithm. And the standard deviation of the learning speed of CVNN is smaller than that of the RVNN. Hence, the CVNN and the related algorithm are natural for learning complex-valued patterns. The main reason for the problem in finding a nonlinear activation function in CVNN design is the conflict between the bountifulness and the differentiability of complex functions in the whole complex plane. Liouville's theorem states that all bounded function must be a constant in complex plane (C), where an entire function is differentiable at every point $z \in C$. Therefore, we cannot find a differentiable complex nonlinear activation function that is bounded everywhere on the entire C. A complex network can use unbounded and analytic output functions and still converge faster with small connection weights and initial values than one with larger connection weights or initial values.

In general, many real-valued mathematical objects are well understood and hence provide the motivation to develop a neural network structure that uses well-defined fully complex nonlinear activation functions. This approach has been shown to be adequate in providing equalization of nonlinear satellite communication channels. In summary, one can say that the RVNN is well studied and well established. In CVNNs, one of the main problems is the selection of an activation function. The split approach typically employs a pair of real-valued function $f_R(z) = \tanh(x), x \in R$, as a nonlinear complex AF as shown in the following equation:

$$f(z) = f_R(Re(z)) + i\, f_I(Im(z)). \tag{9.21}$$

The unbounded nature of nonlinear differentiable functions in complex plane triggered the definition of desirable properties for the complex AFs. The five desirable properties of a fully complex AF are as follows:

1) $f(z) = u(x, y) + i\, v(x, y)$ is nonlinear in x and y
2) Function f(z) is bounded
3) The partial derivatives u_x, u_y, v_x, and v_y exist and are bounded
4) Function f(z) is not entire
5) $u_x v_y \neq v_x u_y$

By Liouville's theorem, the second and the fourth conditions are redundant. In other words, a bounded nonlinear function in C cannot be entire. The last property $u_x v_y \neq v_x u_y$ was originally imposed to guarantee continuous learning. That is, to prevent a situation where a nonzero activation function input forces the gradient of the error function (with respect to the complex weight) to become zero. Hence, a CVNN with proposed activation functions saturates in four regions and thus can achieve higher classification ability. It is expected that a CVNN of given size can perform more complicated functions than a real-valued one. In fact, a CVNN can solve all the two-input Boolean functions too.

9.4.3 Complex-Valued Activation Functions

An AF is for limiting the amplitude of the output of a neuron. Enabling in a limited range of functions is usually called squashing functions. It squeezes the permissible amplitude range of the output signal to some finite value. There are many AFs used in different studies. Some of the most commonly used AFs to solve nonlinear problems are bipolar sigmoid, hyperbolic tangent, and squash.

9.4.3.1 Hyperbolic Tangent Function

Hyperbolic tangent function is proven as the most powerful activation function in neural network. It is similar to sigmoid function with outputs ranging between -1 and 1 and having a broader output space than a linear

AF. It is a function that can easily represent the transition from one state to another. This function is defined as the ratio between the hyperbolic sine and the cosine functions expanded as the ratio of half-difference and half-sum of two exponential functions. In many actual situations, combinations of e^x and e^{-x}arise fairly often. The hyperbolic tangent activation function (Figure 9.17(a)) for real value is written as follows:

$$\sigma(z) = \tanh(z) = \frac{\sinh(z)}{\cosh(z)} = \frac{e^z - e^{-z}}{e^z - e^{-z}}. \tag{9.22}$$

The main feature of this function is that its scaled output property will assist the neural network to produce good results. Hence, the results will converge to the global solutions. Hyperbolic tangent activation function for complex values is given by

$$\sigma(z) = \frac{\tanh(Re(z)) + j\tan(Im(z))}{1 + j\tanh(Re(z))\tan(Im(z))}. \tag{9.23}$$

9.4.3.2 Bipolar Sigmoid Function

This function is similar to the sigmoid function and it goes well for applications that produce output values in the range between -1 and 1. The bipolar sigmoid activation function (Figure 9.17(b)) for real value is written as follows:

$$\sigma(z) = \frac{1 - e^{-z}}{1 + e^{-z}}. \tag{9.24}$$

The reason for the popularity of these sigmoid functions is because they are easily differentiable, easy to distinguish, and minimize the computation capacity for training. It provides better convergence speed compared to sigmoid activation function. Bipolar sigmoid activation function for complex values is given by

$$\sigma(z) = \frac{1 - e^{-2Re(z)}}{1 + e^{-2Re(z)} + 2e^{-Re(z)}\cos(Im(z))} + j\frac{2e^{-Re(z)}\sin(Im(z))}{1 + e^{-2Re(z)} + 2e^{-Re(z)}\cos(Im(z))}. \tag{9.25}$$

9.4.3.3 Squash or Elliot Function

It is mathematically shown that Elliot activation function has a higher speed of approximation. This function was first introduced in 1993 by D.L. Elliot

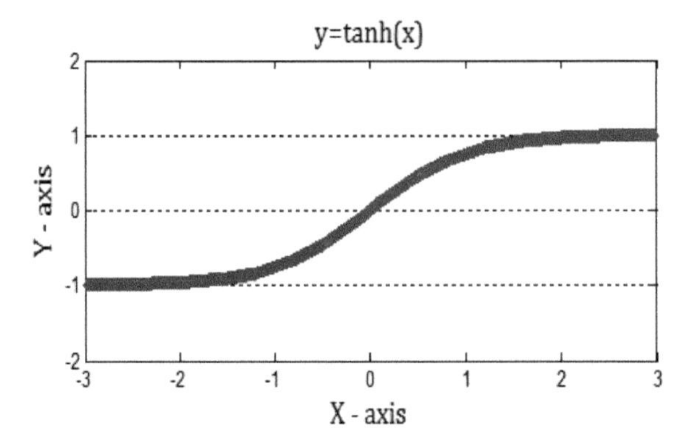

(a) Hyperbolic tangent function

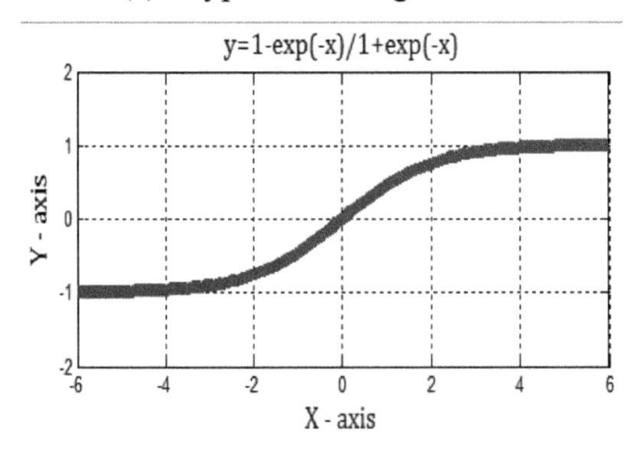

(b) Bipolar sigmoid function

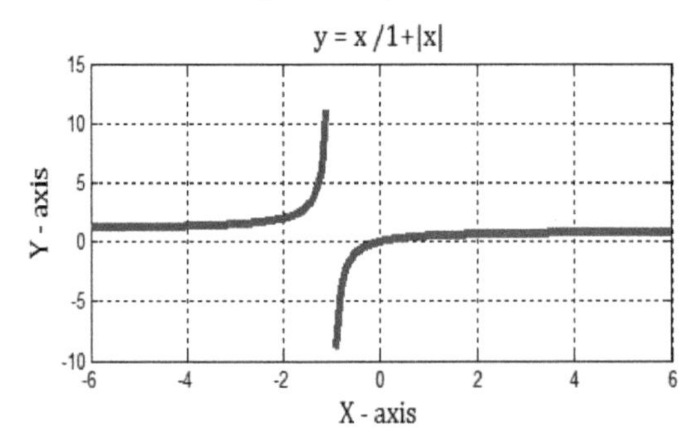

(c) Squash or Elliot function

Figure 9.17 Different activation functions.

and it closely approximates the sigmoid or hyperbolic tangent functions for small values. This function has an ability to compute on modest computing hardware as it does not involve any exponential or trigonometric functions. The squash activation function (Figure 9.17(c)) for real value is written as follows:

$$\sigma(z) = \frac{z}{1 + |z|}. \tag{9.26}$$

The drawback of this function is that it only flattens out for large inputs, so a small number of outputs produced will be trapped into the local minima and might require more training iterations. The output range for this function is between -1 and 1. Squash activation function for complex values is given by

$$\sigma(z) = \frac{z}{1 + \sqrt{\mathrm{real}(z)^2 + \mathrm{imag}(z)^2}}. \tag{9.27}$$

We have chosen the above three functions to investigate whether the robustness of these functions will apply to optimize the weights of an array antenna to steer the beam toward the desired directions. We selected the following desired antenna beam, defined by the function

$$f(\varphi) = \mathrm{sinc}(\varphi - \varphi_0). \tag{9.28}$$

We optimized the weights by taking the distance between two adjacent array elements as half wavelength, for both four and six elements, using SNWOM method. The optimized radiation patterns are compared with the desired radiation patterns. We observed the close match between desired and optimized radiation patterns. We used hyperbolic tangent, bipolar sigmoid, and squash functions as activation function for six-array elements and the results are depicted in Figures 9.18–9.20, respectively. Each was tested for maximum beam direction pointing in the following angles: $\varphi = \pi/3$, $\pi/2$, $2\pi/3$, and $5\pi/3$.

We also have tested the hyperbolic tangent, bipolar sigmoid, and squash functions as activation function for four-array elements and the results are depicted in the Figures 9.21–9.23, respectively. Each was tested for maximum beam angles of $\varphi = \pi/2$, π, $3\pi/2$, and 2π.

Although the input data can be scaled, there is no limit over the values of the complex weights that it can take, making it difficult to implement. To overcome this problem split activation function (Equations (9.23), (9.25), and (9.27)) are used for training the perceptron ANN. Training is done with a given set of input and output data to learn a functional relationship between

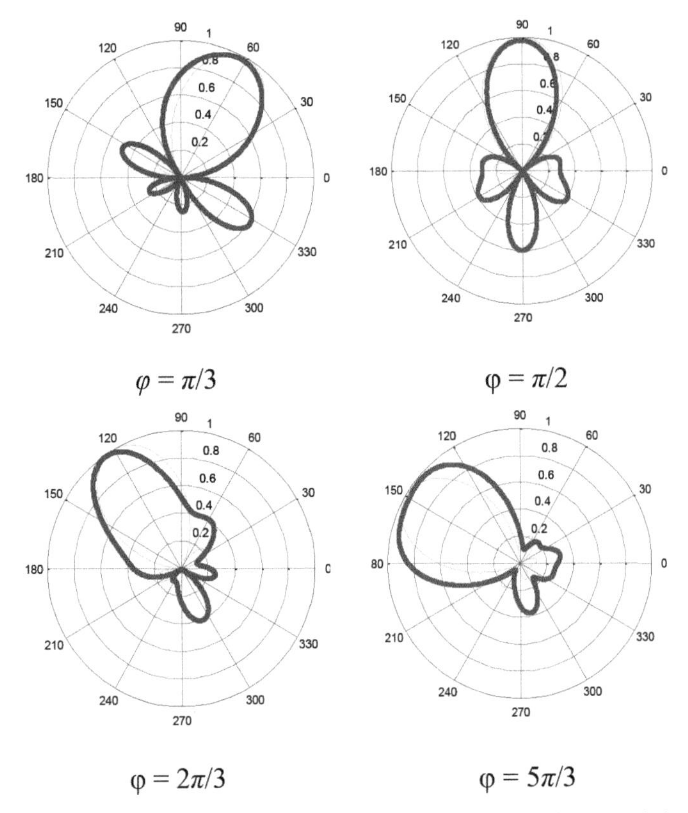

Figure 9.18 Comparison of radiation pattern between optimized beam and desired beam obtained by SNWOM [30].

input and output. The weights are initially set to small values. The outputs are obtained for these random input values. The error between actual output and the desired output is calculated. This error is back-propagated and the weights are updated. Then for these new values of weights, outputs are once again calculated. These actual calculated outputs are once again compared with the target outputs and the error is calculated, which is again back-propagated and the weights are once again updated. This iterative process is continued till the error becomes less than the minimum defined.

We observed the close match between desired and optimized radiation beam patterns. Although most of the optimized radiation antenna beam patterns have good match with the desired radiation patterns, some of the patterns have moderate match only. Further, the variation of match

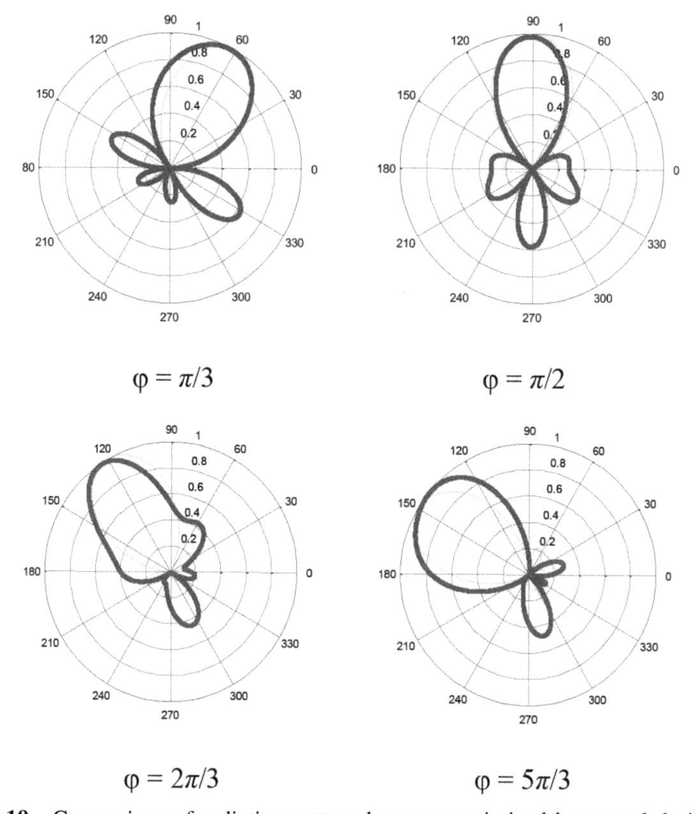

Figure 9.19 Comparison of radiation pattern between optimized beam and desired beam obtained by SNWOM.

is independent of the number of elements. In addition, it is the common observation that variation of match is dependent on the direction of the optimized antenna beam. However, some of the patterns (e.g. in Figure 9.23, angles $\pi/2$and $3\pi/2$) were expected to be matched in similar, but in opposite directions, they did not completely match.

Since only a single perceptron ANN is used for beamforming, the technique is fast and could provide the best set of smart antenna patterns within milliseconds. Even though the precision of the SNWOM depends on the array element (e.g. electric dipole antenna elements) placement and the characteristics of the desired beam selected, it is a fast, efficient, and simple method for the weight optimization compared to the previously proposed neural network based adaptive beamforming methods. The perceptron ANN

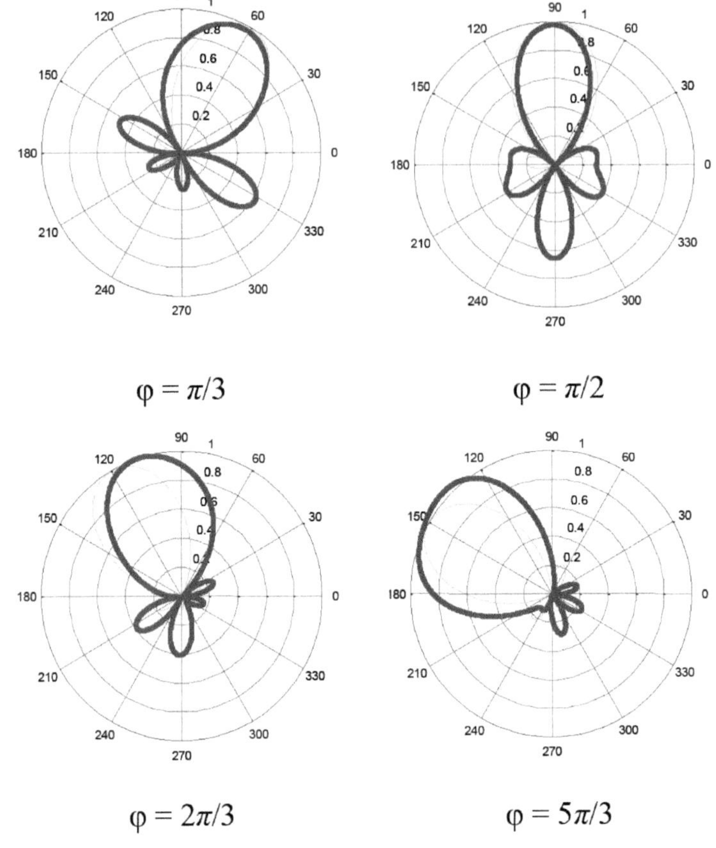

Figure 9.20 Comparison of radiation pattern between optimized beam and desired beam obtained by SNWOM.

beamforming method proposed may work with any chip-based MIMO techniques, including transmit beamforming, spatial multiplexing, space–time block coding, and cyclic delay diversity.

We have compared the perceptron ANN with LMS method for the formation of a desired smart antenna beam using antenna-based beamforming as opposed to chip-based beamforming. In chip-based beamforming, multiple beams are created which should constructively add together at the receiver (mobile station), thus requiring the receiver to send signals to the base station transmitter to steer the beam. While the receiver may have multiple antennas, an ambiguity arises as to the specific antenna that had sent a signal

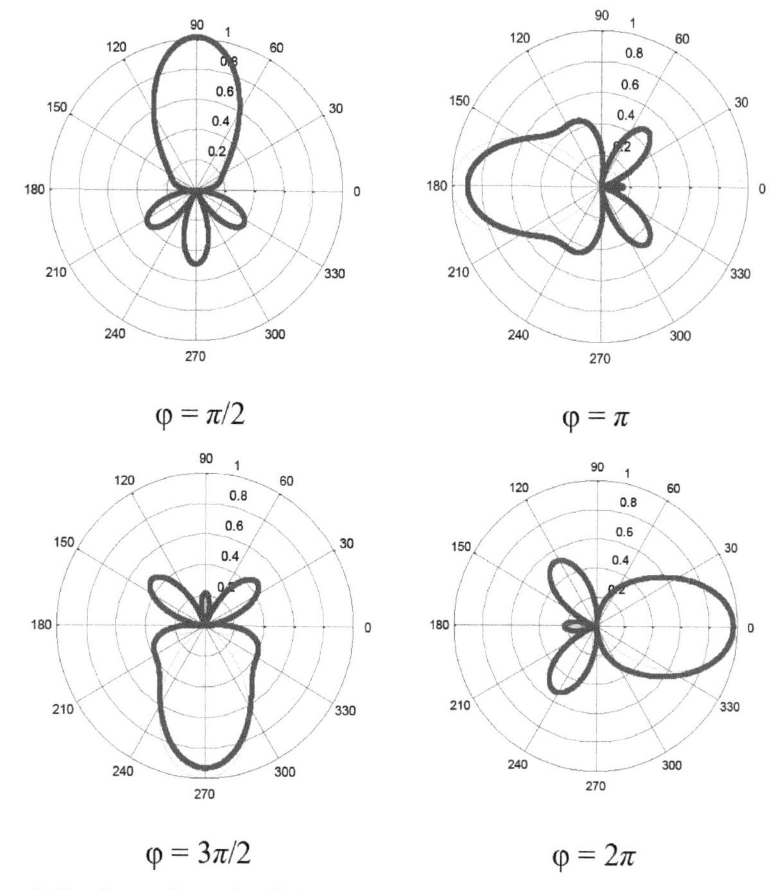

Figure 9.21 Comparison of radiation pattern between optimized beam and desired beam obtained by SNWOM [30].

to the transmitter. In the antenna-based beamforming, reported herein, the beamformer may handle both signal receiver and a single cluster of receivers in one geometrical location, multiple clusters, or antennas.

9.5 MATLAB^{TM} Program

Given below is the program to calculate the smart antenna array weights using the SNWOM algorithm in Section 9.5.1 and to plot the radiation patterns presented in Section 9.5.2. The program supplier below may be used to generate Figure 9.21, 9.22, or 9.23 using the hyperbolic tangent, bipolar

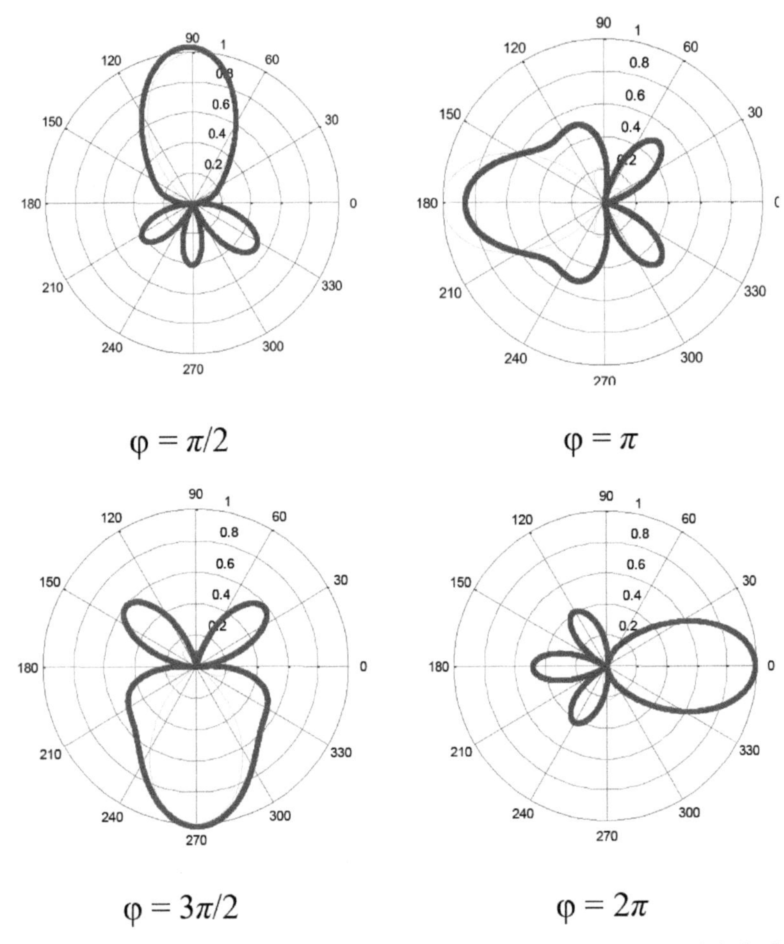

Figure 9.22 Comparison of radiation pattern between optimized beam and desired beam obtained by SNWOM.

sigmoid, or the squash functions as the activation function. The smart antenna array simulated in this program has four elements.

To produce the figure:

1. Create an *m* file of the program listed in Sections 9.5.1 and 9.5.2, respectively.
2. Run the SNWOM algorithm from the program in Section 9.5.1. You need to specify the theta values using the variable named "Theta" in the third line of the program. This output of this program is the four

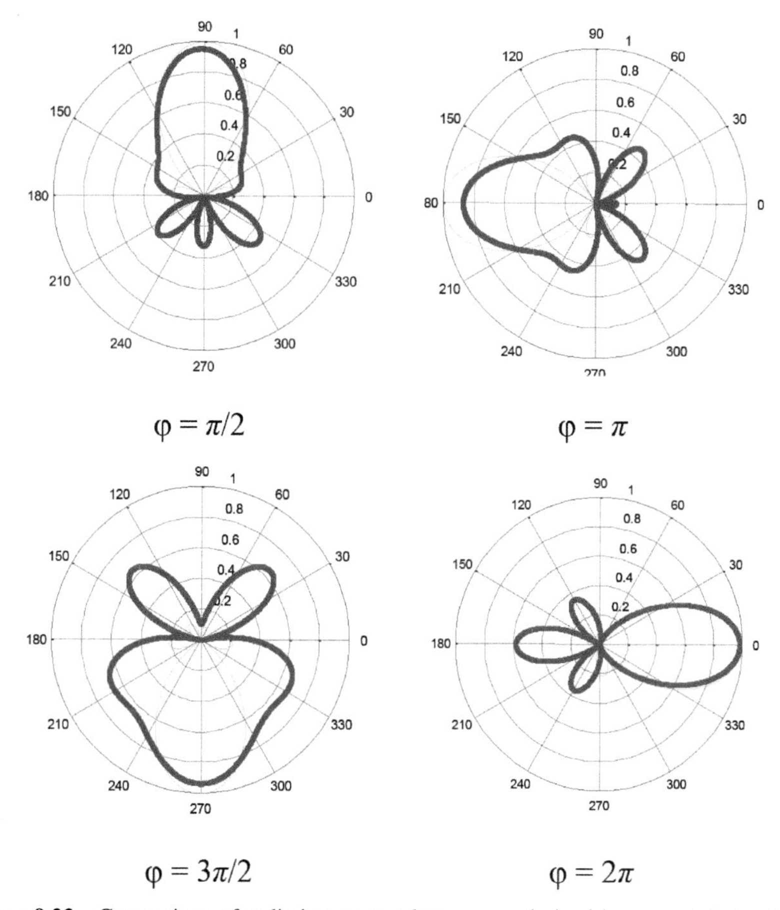

Figure 9.23 Comparison of radiation pattern between optimized beam and desired beam obtained by SNWOM.

weights, stored in a column vector and also stored in a text file in the same directory as the *m* files.

3. Thereafter, run the plotting program in Section 9.5.2. The program should open a new figure window showing two polar plots in the red for the SNWOM plot, while the blue plot shows the desire function.

9.5.1 MATLAB^{TM} Program of the SNWOM Algorithm

% This program is used to plot the calculate
weights for a % 4-element array antenna to generate

```
Figure 9.21, 9.22
% and 9.23.
clear;
numIn = 4;     % number of inputs
theta= pi/2;      % desired angle
bias = 0.002;      % bias
coeff = 0.003;     % learning rate
weights = [-0.001;-0.001;-0.001;-0.001];  % initial
weights
weightsT = [-0.001;-0.001;-0.001;-0.001];
count = 0;
% training is done for different angles in radian
for r=0:1:9
   for d=0:10:360
      fi=d*pi/180;
      % initialize the inputs based on the
      training angle
      inputs = [exp(-1j*pi/sqrt(2)*cos(fi));
      exp(-1j*pi/sqrt(2)*sin(fi));exp(1j*pi/sqrt
      (2)*cos(fi)); exp(1j*pi/sqrt(2)*sin(fi))];
      %desired function
      desired_out = sinc(fi-theta);

      learn=0;
      while (learn==0)
      % calculate linear weighted net output
      y =bias;
         for k = 1:numIn
            y = y + inputs(k)*weights(k);
         end
      % calculate the final output using the
      activation function
      outcome=(1-exp(-2*real(y)))./(1+exp(-2*real
      (y))+ ...
      2*exp(-real(y))*cos(imag(y)))+ ...
      1j*(2*exp(-real(y))*sin(imag(y)))./(1+exp
      (-2*real(y))+ ...
      (2*exp(-real(y))*cos(imag(y))));
      % Bipolar sigmoid function
```

```
%         outcome = y./(1+sqrt(real(y)^2+imag(y)
^2)); % Squash activation function.
% calculate the error
      delta = (desired_out-outcome);
      error=(abs(delta)/desired_out)*100;
         if (abs(error) < 0.001)    % stopping
         criteria
            weightsT(1) = weights(1);
            weightsT(2) = weights(2);
            weightsT(3) = weights(3);
            weightsT(4) = weights(4);
            count=count+1;
            learn=1;
         elseif (count >10000)      % stopping
         criteria
            learn=1;
            weights(1) = weightsT(1);
            weights(2) = weightsT(2);
            weights(3) = weightsT(3);
            weights(4) = weightsT(4);
         else   % adjust the weights
            weights(1) = weights(1)+coeff*
            inputs(1)*delta;
            weights(2) = weights(2)+coeff*
            inputs(2)*delta;
            weights(3) = weights(3)+coeff*
            inputs(3)*delta;
            weights(4) = weights(4)+coeff*
            inputs(4)*delta;
            count=count+1;
         end
      end
   end
fprintf('.');
end
%write the result into a text file after separate
the real part and complex part
% open file
```

```
fad = fopen('mydata4e.txt','w+');       % 'w+'
means "write text"

if (fad < 0)
   error('could not open file "mydata4e.txt"');
end
wfile = fopen('weights.txt', 'w')% fprintf('fi
%d\n',t);
fprintf('\nconverge %d\n',count);
fprintf(wfile,'   W1                  W2
               W3                  W4   \n');
% Write the weights into the file
for a=1:4
   fprintf(wfile, '%5.4g + %5.4gj   ',[real(
   weights(a)) imag(weights(a))]);
end

% close the file
fclose(wfile);
fprintf('\n');
```

9.5.2 MATLAB Program for the Plotting the Radiation Pattern

Below is the code to plot the radiation pattern for a four-element uniform linear array. Create a new *m* file to run this file as the frequency for plotting is different from the weight generation program.

```
% This program is used to plot the radiation
pattern for a % 4-element array antenna to generate
Figure 9.21, 9.22
% and 9.23.

% Frequency for plot

fi=0:pi/100:2*pi;

% Ploting radiation pattern of the desired function
clf
y=sinc(fi-theta);
```

```
polar(fi,abs(y),'b');
hold on

% The relationship for the 4-element antenna
straight line
z=(weights(1)'.*exp(-1j*pi/sqrt(2)*cos(fi-theta))
+weights(2)'.*exp(-1j*pi/sqrt(2)*sin(fi-theta))
+weights(3)'.*exp(1j*pi/sqrt(2)*cos(fi-theta))
+weights(4)'.*exp(1j*pi/sqrt(2)*sin(fi-theta)));

% Plotting the radiation pattern
polar(fi,abs(z),'r*')
```

References

[1] Southall, H.L. Simmers, J.A. and O'Donnell, T.H. "Direction finding in phased arrays with a neural network beamformer," *IEEE Transactions on Antennas and Propagations*, vol. 43, pp.1369-1374, 1995.

[2] El Zooghby, A.H. Christodoulou, C.G. and Georgiopoulos, M. "Performance of radial basis function network for direction of arrival estimation with antenna arrays," *IEEE Transactions on Antennas and Propagations*, vol. 45, pp.1611-1617, 1997.

[3] Mozingo and Miller, Introduction to Adaptive Arrays. New York: Wiley, 1980.

[4] El Zooghby, A.H. Christodoulou, C.G. and Georgiopoulos, M. "Neural Network-Based Adaptive Beamforming for One and Two- Dimensional Antenna Array," *IEEE Transactions on Antennas and Propagations*, vol. 46, pp.1891-1893, 1998.

[5] Wang, N.Y. Agathoklis, P. and Antoniou, A. "A new DOA Estimation Technique Based on Sub array Beam forming," *IEEE Transactions on Signal Processing,* vol. 54, no.9, pp. 3279-3290, 2006.

[6] Jean-Luc Fournier, Diane Titz, Fabian Ferrero, Cyril Luxey, Eric Dekneuvel, and Jacquemod, G. "Phased Array Antenna Controlled by Neural Network FPGA," Loughborough Antennas and Propagation Conference, Loughborough, UK, pp. 1-5, 2011.

[7] Bodhe, S.K. Hogade B.G. and Shailesh D. Nandgaonkar, "Beamforming Techniques for Smart Antennas using Rectangular Array Structure," *International Journal of Electrical and Computer Engineering (IJECE)*, vol.4, no.2, pp. 257-264, 2014.

[8] Ronny HÍansch and Olaf Hellwich, "Classification of Polarimetric SAR data by Complex Valued Neural Networks," proceedings of ISPRS workshop 2009, vol. 38, Hannover, Germany, 2009.
[9] Kim, M. S. and Guest, C. C. "Modification of Back-propagation for complex-valued-signal processing in frequency domain,"*International Joint Conference on Neural Networks*, (San Diego, CA), pp. III-27 – III-31, June 1990.
[10] Taehwan kim, Tulay Adali, "Fully complex multi-layer perception network for non-linear signal processing," *Journal of VLCI signal processing*, vol. 32, pp. 29-43, 2002.
[11] Howard E. Michel, Abdual Ahad, S. Awwal and David Rancour, "Artificial Neural Networks using Complex number and Phase encoded weights – Electronic and Optical Implementations," *International joint conference on Neural Networks*, 2006.
[12] Hamid,Jalab, A., Rabha and Ibrahim, W. , "New activation functions for complex-valued neural network", *International Journal of Physical Sciences*, vol. 6(**??**), pp 1766-1772, 2011.
[13] Widrow, B., McCool, J. and Ball, M., The Complex LMS algorithm, Proceedings of the IEEE, 1975.
[14] Kim M. S. and Guest, C. C., Modification of back propagation for complex-valued signal processing in frequency domain, Int. Joint Conference on Neural Networks, San Diego, vol. 3, pp. 27-31, June 1990.
[15] Georgiou, G. M. and Koutsougeras, C., Complex domain back propagation, IEEE Transaction On Circuits and Systems – II: Analog and Digital Signal Processing, Vol.39, No. 5. pp. 330–334, 1992.
[16] Deville, Y., A neural network implementation of complex activation function for digital VLSI neural networks, Microelectronics Journal, vol. 24, pp. 259-262.
[17] Gangal, A. S. Kalra, P. K. and. Chauhan, D. S., Performance Evaluation of Complex Valued Neural Networks Using Various Error Functions, World Academy of Science, Engineering and Technology, 2007.
[18] Prashanth, A., Investigation on complex variable based back propagation algorithm and applications, Ph.D. Thesis, IIT, Kanpur, India, March 2003.
[19] Uncini, A., Vecci, L., Campolucci, P. and Piazza, F., Complex Valued Neural Networks with Adaptive Spline Activation Functions, IEEE Trans. on Signal Processing, vol. 47, no. 2, 1999.

[20] Kim, T. and Adali, T., Fully Complex Back-propagation for Constant Envelop Signal Processing, in Proceedings of IEEE Workshop on Neural Networks for Sig. Proc., Sydney, pp. 231–240, Dec. 2000.

[21] Silverman, H., Complex Variables, Houghton, Newark, USA, 1975.

[22] Suttisinthong, N., Seewirote, B., Ngaopitakkul, A. and Pothisarn, C., Selection of Proper Activation Functions in Back-propagation Neural Network algorithm for Single-Circuit Transmission Line, Proceedings of the International Multi Conference of Engineers and Computer Scientist, Vol II, IMECS 2014, March 12 - 14, 2014.

[23] Sibi, P., Jones S. A. and Siddarth, P., Analysis of different activation functions using back propagation neural networks. Journal of Theoretical and Applied Information Technology, Vol. 47, pp. 1264-1268, 2013.

[24] Hoole, P.R.P., Smart Antennas and Signal processing for Communication, Biomedical and Radar Systems, WIT Press, UK, 2001.

[25] Wang, X. Hoole, P. R. P. and Gunawan, E., "An electromagnetic-time delay method for determining the positions and velocities of mobile stations in a GSM network," J A Kong (Editor), *Progress In Electromagnetics Research*, Vol. 23, 165-186, 1999.

[26] K. Pirapaharan, Kunsei. H, Senthilkumar. K.S, Hoole. P.R.P, and Hoole. S.R.H, “A Single Beam Smart Antenna for Wireless Communication in Highly Reflective and Narrow Environment,” in Proceedings of International Symposium on Fundamentals of Electrical Engineering (ISFEE), 2016, pp. 1-5.

[27] K.Pirapaharan, P.R.P. Hoole, Norhuzaimin Julai, Al-Khalid Hj Othman, Ade S W Marzuki, K.S. Senthilkumar, , and S.R. H. Hoole, , Polygonal Dipole Placements for Efficient, Rotatable, Single Beam Smart Antennas in 5G Aerospace and Ground Wireless Systems Journal of Telecommunication, Electronic and Computer Engineering (JTEC), 2017

[28] K.S. Senthilkumar, K.Pirapaharan, G.A.Lakshmanan, P.R.P Hoole, S.R.H. Hoole, Accuracy of Perceptron Based Beamforming for Embedded Smart and MIMO Antennas, Proc. International Symposium on Fundamentals of Engineering, IEEE Xplore Digital Library, Jan 2017,

[29] K.S. Senthilkumar, Lorothy Singkang, P.R.P Hoole, Norhuzaimin Julai, S. Ang, Shafrida Sahrani, Kismet Anak Hong Ping, K. Pirapaharan,

S.R.H. Hoole, A Review of a Single Neuron Weight Optimization Model for Adaptive Beam Forming Journal of Telecommunication, Electronic and Computer Engineering (JTEC), 2017

[30] K.S. Senthilkumar, K. Pirapaharan, Norhuzaimin Julai, P.R.P Hoole, Al-Khalid Hj Othman,, R. Harikrishnan, S.R.H.Hoole, Perceptron ANN Control of Array sensors and transmitters with different activation functions for 5G wireless systems, Proc. IEEE Int Conf on Signal Processing and Communication, India, July 2017

10

Advanced Wireless Systems: A Comprehensive Survey

K. Pirapaharan[1], P.R.P. Hoole[2], and S.R.H. Hoole[3]

[1]Faculty of Engineering, University of Jaffna, Sri Lanka
[2]Wessex Institute of Technology, Southampton, United Kingdom
[3]FIEEE as Professor of Electrical Engineering (Retired), Michigan State University, USA

Abstract

In this chapter is presented the historical information of wireless technology and its evolution in the recent years. The 5G architecture and its extensive multidimensional applications are described. The advantages of 5G are addressed. The physical layer design requirements in terms of millimeter-wave channel modeling, adaptive beamforming, and massive multiple-input multiple-output (MIMO) system implementation are elaborated. In addition, the media access control layer restoration requirements to support the modification in the physical layer, adaptive special beamforming, multiple access protocols, and simultaneous multicell operation requirements are deliberated. Further, the massive MIMO system requirements to support 5G technology are discussed. The future 6G wireless system and its key features and potential applications are presented.

Keywords: Review of 5G, evolvement of wireless systems.

10.1 Introduction

Technological resolution in wireless communications is quickly evolving to provide ubiquitous communication access to people wherever they are. The exponential progress of wireless technologies, services, and business applications has resulted in the wide-scale deployment and usage of wireless and mobile networks. The development of wireless access technologies has achieved its fifth generation (5G) of development. Wireless access technologies have gone through different evolutionary paths for the performance and efficiency in highly mobile environments. The first generation (1G) has fulfilled the basic mobile voice with analog signals, while the second generation introduced capacity and coverage through digital signals, followed by the third generation (3G), which successfully targeted broadband data at higher speeds using code division multiplexing. The fourth generation (4G) provided efficient access to a wide range of telecommunication services by means of incorporating space division multiplexing that enhanced the efficient use of the available frequency spectrum.

Wireless communications today is in its best ever growth period in history due to the enabling technologies that permit universal deployment, miniaturization, digital signal processing, networking, and switching. Historically, development in the mobile communications area came gradually and in the past was coupled closely to technological enhancements that hailed from much effort by engineers. The aptitude to provide wireless communications to the whole population was developed by Bell Labs in the late 1960s and 1970s [4–6]. The wireless communications era was born with the invention of highly dependable, solid-state radio frequency (RF) devices in the 1970s [1].

The recent rapid growth in cellular radio devices throughout the world is directly attributable to the new developments in the 1970s. The exceptional development in the number of wireless subscribers in the late 1990s, combined with the novel business approach of purchasing private mobile radio licenses for bundling as a nationwide commercial cellular service has increased today's subscriber base for cellular and personal communication systems.

Today, wireless communication has arguably become the most active area of technology development. This development is due to the transition of what has been largely a medium for supporting voice telephony into a medium for supporting other services, such as the transmission of video, images, text, and data. Wireless communications have become the most robust, viable voice and data transport mechanism. This has actually led to the development of newer wireless systems and standards for many other types of telecommunications traffic besides mobile voice telephone calls [7].

The explosive development of wireless communications continues to drive the development of cellular networks with advanced services. This chapter highlights the key developments and standards of the existing wireless generations and the upcoming ones while emphasizing developments in the physical layer architecture, especially adaptive antennas. This chapter is systematized into providing details such as the history and progresses of wireless communications, the variety of services available, multiplexing technologies, and the enhancing technologies in physical and datalink layers.

10.2 Evolution of the Wireless Technology

10.2.1 The Zero Generation

Mobile radio telephone systems referred to as zero generation (0G), or *pre-cellular mobile telephony technology*, were introduced in the 1970s [2]. In the early 1970s, wireless service offerings were a rare and high-priced way of communication. The first wireless service in the U.S. was *Improved Mobile Telecommunications Service (IMTS)*, which consisted of a 100-watt base station (BS) centrally located in a service area. In order to complete a call, either operator assisted or manual selection of an available frequency was needed. The rotary telephone was mounted on the dashboard of the vehicle and contained 12 buttons for manual selection of an available RF [7]. This early service was a party-line service in which to search for an available RF, the user had to listen to an accessible channel before starting a call. Since the spectrum was a restricted resource, early subscriptions to mobile communications needed a customer to be placed on a waiting list for service [3]. The service request was generally granted to those in greatest need of mobile communications. MTS, IMTS, and AMTS are headed under 0G.

This generation was followed by 0.5G which was a group of technologies with improved features over 0G. HCMTS and Autotel, or PALM, and ARP come under this generation [2].

10.2.2 The First Generation

Also known as the *Analog Radio Interface Generation*, the 1st generation began in the early 1980s. It was analog, circuit switched, and carried only voice traffic. The core difference between the existing 0G systems and 1G was the discovery of cellular technology which therefore was also known as *First Generation Analog Cellular Telephone*. In this generation, the network consists of many cells and each cell is covered by a radio network with one transceiver, so the same frequency can be reused many times which allowed large spectrum usage and thus increased the system capacity.

The different standards of 1G which were used worldwide are: advance mobile phone service (AMPS), total access communication system (TACS), extended total access communication system (ETACS), Nordic Mobile Telephone-450 (NMT-450), Nordic Mobile Telephone-900 (NMT-900), radio telephone mobile system (RTMS), Nippon telephone and telegraph (NTT), narrowband total access communications system (NTACS), and Japanese total access communication system (JTACS).

The 1G systems suffered from the limitations of low capacity, unreliable handoff, poor voice links, and very poor security since anybody with an all-band radio receiver can listen to the conversation. Further details are:

1) The mode of modulation used is amplitude/frequency modulation.
2) The type of communication allowed was full duplex (FD) which allows both parties to communicate at the same time.
3) The relevant device was equipped with direct dialing; hence, operator help was no more required to connect to a call.
4) Channel access method used was frequency division multiple access or FDMA.

10.2.3 The Second Generation

In the 1990s, the "second generation" (2G) mobile phone systems such as GSM (global system for mobile communications), IS-136 (time division multiple access (TDMA)), iDEN, and IS-95 (code division multiple access (CDMA)) was launched. 2G networks employ digital modulation and advanced call processing capabilities. These systems also provided dedicated voice and signaling trunk between MSC's and between each MSC and PSTN. These networks provided services such as paging, facsimile, and high data rate network access. This generation of systems catered to the needs of all four information types – text, picture, data, and voice.

This generation supports data and voice services. Two types of digital modulation schemes are used: TDMA and CDMA [3]. The 2G mobile phone networks provide services such as text messages, picture messages and multimedia messages (MMS). It converts the voice into digital pulses/codes, which give a clear signal that can be encrypted for security. 2G phone systems were run through digital circuit switched transmission, and a more advanced and fast phone-to-network signaling. The bitrate offered was higher with better error detection.

The 2.5G is a group of bridging technologies between 2G and 3G wireless communications. The basic change in it is that the system is upgraded to data services as well, which means that the share of voice and data is distributed within GSM arch. It involves digital communication allowing e-mail and simple web browsing. In addition to voice, 2.5G supports higher data speeds. The term 2.5G also applies to technology such as wireless application protocol (WAP), which uses a version of the web to fit into a mobile phone's slow data rate and small screen. 2.5G networks include enhanced data rates (EDGE) and general packet radio service (GPRS).

The 2.75G is used to refer to EDGE (enhanced data rates for GSM evolution) networks. It supports high speed data. The main advantage of using this particular technology is that it is feasible for enhanced data rates for GSM evolution. Its features are:

1) Narrow band bandwidth
2) Circuit switching for voice and packet switching for data
3) Multiple users on a single channel
4) Channel access methods used were TDMA and CDMA

10.2.4 The Third Generation

The 3G of wireless communication technologies is used to support broadband data rate, voice, data, and multimedia communications over wireless networks. The performance of GSM is improved by adding more functionality and provides value to the existing GSM network. 3G technology provides high speed, fast data rate capacity, and good quality of service. This system provides various facilities that support multimedia applications such as full-motion video, video conferencing, and internet access. Technologies like universal mobile telecommunication system, wideband CDMA (WCDMA), CDMA 2000, and TD-SCDMA. 3G technology starts from 2001 with having data capacity of 384 kbps. The main feature is that it provides high speed voice, data, and video services at 1.6–2.5 GHz frequency.

The 3.5G technologies provide services like mobility with greater speed over 3G. 3G comprises additional technologies along with 3G wireless/mobile technologies. The 3.5G generation starts from 2003 with data capacity 2 Mbps. Technology used in this generation is GSM/3GPP.The main feature provided in this generation is its high speed voice, data, and video services. Packet switching with frequency at 1.6–2.5 GHz is provided having horizontal handoff services. The main networks used in CDMA multiplexing are GSM TDMA. Technologies like EV-DO, HSDPA, HSUPA generally called HSxPA are involved in this generation. The enhanced version of this HSxPA emerged with the name of HSPA+. Its features are:

1) Multimedia capabilities such as video conferencing, mobile TV, and GPS
2) Broadband bandwidth
3) High security
4) International roaming

10.2.5 The Fourth Generation

The 4G refers to the next generation of wireless communication or the heterogeneous networks (HetNets). The main aim of 4G is to congregate all the technologies into one another with simplified structure, wherein any user can use his services even on the phone line, and similarly he can use his broadband services on mobile. In order to support roaming across heterogeneous wireless networks and packet-switched wireless communications, there is a growing interest in the design and development of 4G wireless networks [5], which will allow users to move from one type of wireless network to another using multinetwork devices or interconnected wireless networks.

The main technologies of 4G are long-term evolution (LTE), worldwide interoperability for microwave access (Wi-Max), ultra-mobile broadband (UMB), and EV-DO (Rev. C).

The factors that distinguish the 4G networks are roaming across networks, IP interoperability, and higher speeds. The 4G systems will encompass all systems from various networks, public to private, operator-driven broadband networks to personal areas, and *ad hoc* networks. The 4G systems will be interoperable with 2G and 3G systems, as well as with digital (broadband) broadcasting systems. 4G systems will have broader bandwidth, higher data rate, and smoother and quicker handoff, and will focus on ensuring seamless service across a multiple of wireless systems and networks.

The 4G technologies will consist of direct channel condition estimations of several users to distribute transmission load, better access methods than 3G, adaptive antenna, adaptive coding and modulation, adaptive channel/code allocation, and scheduling among sectors and users. Applications of 4G will be virtual presence, virtual navigation, tele-medicine, tele-geo-processing applications, and online education.

4G wireless systems are packet-switched wireless systems with wide area coverage and high throughput. They are designed to be cost effective and to provide high spectral efficiency. Other key features/functionalities are:

1) Wearable devices
2) Ultra-broadband bandwidth
3) Integrated services
4) Global mobility
5) Multiple-input multiple-output (MIMO) technology
6) High speed real-time streaming

10.2.6 The Fifth Generation

An average mobile user is expected to download around 1 terabyte of data annually in 2020 and beyond [8]. Researchers are also exploring new applications such as Internet of Things (IoT), Internet of Vehicles (IoV), device to device (D2D) communications, e-healthcare, machine to machine (M2M) communications, and financial technology (FinTech). Supporting this mammoth and quick increase in data usage and connectivity is an extremely frightening task in the present 4G LTE cellular systems. The LTE cellular network is exploring avenues of different research and development such as MIMO, small cells, coordinated multipoint (CoMP) transmissions, HetNets, and multiple antennas to enhance capacity and data rates. However, it is unlikely to sustain this ongoing traffic explosion in the long run [8].

Thus, the increasing capacity of wireless communications is going to face the new challenges of high frequency bandwidth. The key essence of the new generation 5G networks lies in exploring the unused, high frequency millimeter-wave (mm-wave) band, ranging in 3–300 GHz. The combined effect of emerging mm-wave spectrum access, hyper-connected vision, and new application-specific requirements has triggered the next major evolution in wireless communications – the 5G. Wireless communications envision a much higher magnitude of increased data rates, bandwidth, convergence, and connectivity, with a massive reduction in round trip latency and energy consumptions [9]. It points out that the first standard matured in 2020. Group Special Mobile Association (GSMA) is working with its partners toward the ultimate shaping of 5G communication.

The 5G of wireless mobile communications refers to worldwide wireless web (WWWW). It is the wireless internet network that is supported by technologies like orthogonal frequency division multiplexing (OFDM), MC-CDMA, LAS-CDMA, UWB, Network-LMDS, and IPv6. IPv6 will be the basic protocol for running 5G systems. The physical and datalink layer will define the wireless technology. This wireless technology is like an open wireless architecture.

The 5G systems maintain virtual multiwireless networks. Therefore, the network layer is divided into two sub-layers: the upper network layer for the mobile node and the lower network layer for the interface. This is the preliminary framework for the internet, where all the routing will be based on IP addresses which will be different in each IP network worldwide. In the existing wireless radio interface, a higher bit rate is a big loss; to control this loss, the 5G systems use *open transport protocol* (OTP). The application layer is for quality of service management over different types of networks.

Merging the different research initiatives by industries and academia, the following major requirements of the new generation 5G systems are identified [9–11]:

1) 1–10 Gbps data rates in real networks
2) 1 ms round trip latency
3) High bandwidth in unit area
4) Enormous number of connected devices
5) Perceived availability of 99.99% of what
6) Almost 100% coverage for "anytime anywhere" connectivity
7) Reduction in energy usage by almost 90%
8) High battery life

The complete replacement of the 2G, 3G, and 4G wireless systems by 5G systems will take a long time from the time of introduction of 5G in 2020. Therefore, the 5G architecture must be capable of incorporating the existing 2G to 4G architectures.

When compared, therefore, to 4G system which operates in the available spectrum around 3 GHz, the advantages and disadvantages of the 30 GHz spectrum 5G system are as follows. The advantages of the 5G wireless over 4G wireless system include the following:

1) Speed: about 20 Gbps compared to about 1 Gbps for 4G. With 3G, you would be able to download an average HD movie in about 25 hours, with 4G, it would be less than 10 minutes, and with 5G, it would be as less as 4 seconds. The theoretical maximum speed is 20 Gbps, which is 20 times faster than Google Fibre.
2) Reduced latency: less than 1 ms compared to about 10 ms for 4G. Perhaps more important than speed, 5G solves the problem of latency

which is the delay between commands and responses between the server, also called response times. Scientists are working on low earth orbit satellites also called LEOs which can help us achieve this. 5G would have 50 times faster response than 4G and this is particularly important not in smartphones but in connected machinery and self-driving cars. Even more important is the health field, where doctors can perform real-time surgery.

3) Efficient and uniform platform: with 5G bringing all devices together, it will be much easier to support and provide services. This would mean that the process would be extremely efficient. The connection density of the 5G system will be around 1 million connections per km^2, whereas for 4G, it is around 0.1 million per km^2.
4) Concentrated networks and personalized internet: there can be a requirement for providing increased internet capacity as there can be a massive flux of people using the internet. With 5G, organizers can pay for increased bandwidths, thereby improving the people's online experience. The 5G data traffic is expected to be about 50 Exabytes/month compared to about 7.2 Exabytes/month for 4G.

The disadvantages of the 5G system include:

1) 5G is more costly compared to other mobile network technologies because many technical/official engineers are required to install and maintain it.
2) The risk of overcrowding the frequency range of the 5G wireless spectrum is greater as more devices are connected to one channel.
3) 5G network technology will take more time for security and privacy issues.
4) Coverage indoor distance up to 2 and 300 m outdoors can be achieved due to greater losses at higher frequencies as 5G mm-wave influences from such losses (rain losses, attenuation due to rain, etc.).
5) 5G infrastructure comes at a high cost and greater health and safety concerns.

10.3 5G Architecture

With the requirements of sub-millisecond latency and bandwidth limitation in the traditional wireless spectrum, cellular networks are now dignified to break the BS centric network architecture. The increase in demand by wireless industry motivated the advancement toward much smaller cell disposition from the initial macro-hexagonal coverage. Researchers these days are focused on ways to design user centric networking. The evolution of 5G architecture will end up with ultra-wideband mm-wave channels. To meet the requirements, massive MIMO-based physical layer deployment with adaptive beamforming is required in the 5G evolution.

Future networks are expected to connect diverse nodes in different proximity. Small, micro-, pico-, and femto-cell deployment is already underway depending on the user density. Thus, dense 5G networks will have high co-channel interference that will gradually render the current air interface obsolete. This pushes in the concept of sectorized and directional antennas, as opposed to the age-old omni-directional antennas. Therefore, space division multiple access (SDMA) and efficient antenna design are of utmost necessity. The decoupling of user and control planes, along with seamless interoperability between various networks, is expected to strengthen the foundation for 5G systems.

10.3.1 Radio Network Evolution

Overall layout of 5G wireless networks breaks the rule of the BS centric cellular concept and moves toward a device centric topology [12]. The 5G network proposes the use of higher frequencies for communications. The propagation and penetration of mm-wave signals in the outdoor environment is quite limited [13]. Thus, the node layout cannot follow traditional cellular design or even any definite pattern. For instance, ultra-dense deployment is necessary in areas requiring high data rates. Line of sight (LOS) communication is undisputed in performance over non-line of sight (NLOS) communications. Alternatively, reflected, scattered, and diffracted signals still might have sufficient energy, which needs to be explored when LOS is completely blocked [14].

5G cellular technology needs to work with an enormous number of users, variety of devices, and diverse services. The primary concern, therefore, is the integration of 5G BSs with the legacy cellular networks [16]. Large beamforming gains extend the coverage, while reducing interferences and improving link quality at the cell edges. This feature enables mm-wave BS grids to provide low latency and cost-effective solutions [15]. Thus, radio networking in 5G communications is very different from legacy networks. Evolutions in radio would also change the schematics of the air interface.

10.3.2 Advanced Air Interface

Small radio wavelengths of mm-wave propagation demand small antenna sizes. This enables the use of large number of smaller antennas. Controlling the phase and the amplitude of a signal using array antennas helps in enhancing electromagnetic waves in the desired directions while cancelling in all other directions [16]. This necessitates the introduction of directional air interfaces. Highly directional radiation patterns could be secured by using adaptive beamforming techniques, resulting in the introduction of SDMA [17]. Effective SDMA improves frequency reuse for beamforming antennas at both the transmitter and the receiver [18].

10.3.3 Next Generation Smart Antennas

Successful deployment of 5G networks depends on the effective antenna array design. This exploits the advantage of a change in the air interface. The multibeam smart antenna array system should be used to realize SDMA capabilities. Smart antennas help in interference mitigation, while maintaining the optimal coverage area and transmission power reduction of both the mobile handset and the BS [19]. Further, for the same physical aperture size, more energy can be transmitted at higher frequency by the use of narrow beams [20]. Smart antenna implementation enables the same channel to be used by different beams. This reduces one of the major problems of wireless communications: co-channel interference. Use of beamforming antennas, with a fractional loading factor, further dilutes the co-channel interference problem [21]. Application of highly directional beams does not necessarily require any fractional loading. Infrastructure expenses and complex operations impede indiscriminate use of directional antennas. Therefore, a smart antenna design, optimized over directional gains, cost, and complexity, is very important for development of 5G wireless communications.

10.3.4 Heterogeneous Approach-HetNets

Another way to handle the wireless traffic explosion, expected in 5G communication, is through the deployment of large number of small cells giving rise to HetNets. HetNets are typically composed of small cells, having low transmission power, besides the legacy macrocells. By deploying low power, small BSs, network capacity is improved and the coverage is extended to cover holes. Moreover, the overlap of all small, pico-, and femto-cells with the existing macrocells leads to improved and efficient frequency reuse. Deployment of HetNets calls for a coordinated operation between traditional macrocells and small cells for mutual interference reduction [22].

10.4 Physical Layer Design Issues

Blending 5G architecture with the existing wireless systems requires a novel approach to make the process smooth and fast. Hence, it is critical to understand the physical layer technologies and integrate them for maximum performance and minimum overhead. Previously, we explored the physical layer concepts such as the understanding of the channel model, adaptive beamforming, and massive MIMO system technology.

10.4.1 mm-Wave Wireless Channel Model

The emerging mm-wave frequencies raise many new challenges in mobile wireless communications. The primary challenge is the non-availability of any standard channel model. Technical understanding of channel behavior presents new architectural techniques and different multiple access and novel methods of air interface [13]. Moreover, the biological safety at mm-wave frequencies is also questionable and still subject to study [23].

Propagation losses are expected to be high at higher frequencies due to high path losses. However, it is true only for the path loss at a particular frequency prevailing with two isotropic antennas. Shorter wavelengths enable dense packing of smaller antennas in small areas, thereby challenging the use of isotropic antennas for future 5G networks. In addition, mm-wave links are capable of casting very narrow beams. Furthermore, recent research [24] has also demonstrated that directional transmission of narrow beams reduces interferences and spatial multiplexing capabilities for cellular applications. Hence, mm-wave links are expected to perform perfectly. However, mm-wave link performance depends on many other factors as well, like distance between the nodes, link margin of the radios, and multipath diversity. Understanding diffraction, penetration, scattering, and reflection of mm-waves in different possible environments lays the foundation for 5G network deployment.

Signal outage investigation and comparison of reflection coefficients for building materials like tinted glass, clear glass, dry wall, doors, cubical, and metallic elevators revealed that the common outdoor building materials present high penetration resistance to mm-waves [13]. Moreover, indoor environment structures such as dry wall, white board, clutter, and mesh glass are also found to significantly impact attenuation, multipath components, and free-space path loss [25]. Indoor channel impulse responses confirm that human bodies create considerable obstruction to mm-wave propagation. Movement of people generates shadowing effects, which could be mitigated by a larger antenna beam width and the introduction of angular diversity [26]. From the available propagation results [13], [24], we can conclude that outdoor mm-wave signals are mostly confined to the outdoor. Very little signal penetrates indoors through glass doors, open doors, and windows. The indoor–outdoor isolation emphasizes the need for different nodes to serve different coverage sites [13], [24]. Further, separation of indoor and outdoor traffic relaxes the overhead associated with radio resources allocation and transmission power consumption. Interestingly, small cell architecture is already under deployment in dense urban areas. Thus, application of LOS propagation in a small cell environment looks promising for mm-wave communications. Ensuring LOS would require massive antenna deployment without any pre-defined pattern. An example of random, dense, and site-specific LOS communications is given in [60].

Carrier frequency and mobility characterize the Doppler effect. Received incoming waves have different shift values, thus resulting in a Doppler spread [17], [24]. Doppler-induced time-selective fading is easily alleviated by packet sizing and suitable coding over the coherence time of the channel [14]. Moreover, reduced angular spread in narrow beam transmissions, inherent to mm-wave propagation, further reduces Doppler spread [17], [24]. Therefore, it is unlikely that Doppler effects could raise any significant challenge in 5G networks.

10.4.2 Adaptive Beamforming

Design of smart antennas is vital for effective mm-wave communications since directional beams are the vibrant components of emerging 5G networks. The mm-wave beamforming algorithm is indispensable to focus energy in the desired direction to meet the air interface desires of 5G networks. Different configurations of antenna arrays and sub-arrays, with designated beamforming weights, steer and control the beams. Beamforming is possible in the analog, digital, or RF front end [15]. Beamforming weights are applied in the digital or analog domain to create directive beams [20]. In digital beamforming, coefficients are applied and multiplied per RF chain, over modulated baseband signals, before or after fast Fourier transformation (FFT) at the transmitter or receiver, respectively; whereas, the analog beamforming is done by applying coefficients to the modified RF signals in the time domain itself. Digital beamforming offers better performance at increased complexity and cost. On the other hand, analog beamforming is a simple and effective method with less flexibility. The hybrid beamforming architecture provides sharp beams with phase shifters at the analog domain and flexibility of the digital domain [20]. For a large antenna array, it is expensive and complicated to use separate transceivers for every antenna element. It not only causes a rise in component cost but also increases power consumption. Challenges in 5G communications could be resolved with MIMO, RF, and hybrid beamforming architectures [27].

For mm-wave frequencies, the efficiency of RF components is usually poor, thus imposing the power amplifiers to operate only at maximum power. Hence, the control of the array is done by phase shifters. There is also a proposition to use narrow beams for data and broader beams for control channels [28]. The work in [29] provides a new beamforming algorithm for sum-rate maximization in virtual cell networks. The solution achieves a balance between the desired signal maximization and interference minimization.

It is implied that beamforming by the steerable and extremely directional antenna is important for further 5G development. Consequently, it raises momentous new issues in complex communication protocol design. Steerable beams can be used at the mobile hand set and BS for RF communication and backhaul coordination. Early work on mm-wave antenna pointing protocols used pseudo-noise sequences. With narrowband pilot signals and multipath angular spreads, antenna pointing directions could efficiently be determined [14].

Authors in [30] have proposed singular value decomposition (SVD) based transmission preceding and receive combining method. It is employed for training antenna coefficients in a multistage iterative fashion. For a system with a lower number of RF chains and large number of antennas, this training method is very effective [31].

Beam coding is another beam training technique which assigns a unique signature code to every beam angle, in a training packet, and helps in fast estimation of the best angle pair. Moreover, the technique shows robustness in an NLOS environment, critical for future mm-wave communications [31].

For outdoor mobile channels, knowledge of time-varying angle of arrival (AOA) and Doppler spread characteristics is necessary. As compared to

LOS, antenna pointing for the NLOS scenario produces higher path loss and multipath delay spread. Adaptive beam arrays for narrow beam steerable antennas demonstrate that links could be created by illuminating surrounding objects in an NLOS antenna [32]. Knowledge of AOA is also useful for finding an alternative path for the NLOS scenario. The conventional method is to identify an alternative path by ranking signal strengths of all training beam pairs. In order to achieve lower redundancy, with energy and bandwidth conservation, AOA information is also used by directional self-pursuing protocol (DSP). Directional antenna and on-demand route discovery for the shortest path are special cases of DSP for enhancing the efficiency and reliability of wireless broadcasts.

The mm-wave MIMO systems are constrained by the need for RF beamforming to overcome poor link budget. The challenge can be mitigated by using switched narrow beams to employ fixed antenna patterns for transmitting or receiving from specific directions. A sectorized antenna model is considered ideal for these systems. These arc creating antennas provide high gain over a confined range of azimuths. The range of each transmitting node is divided into overlapping sectors. For transmission or reception, the node is configured to switch on one or several sectors. The jointly covered transmission range is usually more effective than omni-directional mode. It also demands less hardware requirements while the beam combining protocol and SDMA could be employed along with FDMA or TDMA techniques to increase spectrum capacity and frequency reuse.

10.4.3 Massive MIMO Systems

Active phased array antenna (APAA) based massive MIMO, with the help of signal processing techniques, provides BS with a huge number of antennas. The grid of antennas is capable of directing horizontal and vertical beams with full dimension MIMO. Massive MIMO significantly enhances spectral and energy efficiency. Every single antenna is positioned to achieve directivity in transmission. Coherent superposition of wave fronts is the underlying principle of massive MIMO technology. Emitted wave fronts add constructively at the intended location and reduce strength everywhere else. Hence, the spatial multiplexing at massive MIMO enabled BS increases capacity by several magnitudes [33]. Such a design requires effective algorithms for massive MIMO systems with advanced modulation techniques [34]. Increase in number of antennas cannot render a highly correlated channel vector as orthogonal. Time division duplexing (TDD) is the preferred choice for massive MIMO systems to avoid the complexity associated with channel estimation and channel sharing in frequency division duplexing (FDD). More investigations into frequency correction algorithms, such as direction of arrival-based frequency correction, covariance matrix, and spatio-temporal correlation, would enable the use of FDD in massive MIMO systems [34].

Massive MIMO deployment schemes, distributed antenna arrays, and directional antenna arrays are proposed in [33] for future BS designs. Moreover, inexpensive and low-power components can be used to build a massive MIMO system. Channel state information associated with a large number of BS antennas for massive MIMO systems and coordination among different cells includes a huge amount of information exchange overhead. This hampers system performance with limited-capacity backhaul links. In order to achieve better energy focus and reduced spatial interference, a new massive MIMO design by integrating an electromagnetic lens with large antenna array is also proposed [35].

The energy efficiency of small cell networks (SCN) is found to be larger than that of a massive MIMO system. Reduction in the size of antenna array and related electronic circuitry makes the small cells, with low frequency mm-wave transmission, a suitable candidate for a massive MIMO system. Hence, an efficient combination of the two technologies is expected to give better results [34]. Moreover, the spatial and temporal freedoms of massive MIMO systems can help in managing residual self-interference (SI).

FD is a new physical model by receiving and transmitting on the same frequency channel simultaneously which offers double spectral efficiency. Crosstalk between the transmitter and the receiver as well as the internal interference hinders the approval of simultaneous communication on the same frequency channel. However, recent developments in RF and beamforming antenna design technologies encourage FD transmission in the same frequency band [37]. Further, advances in MIMO systems, especially the present effective methods of SI reduction in the spatial domain, enable successful FD transmission in spite of interferences [38]. FD promises to double the capacity and improve the feedback and latency mechanism, while maintaining the security in the physical layer [37]. Simultaneous scheduling of uplink and downlink on the same resource block causes every FD transmission to suffer from SI challenges not only from within but also from neighboring cells as well. Thus, a reduction of SI is the major challenge to be addressed in the implementation of FD. SI cancellation categories are broadly classified as passive and active [38]. Passive SI cancellation exploits directional antennas, absorptive shielding, and cross-polarization to isolate the transmitter and receiver. The active technique utilizes information of a node's transmission signal to cancel the interference [38]. 5G networks with beamforming technology, massive MIMO deployment, centralized architecture, and small cell design appear to be conducive for FD realization. Intelligent device scheduling with suitable rate and power assignment can enable high capacity gains from FD operations [39].

10.5 MAC Layer Upgrading Requirements

10.5.1 MAC Layer Restoration to Meet the Modifications in Physical Layer

Modifications in the physical layer necessitate adjustments in the media access control (MAC) layer. The restoration is required in MAC protocol, multiple access, multiplexing, and frame structure along with the upcoming concepts of random access channel (RACH), cognitive radio, and FD reception and transmission.

10.5.2 Spatial Beam Patterns

SDMA is an attractive fit for adaptive antennas, beamforming, and device centric 5G architecture. Beamforming coefficients need to be trained beforehand for achieving the desired spatial beam patterns for the successful implementation of SDMA [40]. In order to support SDMA, BSs are required to transmit simultaneously and receive multiple beams in different directions [15].

Accurate computations of channel matrices, beamforming vectors, and feedback mechanisms are very crucial for effective SDMA implementation. However, for large number of antennas and small RF chains, traditional estimation and feedback procedures are inadequate. Thus, for smaller number of RF chains, an antenna training protocol is proposed in [41].

A new MIMO SDMA-OFDM system model is proposed in [42]. BSs receiver exploits array antenna of "P" elements, while each of the "L" simultaneous mobile users employs a single transmission antenna. The complex signal vector is a function of the *n*th sub-carrier of the *k*th OFDM symbol, received at the *p*th element of the antenna array [42].

10.5.3 Directional MAC Protocols

An MAC protocol that exploits spatial features could effectively increase the network capacity. Major MAC layer protocols, relevant to 5G wireless networks are:

1) Channel time allocation (CTA)
2) Multihop MAC protocol (MMAC)
3) Directional MAC protocol (DMAC)
4) Directional network allocation vector (DNAV)
5) Simultaneous transmission and reception (STR)
6) Request to send/clear to send (RTS/CTS)
7) Directional carrier sense multiple access with collision avoidance (D-CSMA/CA)

In multiuser MAC, protocols are proposed based on the Markov chain model of multiuser SDMA. Interference reduction in directional and adaptive beamforming aids BSs in simultaneous communications with many users in the same multiuser group [43]. With physical layer technologies nearing the Shannon capacity, concurrent transmissions and receptions at the same time and frequency effectively double the spectral efficiency and throughput [43].

TDMA with time portioned in super frames is also possible for 5G communications. Super frames are composed of many time slots named as CTA [44]. In every CTA, many local links communicate concurrently to achieve the spatial reuse. Directionality can also be added to the MAC protocol, which sends RTS and CTS. Advantages and disadvantages of RTS/CTS transmissions in both omni-directional and directional operation are discussed [16], [44]. The nodes need to transmit in the same direction from where it received the CTS/RTS. The DNAV table helps in tracking directions where the node must (or must not) initiate a transmission. DNAV is integral to the DMAC protocol, in which the upper layer is assumed to be aware of its neighbors [45].

10.5.4 Multiple Access Techniques for 5G

All mm-wave channels are characterized with a broad clear bandwidth and small delay spread. OFDM is favored as the multiple access technique since it is efficient and less sensitive to time offsets [46]. Flexibility, support for multiple bandwidth, and simple equalizer design are some of the major advantages of OFDM. It is also proposed to have OFDM and single carrier FDM for mm-wave mobile broadband (MMB) systems [17]. MMB frame structure and simulation analysis are also presented in [17] and [24]. Lower channel delay spread helps in reducing frequency selective fading in OFDM [42]. It is believed that OFDM might be an effective choice for mm-wave broadband networks subject to the condition that technical considerations are not ignored [15]. However, diverse user requirements, tactile Internet, and ultra-low latency demands make synchronization and orthogonality the challenges.

Interleave division multiple access (IDMA), for generating signal layers, needs further investigations [47]. Similarly, to leverage synchronization and orthogonality, sparse code multiple access (SCMA) and non-orthogonal multiple access are also proposed for consideration in 5G communication [48]. IDMA is a special case of CDMA. Instead of considering a spread sequence specific to the user, IDMA uses specific interleaves for user segregation. Interleaves generally utilize a less complex, iterative multiuser identification at the receiver [49]. SCMA combines quadrature amplitude modulation (QAM) symbol mapping and spreading. Multidimensional code words from the SCMA codebook are directly mapped over incoming bits. SCMA has less complexity but better performance compared to low density version of CDMA [50].

Novel concepts such as generalized frequency division multiplexing (GFDM) or filter bank multicarrier (FBMC) are also the candidates to overcome the challenges of 5G system. GFDM's flexibility and block structure help in the fulfillment of low latency requirements of 5G systems [47]. Researchers have also proposed GFDM implementation by integrating FFT/IFFT algorithms. FBMC is another key enabling technology for emerging 5G MAC. FBMC is naturally non-orthogonal and does not require complex synchronization. Thus, it offers to reduce signaling overheads which in turn would improve latency and enhance user experience in irregular traffic environment.

10.5.5 Other Methods

RACH and multicell operations which are important for future communications are not compatible with orthogonality. Further, sporadic traffic generation from M2M and IoT applications cannot be avoided. Such devices have to incorporate bulky synchronization procedures of random access, specifically designed for orthogonality. Hence, non-orthogonal waveforms are proposed to carry sporadic traffic for asynchronous signaling over the RANCH [47].

Cognitive radio is one of the promising technologies aimed at improved resource utilization. It prescribes the existence of both licensed and unlicensed radio nodes on the same bandwidth. Dynamic spectrum allocation algorithms are of prime importance in cognitive networks. Routing and resource allocation for mesh networks using cognitive radio techniques are analyzed in [51]. The design scheme is proposed for higher traffic load and lower delay. Both characteristics are fundamental to 5G systems. Hence, further developments in cognitive radio are expected to support and enhance emerging 5G networks.

Efficient and novel MAC protocols are critical to achieve fully the capabilities of FD design. FD enables nodes to transmit and receive a designated packet simultaneously at the same frequency. A-Duplex establishes packet-alignment-based dual link between two different half duplex clients and FD access point (AP). Therefore, FD and half duplex may coexist in the same application environment. An MAC protocol with RTS/full duplex clear to send (FCTS), supporting both bidirectional and unidirectional techniques, is proposed in [52]. FD is expected to play a crucial role in accomplishing low latency requirements of 5G networks [25]. Smart device scheduling and suitable rate/power allocation enable high capacity gain from FD operations [39].

With the development in physical and MAC layer technologies (such as mm-wave spectrum, multiple antennas, small cells, adaptive beamforming, massive MIMO, SDMA, RACH, and cognitive radio), 5G networks are expected to uplift itself through a big paradigm shift in the communications industry along with the announcements of novel applications.

10.6 MIMO

MIMO systems that improve communication performance between end-to-end systems are composed of multiple antennas at the transmitter and receiver ends. The MIMO technique has the capacity to increase channel rate with the use of multiple antennas, thus increasing spatial diversity gain and improving quality of service without increasing the transmitted power of antennas [66]. MIMO is effectively a radio antenna technology as it uses multiple antennas at the transmitter and receiver to enable a variety of signal paths to carry the data, choosing separate paths for each antenna to enable multiple signal paths to be used [64].

One of the core ideas behind MIMO wireless systems space–time signal processing in which time is complemented with the spatial dimension inherent in the use of multiple spatially distributed antennas, i.e. the use of multiple antennas located at different points. Accordingly, MIMO wireless systems can be viewed as a logical extension to the smart antennas that have been used for many years to improve wireless.

It is found that between a transmitter and a receiver, the signal can take many paths. Additionally, by moving the antennas even a small distance, the paths used will change. The variety of paths available occurs as a result of the number of objects that appear at the side or even in the direct path between the transmitter and receiver. Previously, these multiple paths only served to introduce interference.

By using MIMO, these additional paths can be used as an advantage. They can be used to provide additional robustness to the radio link by improving the signal-to-noise ratio (SNR) or by increasing the link data capacity.

The two main formats for MIMO are given below:

1) **Spatial diversity:** Spatial diversity used in this narrower sense often refers to transmit and receive diversity. These two methodologies are used to provide improvements in the SNR and they are characterized by improving the reliability of the system with respect to the various forms of fading.
2) **Spatial multiplexing:** This form of MIMO is used to provide additional data capacity by utilizing the different paths to carry additional traffic, i.e. increasing the data throughput capability.

One of the key advantages of MIMO spatial multiplexing is the fact that it is able to provide additional data capacity. MIMO spatial multiplexing achieves this by utilizing the multiple paths and effectively using them as additional "channels" to carry data. The maximum amount of data that can be carried by a radio channel is limited by the physical boundaries defined under Shannon's law.

MIMO antenna systems are used in modern wireless standards, including in IEEE 802.11n, 3GPP LTE, and mobile WiMAX systems. The technique supports enhanced data throughput even under conditions of interference, signal fading, and multipath. The demand for higher data rates over longer distances has been one of the primary motivations behind the development of MIMO OFDM communications systems.

Shannon's law defines the maximum rate at which error-free data can be transmitted over a given bandwidth in the presence of noise. It is usually expressed in the form:

$$C = \text{BW} * \log_2 (1 + \text{SNR}) \tag{10.1}$$

where C is the channel capacity in bits per second, BW is the bandwidth in Hertz, and SNR is signal-to-noise ratio.

Equation (10.1) shows an increase in a channel's SNR results in marginal gains in channel throughput. As a result, the traditional way to achieve higher data rates is by increasing the signal bandwidth. Unfortunately, increasing the signal bandwidth of a communications channel by increasing the symbol rate of a modulated carrier increases its susceptibility to multipath fading. For wide bandwidth channels, one partial solution to solving the multipath challenge is to use a series of narrowband overlapping sub-carriers. Not only does the use of overlapping OFDM sub-carriers improve spectral efficiency, but the lower symbol rates used by narrowband sub-carriers reduces the impact of multipath signal products.

MIMO communication channels provide an interesting solution to the multipath challenge by requiring multiple signal paths. In effect, MIMO systems use a combination of multiple antennas and multiple signal paths to gain knowledge of the communications channel. By using the spatial dimension of a communications link, MIMO systems can achieve significantly higher data rates than traditional single-input single-output (SISO) channels. In a 2×2 MIMO system, signals propagate along multiple paths from the transmitter to the receiver antennas.

Using this channel knowledge, a receiver can recover independent streams from each of the transmitter's antennas. A 2×2 MIMO system produces two spatial streams to effectively double the maximum data rate of what might be achieved in a traditional 1×1 SISO communications channel.

The maximum channel capacity of a MIMO system can be estimated as a function of N spatial streams. A basic approximation of MIMO channel capacity is a function of spatial streams, bandwidth, and SNR and is shown in the following equation:

$$C = N * \mathrm{BW} * \log_2 (1 + \mathrm{SNR}) . \tag{10.2}$$

Given the equation for MIMO channel capacity, it is possible to investigate the relationship between the number of spatial streams and the throughput of various implementations of SISO and MIMO configurations.

As an example, the IEEE 802.11g specs prescribe that a wireless local area network (WLAN) channel uses a SISO configuration. With this standard, the maximum coded data rate of 54 Mbps requires the use of a 64-QAM modulation scheme and a code rate of 3/4. As a result, the encoded bit rate is 72 Mbps (4/3 $\times$ 54 Mbps). With minimum transmitter error vector magnitude (EVM) at −25 dB, an SNR of 25 dB can be estimated as the requirement for a 64-state QAM (64-QAM) scheme. While EVM and SNR are not equivalent in all cases, we can assume that the magnitude error of a symbol will dominate the signal error as the SNR approaches its lower limit. The maximum data rate of IEEE 802.11g maps closely with the maximum channel capacity dictated by the Shannon–Hartley theorem. According to this theorem, a Gaussian channel with an SNR of 25 dB should produce an encoded data rate of 94 Mbps in a 20-MHz channel bandwidth.

In contrast, Equation (10.2) would suggest that a MIMO channel with four spatial streams should be capable of four times the capacity of the SISO channel. A 20 MHz channel with an SNR of 25 dB and four spatial streams should have an encoded bit rate of 4×94 Mbps = 376 Mbps. This estimation maps closely with the expected data rates of the draft IEEE 802.11n physical layer specs. IEEE 802.11n is designed to support MIMO configurations with as many as four spatial streams. At the highest data rate, bursts using a 64-QAM modulation scheme with a 5/6 channel code rate produce a data rate of 288.9 Mbps and an uncoded bit rate of 346.68 Mbps. At the highest data rate, the IEEE 802.11n channel with four spatial streams produces a data rate that is comparable to the theoretical limit of 376 Mbps.

It can be observed that the bit rate of a 4×4 (four spatial stream) MIMO configuration exceeds that of the Shannon–Hartley limit at all data rates, making MIMO systems attractive for higher data throughput. While MIMO systems provide users with clear benefits at the application level, the design and test of MIMO devices is not without significant challenges.

10.6.1 Benefits of MIMO Technology

Multiple antenna configurations can be used to overcome the detrimental effects of multipath and fading when trying to achieve high data throughput in limited-bandwidth channels. MIMO antenna systems are used in modern wireless standards, including in IEEE 802.11n, 3GPP LTE, and mobile WiMAX systems. The technique supports enhanced data throughput even under conditions of interference, multipath, and fading. The demand for higher data rates over longer distances has been one of the primary motivations behind the development of MIMO OFDM communications systems.

10.6.2 Superior Data Rates, Range, and Reliability

Systems with multiple antennas at the transmitter and receiver – also referred to as MIMO systems – offer superior data rates, range, and reliability without requiring additional bandwidth or transmit power. By using several antennas at both the transmitter and receiver, MIMO systems create multiple independent channels for sending multiple data streams.

A 4×4 MIMO system supports up to four independent data streams. These streams can be combined through dynamic digital beamforming and MIMO receiver processing to increase reliability and range.

The number of independent channels and associated data streams that can be supported over a MIMO channel is equivalent to the minimum number of antennas at the transmitter or receiver. Thus, a 2×2 system can support at most two streams, a 3×3 system can support three streams and a 4×4 system can support four streams. Some of the independent streams can be combined through dynamic digital beamforming and MIMO receiver processing which results in increased reliability and range.

10.6.3 Other Methods Downlink MIMO

For the LTE downlink, a 2×2 configuration for MIMO is assumed as baseline configuration, i.e. two transmit antennas at the BS and two receive antennas at the terminal side. Configurations with four antennas are also being considered.

Different MIMO modes are envisaged. It has to be differentiated between spatial multiplexing and transmit diversity, and it depends on the channel condition which scheme to select.

10.6.4 Spatial Multiplexing

Spatial multiplexing allows to transmit different streams of data simultaneously on the same downlink resource block(s). These data streams can belong to one single user (single-user MIMO/SU-MIMO) or to different users (multiuser MIMO/MU-MIMO). While SU-MIMO increases the data rate of one user, MU-MIMO allows to increase the overall capacity. Spatial multiplexing is only possible if the mobile radio channel allows it.

The principle of spatial multiplexing, exploiting the spatial dimension of the radio channel, allows to transmit the different data streams simultaneously.

$$\mathbf{H} = \begin{bmatrix} h_{11} & h_{12} & \dots & h_{1Nt} \\ h_{21} & h_{22} & \dots & h_{2Nt} \\ \vdots & \vdots & \vdots & \vdots \\ h_{Nr1} & h_{Nr2} & \dots & h_{NrNt} \end{bmatrix}. \tag{10.3}$$

Each transmitting antenna transmits a different data stream. Each receive antenna may receive the data streams from all transmit antennas. The channel (for a specific delay) can thus be described by the channel matrix **H**, as given in Equation (10.3).

In this general description, *Nt* is the number of transmit antennas and *Nr* is the number of receive antennas, resulting in a 2×2 matrix for the baseline LTE scenario. The coefficients h_{ij} of this matrix are called channel coefficients from transmitting antenna *i* to receive antenna *j,* thus describing all possible paths between transmitter and receiver sides. The number of data streams that can be transmitted in parallel over the MIMO channel is given by min $\{Nt, Nr\}$ and is limited by the rank of the matrix **H**. The transmission quality degrades significantly in case the singular values of matrix **H** are not sufficiently strong. This can happen in case the two antennas are not sufficiently de-correlated, for example, in an environment with little scattering or when antennas are too closely spaced. In LTE, up to two code words can be mapped onto different so-called layers. The number of layers for transmission is equal to the rank of the matrix **H**. There is a fixed mapping between code words to layers [66].

Pre-coding on the transmitter side is used to support spatial multiplexing. This is achieved by applying a pre-coding matrix ***W*** to the signal before transmission.

The optimum pre-coding matrix ***W*** is selected from a pre-defined "codebook" which is known at eNodeB and user equipment (UE) side. Unitary pre-coding is used, i.e. the pre-coding matrices are unitary: ***WHW***= **I**. The UE estimates the radio channel and selects the optimum pre-coding matrix. The optimum pre-coding matrix is the one that offers maximum capacity. The UE provides feedback on the uplink control channel regarding the preferred pre-coding matrix (pre-coding vector as a special case). Ideally, this information is made available per resource block or at least group of resource blocks since the optimum pre-coding matrix varies between resource blocks.

10.6.5 Transmit Diversity

Instead of increasing data rate or capacity, MIMO can be used to exploit diversity. Transmit diversity schemes are already known from WCDMA release 99 and will also form part of LTE as one MIMO mode. In case the channel conditions do not allow spatial multiplexing, a transmit diversity scheme will be used instead, so switching between these two MIMO modes is possible depending on channel conditions. Transmit diversity is used when the selected number of streams (rank) is one [67].

10.6.6 Uplink MIMO

Uplink MIMO schemes for LTE will differ from downlink MIMO schemes to take into account terminal complexity issues. For the uplink, MU-MIMO can be used. Multiple user terminals may transmit simultaneously on the same resource block. This is also referred to as SDMA. The scheme requires only one transmit antenna at UE side which is a big advantage. The UEs sharing the same resource block have to apply mutually orthogonal pilot patterns [65]. To exploit the benefit of two or more transmit antennas but still keep the UE cost low, antenna sub-set selection can be used. In the beginning, this technique will be used, e.g. a UE will have two transmit antennas but only one transmit chain and amplifier. A switch will then choose the antenna that provides the best channel to the eNodeB.

10.7 Impact of 5G Wireless Systems on Human Health

Using millimeter waves and higher frequencies, 5G requires widespread network of transmitting devises. Shorter wavelengths (higher frequencies) are more powerful in terms of energy. Even though microwave and millimeter wavelength radiation is a non-ionizing radiation, it produces tissue heating at high exposure levels that can cause damages in temperature-sensitive biological tissues.

5G cell antenna devices are installed very close to one another, which will result in constant exposure of population to mm-wave radiation. Use of 5G technology will require the tools of adaptive antennas capable of beamforming and massive MIMOs. Also, BSs will be more closely packed to provide complete coverage. Use of small cells ranging from of 20 to 50 m achieves complete coverage. The cell radius of 20 m results in about 800 BSs per km^2. This contrasts with 3G or 4G technologies which use large or macrocells, with each cell covering a radial range of 2 km.

The ongoing investigations into non-ionizing frequencies appear to indicate potentially harmful, non-thermal, biological effects on humans, animals, and plants, even when exposed to levels that are below the official threshold level. The issue of possible environmental and health effects of electromagnetic fields is considered to have clear parallels with other current issues such as licensing of medication, chemicals, pesticides, heavy metals, and genetically modified organisms.

Non-ionizing radiation, which includes 5G mm-wave radiation, is considered as harmless in general due to its signal strength which is regulated by the respective regulatory bodies. However, the pulsation to which the whole population will be exposed due to the dense network of antennas and the estimated enormous number of simultaneous connections may cause environmental and health effects. Consequently, even though 5G can be weak in terms of power, its massive pulsing electromagnetic waves seem to increase the biologic and health impacts of exposure.

10.8 Next Generation Wireless Systems

It is believed that conventional text, voice and video mobile communications will still be the most important application of next generation (6G) in the 2030s, though other application scenarios will become ubiquitous and increasingly significant. Consequently, the 6G network should be human centric, rather than machine centric, application centric, or data centric [68]. Following this rationale, high security, secrecy, and privacy are key features of 6G. Furthermore, user experience would be adopted as a pivotal metric in 6G communication networks [69].

6G communications should be human centric, which implies that conventional mobile communications will still characterize 6G, in which the classic cellular phone is the major tool for mobile communications. To have a qualitative comparison between 5G and 6G communications, the spectral efficiency in 5G has already been close to its operational frequency spectrum boundary by using massive MIMO, network densification, and mm-wave transmission. As bounded by the Shannon limit, the spectral efficiency in 6G would hardly be improved on a large scale. In contrast, secrecy and privacy in 6G communications should be significantly enhanced by new technologies. In 5G networks, traditional encryption algorithms based on the Rivest–Shamir–Adleman (RSA) public-key cryptosystems are still in use to provide transmission security and secrecy. RSA cryptosystems have become insecure under the pressure of big data and artificial intelligence

(AI) technologies. But novel privacy protection mechanisms are still far from full-fledged in the 5G. Incremental improvements would appear in energy efficiency, intelligence, affordability, and customization. The energy efficiency gain would be accomplished by the maturity of energy harvesting technology and green communications. Intelligence in 6G can be classified into operational, environmental, and service levels, which will benefit from the thrust in AI developments. The improvements on affordability and customization rely on novel networking architectures, promotion, and operational strategies on the market.

Researchers placed great emphasis on network throughput, reliability, latency, and the number of served users in 4G and 5G communications. However, the security, secrecy, and privacy issues of wireless communications have been, to some extent, overlooked in the past decades. To protect data security, classic encryption based on RSA algorithms is being challenged by increasingly powerful computers. Meanwhile, communication service providers have legally collected an enormous amount of user information, and private data leakage incidents have worryingly increased. This becomes an unstable factor in the human-centric 6G network and could lead to a disastrous consequence without proper countermeasures. To solve this problem, it is envisioned that complete anonymization, decentralization, and untraceability can be realized in 6G networks by blockchain technology [70].

From a human-centric perspective, high affordability and full customization should be two important technological indicators of 6G communications. All users will be granted the right to choose what they like in 6G, and such rights should not be diminished by intelligent technologies or unnecessary system configurations. To overcome the daily charging constraint for most communication devices and facilitate continuous communication services, low energy consumption and long battery life are two key factors in 6G communications. To achieve a long battery life, various energy harvesting methodologies would be applied in 6G, which not only harvest energy from ambient RF signals but also from other renewable energy sources. In addition, the high intelligence in 6G will benefit network operations, wireless propagation environments, and communication services, which refer to operational intelligence, environmental intelligence, and service intelligence, respectively. High intelligence introduces a high degree of system complexity, which could rise up the costs to network operators and device manufacturers. All these raised costs will translate into less affordable products for end users. To resolve this tradeoff, technological breakthroughs in intelligent systems are necessary, but more importantly, a new commercial strategy will be important.

References

[1] Rappaport, Theodore S. Wireless communications: principles and practice. Vol. 2. New Jersey: Prentice Hall PTR, 1996.

[2] Bhalla, Mudit Ratana, and Anand Vardhan Bhalla. "Generations of mobile wireless technology: A survey." International Journal of Computer Applications. Vol. 5, No. 4, 2010.

[3] Dekleva, Sasha, J. P. Shim, Upkar Varshney, and Geoffrey Knoerzer. "Evolution and emerging issues in mobile wireless networks." Communications of the ACM 50, no. 6 (2007): 38-43.

[4] Noble D., "The History of Land-Mobile Radio Communications," IEEE Vehicular Technology Transactions, pp. 1406-1416, 1962.

[5] MacDonald V. J., "The Cellular Concept," Bell Systems Technical Journal, Vol. 58, No. 1, pp. 15-43, 1979.

[6] Young W. R., "Advanced Mobile Phone System: Introduction, Background, and Objectives," Bell System Technical Journal, Vol. 58, No. 1, p.7, 1979.

[7] Rajinder Vir, "A Comprehensive Survey of the Wireless Generations," International Journal of Research in Computer Applications and Robotics, Vol.3, No. 9, pp. 45-51, 2015.

[8] Rappaport T.S., Roh.W and Cheun K., "Wireless engineers long considered high frequencies worthless for cellular systems. They couldn't be more wrong," IEEE Spectrum, Vol.51, No. 9, pp. 34-58, 2014.

[9] Mamta Agiwal, Abhishek Roy and Navrati Saxana, "Next Generation 5G Wireless Networks: A comprehensive Survey," IEEE Communications Surveys & Tutorials, Vol. 18, No. 3, pp. 1617-1655, 2016.

[10] Andrews J.G. *et al.*, "What will 5G be?" IEEEJ. Selected Areas of Communications, Vol. 32, No. 6, pp. 1065-1082, 2014.

[11] Chen S. and Zhao J., "The requirements, challenges, and technologies for 5G of terrestrial mobile telecommunications," IEEE Communications Magazine, Vol. 52, No. 5, pp. 36-43, 2014.

[12] Boccardi F, Heath R.W, Lozano A, Marzetta T.L, and Popovski P, "Five disruptive technology directions for 5G," IEEE communications Magazine, Vol. 52, No. 2, pp. 74-80, 2014.

[13] Rappaport T.S. *et al.*, "Millimeter wave mobile communications for 5G cellular: It will work!," IEEE Access, Vol.1, pp. 335-345, 2013.

[14] Rappaport T.S, Gutierrez F, Ben-Dor E, Murdock N.J, Qiao Y, and Tamir J.I, "Broadband millimeter wave propagation measurements and models using adaptive beam antennas for outdoor urban cellular communications," IEEE Transactions on Antennas and Propagation, Vol. 61, No. 4, pp. 1850-1859, 2013.

[15] Pi Z, and Khan F, "System design and network architecture for a millimeter-wave mobile broadband (MMB) systems," in Proc. IEEE Sarnoff Symposium, 2011, pp.1-6.

[16] Korakis T, Jakllari G, and Tassiulas L, "A MAC protocol for full exploitation of directional antennas in ad-hoc wireless networks," in Proceedings of AMC international Symposium on Mobile Ad Hoc networks and computing, 2003, pp.97-108.

[17] Pi Z, and Khan F, "An introduction to millimeter-wave mobile broadband systems," IEEE Communications Magazine, Vol. 49, No. 6, pp. 101-107, 2011

[18] Bae J, Choi Y.S, Kim J.S, and Chung M.Y, "Architecture and performance evaluation of mmWave based 5G mobile communication systems," in Proceedings of International Conference on Information and Communication Technology Convergence (ICTC), 2014, pp. 847-851

[19] Feng Z, and Zhang Z, "Dynamic spatial channel assignment for smart antenna," Wireless Pers. Commun. Vol. 11, No. 1, pp. 79-87, 1998.

[20] Roh W *et al.,* "Millimeter-wave beamforming as an enabling technology for 5G cellular communications: Theoretical feasibility and prototype results," IEEE communication Magazine, Vol. 52, No. 2, pp. 106-113, 2014.

[21] Cardieri P, and Rappaport T.S, "Application of narrow-beam antennas and fractional loading factor in cellular communication systems," IEEE Trans. Veh. Technol., Vol. 50, No. 2, pp. 106-113, 2014.

[22] Wang Z, Li H, Wang H, and Ci S, "Probability weighted based spectral resources allocation algorithm in HetNet under Cloud-RAN architecture," in Proceedings of International Conference on Communication China Workshop, 2013, pp. 88-92.

[23] Wu T, Rappaport T.S, Collins C.M, "Safe for generation to come: Consideration of safety for millimeter waves in wireless communications," IEEE Microwave Magazine, Vol. 16, No. 2, pp. 65-84, 2015.

[24] Khan F, Pi Z, and Rajagopal S, "Millimeter-wave mobile broadband with large scale spatial processing for 5G mobile communication," in Proc. 50th Annual Allerton Conf. Commun. Control Comput. 2012, pp. 1517-1523.
[25] Anderson C.R, and Rappaport T.S, "In-building wideband partition loss measurements at 2.5 and 60 GHz," IEEE Trans. Wireless Communication, Vol. 3, No. 3, pp.922-928, 2004.
[26] Collonge S, Zhharia G, and Zein G.E, "Influence of the human activity on wide-band characteristics of the 60 GHz indoor radio channel," Vol. 3, No. 6, pp.2396-2406, 2004.
[27] Vook F.W, Ghosh A, and Thomas, T.A, "MIMO and beamforming solutions for 5G technology," in Proc. IEEE Microwave Symposium (IMS), 2014, pp. 1-4.
[28] Rajagopal S, "Beam broadening for phased antenna array using multi-beam subarrays," in Proc. IEEE International Conference on Communication, 2012, pp. 3637-3642.
[29] Kim. J, Lee. H.W, and Chong. S, "Virtual cell beamforming in cooperative networks," IEEE Journal of Selected Areas in Communications," Vol. 32, No. 6, pp. 1126-1138, 2014.
[30] Xia. P, Yong. S.K, Oh. J, and Ngo. C, "Multi-stage iterative antenna training for millimeter wave communications," in Proc. IEEE Global Telecomm. Conf. (Globecom), 2008, pp. 1-6.
[31] Tsang. Y.M, Poon. A.S.Y, and Addepalli. S, "Coding the beams: Improving beamforming training in mmwave communication systems," in Proc. IEEE Global Telecomm. Conf. (Globecom), 2011, pp. 1-6.
[32] Ben-Dor. E, Rappaport. T.S, Qiao. Y, and Lauffenburger. S.J, "Millimeter-wave 60 GHz outdoor and vehicle AOA propagation measurements using a broadband channel sounder," in Proc. IEEE Global Telecomm. Conf. (Globecom), 2011, pp. 1-6.
[33] Mehmood. Y, Afzal. W, Ahmad. F, Younas. U, Rashid. I, and Mehmod. I, "Large scaled multi-user MIMO system so called massive MIMO systems for future wireless communication networks," in Proc. Int. Conf. Autom. Comput., 2013, pp. 1-4.
[34] Lu. L, Li. G.Y, Swindlehurst. A.L, Ashikhmin. A, and Zhang. R, "An overview of massive MIMO: Benefits and challenges," IEEE Journal of selected topics in signal processing, Vol. 8, No. 5, pp. 742-758, 2014.
[35] Zeng. Y, Zhang. R, and Chen. N, "Electromagnetic lens-focusing antenna enabled massive MIMO: Performance improvement and cost

reduction," IEEE Journal of Selected Areas in Communications," Vol. 32, No. 6, pp. 1194-1206, 2014.

[36] Cirik. A.C, Rong. Y, and Hua. Y, "Achievable rates of full-duplex MIMO radios in fast fading channels with imperfect channel estimation," IEEE Trans. Signal Process., Vol 62, No. 15, pp. 3874-3866, 2014.

[37] Talwar. S, Choudhury. D, Dimou. K, Aryafar. E, Bangerter. B, and Stewart. K, "Enabling technologies and architectures for 5G wireless," in Proc. MTT-S Int. Microw. Symp. (IMS), 2014, pp. 1-4.

[38] Zheng. G, "Joint beamforming optimization and power control for full-duplex MIMO two-way relay channel," IEEE Trans. Signal Process., Vol. 63, No. 3, pp. 555-566, 2015.

[39] Goyal. S, Liu. P, Panwar. S.S, DiFazio. R.A, Yang. R, and Bala. E, "Full-duplex cellular systems: Will doubling interference prevent doubling capacity?" IEEE Commun. Mag., Vol. 53, No. 5, pp. 121-127, 2015.

[40] Pirapaharan. K, Kunsei. H, Senthilkumar. K.S, Hoole. P.R.P, and Hoole. S.R.H, "A Single Beam Smart Antenna for Wireless Communication in Highly Reflective and Narrow Environment," in Proceedings of International Symposium on Fundamentals of Electrical Engineering (ISFEE), 2016, pp. 1-5.

[41] Xia. P, Yong. S.K, Oh. J, and Ngo. C, "A practical SDMA protocol for 60 GHz millimeter wave communications," in Proc. 42nd Asilomar Conf. Signals Syst. Comput., 2008, pp. 2019-2023.

[42] Jiang. M, and Hanzo. L, "Multiuser MIMO-OFDM for next-generation wireless systems," Proc. IEEE, Vol. 95, No.7, pp. 1430-1469, 2007.

[43] Gong. M.X, Akhmetov. D, Want. R, and Mao. S, "Multi-user operation in mmWave wireless networks," in Proc. IEEE Int. Conf. Commun., 2011, pp.1-6.

[44] Qiao. J, Shen. X, Mark. Q, Shen. Y, He. Y, and Lei. L, "Enabling device-to-device communications in millimeter-wave 5G cellular networks," IEEE Commun. Mag., Vol. 53, No. 1, pp. 209-215, 2015.

[45] Choudhury. R.R, Yang. X, Ramanathan. R, and Vaidya. N.H, "On designing MAC protocols for wireless networks using directional antennas," IEEE Trans. Mobile Comput., Vol. 5, No. 5, pp. 477-491, 2006.

[46] Kim. J, and Kim. I.G, "Distributed antenna system-based millimeter-wave mobile broadband communication system for high

speed trains," in Proc. IEEE Int. Conf. ICT Convergence, 2013, pp. 218-222.

[47] Wunder. G, et al., "5G NOW: Non-orthogonal, asynchronous waveforms for future mobile applications," IEEE Commun. Mag., Vol. 52, No.2, pp. 97-105, 2014.

[48] Zakrzewska. A, Ruepp. S, and Berger. M.S, "Towards converged 5G mobile networks-challenges and current trends," in Proc. IEEE ITU Kaleidoscope Acad. Conf. Living Converged World Impossible Without Stand.?, 2014, pp.39-45.

[49] Niroopan. P, and Chung. Y.H, "A user-spread interleave division multiple access system," Int. J. Adv. Res. Comput. Commun. Eng., Vol. 1, No. 10, pp. 837-841, 2012.

[50] Hosein. N, and Hadi. B, "Sparse code multiple access," in Proc. IEEE 24th Int. Symp. Pers. Indoor Mobile Radio Commun. (PIMRC), 2013, pp. 332-336.

[51] El-Sherif. A.A, Mohamed. A, " Joint routing and resource allocation for delay minimization in cognitive radio based mesh networks," IEEE Trans. Wireless Commun., Vol. 13, No. 1, pp. 186-197, 2014.

[52] Cheng. W, Zhang. X, and Zhang. H, "RTS/FCTS mechanism based full-duplex MAC protocol for wireless networks," in Proc. Global Commun. Conf. (GLOBECOM), 2013, pp. 5017-5022.

[53] Gustavo N, "Radio Link Performance of Third Generation (3G) Technologies of Wireless Networks" Faculty of Virginia Polytechnic Institute and State University, chp.2, pp.3, 2002

[54] Comparison of speed 2G to 4G.Adapted from"3G VS 4G Technology". Retrieved from https://technicgang.com/3g-vs-4g-technology-what-is-the-difference-between-3g-and-4g/amp/

[55] Kadir E. A., Shamsuddin S. M., Rahman T. A., Ismail A. S,"Big Data Network Architecture andMonitoring Use Wireless 5G Technology," Int. J. Advance Soft Compu. Appl, Vol. 7, No. 1, March 2015, pp.9

[56] Nguyen T.,"Small Cell Networks and the Evolution of 5G (Part 1)," 2017. Retrieved from https://www.qorvo.com/design-hub/blod/small-cell-netwroks-and-the-evolution-of-5g.

[57] Nasimi M., Hashim F. and Kyun C.N.,"Characterizing energy efficiency for heterogeneous cellular networks,"IEEE Student Conf. on Research and Dev.,pp. 200, 2012

[58] Chen N., Kadoch M. and Rong B." SDN Controlled mmWave Massive MIMO Hybrid Precoding for 5G Heterogeneous Mobile Systems,"Hindawi Publishing Corp., 2016, pp.4

[59] Gür G."Multimedia transmission over networks fundamentals and challenges,"Bogazici University, pp.717
[60] "5G and EMF explained."Retrieved from https//emfexplained.info/?ID=25916
[61] Ali E., Ismael M., Nordin R. and Nor F. A."Beamforming techniques for massive MIMO systems in 5G:overview, classification, and trends for future research,"Frontier of Information Technology & Electronic Engineering, Vol. 18, Issue 6, pp. 753, 2017
[62] "Massive MIMO."Retrieved from https://techblog.comsoc.org/tag/massive-mimo/
[63] "5G-FD MIMO."Retrieved from https://www.sharetechnote.com/html/5G/5G_MassiveMIMO_FD_MIMO.html
[64] Yang Wen Liang, "Ergodic and Outage Capacity of Narrowband MIMO Gaussian Channels", Department of Electrical and Computer Engineering, The University of British Columbia, Vancouver, British Columbia.
[65] BengtHolter, "On The Capacity of the MIMO Channel-A Tutorial Introduction".
[66] D. W. Bliss, Keith W. Forsythe, and Amanda M. chan "MIMO Wireless Communication", Lincoln Labt. Journal, Vol. 15, No. 1, 2005
[67] X Gu, X-H Peng and G C Zhang, 2006. "MIMO systems for broadband wireless Communications", BT Technology Journal, Vol 24 No 2, April 2006.
[68] Shuping Dang, Osama Amin, Basem Shihada and Mohamed-Slim Alouini. "What should 6G be?", Nature Electronics. Vol. 3, pp. 20-29, 2020.
[69] David, K. & Berndt, H. "6G vision and requirements: is there any need for beyond 5G?", IEEE Veh. Technol. Mag. 13, 72–80 (2018).
[70] Henry, R., Herzberg, A. & Kate, A. "Blockchain access privacy: challenges and directions", IEEE Secur. Priv. 16, 38–45 (2018).

11

Emerging Technologies for 5G/6G Wireless Communication Networks

Ade Syaheda Wani Marzuki[1], Dayang Azra Awang Mat[1], Dayang Nurkhairunnisa Abang Zaidel[1], Kho Lee Chin[1], and Paul RP Hoole[2]

[1]Faculty of Electrical and Electronic Engineering, Universiti Malaysia Sarawak, Malaysia
[2]Wessex Institute of Technology, Southampton, United Kingdom
maswani@unimas.my

Abstract

This chapter gives an overview of the important, present, and future 5/6G technologies, of which the smart antenna is an integral part. These include the cloud-based system architecture, cell densification, increasing operating signals to millimeter (mm) electromagnetic wave techniques, massive multi-input multi-output (MIMO) systems, beamforming, ubiquitous networks, and green communication systems. The fifth generation system requirements are also presented.

Keywords. 5G technologies, mm-wave, green communications, massive MIMO, HetNet, ubiquitous networks.

11.1 Introduction

In recent years, the demand for higher data rates in wireless communication networks is escalating due to the rapid evolution of smart electronic devices and systems. As these technologies proliferate, data traffic volume is estimated to reach 49 exabytes per month in 2021, with approximately 86% of it generated by smartphone users alone [1]. This huge data traffic volume is beyond the current fourth generation (4G) network capacity, particularly when spectrum and energy efficiency and related issues remain unresolved in network design. The current 4G networks have just about reached the theoretical data rate limit [2] and therefore not capable in solving bandwidth scarcity, limited capacity, and high energy consumption issues relating to the increased number of mobile users. The fifth generation (5G) wireless networks, which was standardized in 2020, is expected to accommodate and resolve all these issues efficiently with various new approaches.

The emerging 5G networks are expected to achieve a 1000-fold capacity increase, by offering high data rates and spectral efficiency, massive ubiquitous device connectivity, low energy consumption, better mobility support, and consistent quality of experiences (QoE) provisioning [3]. Moreover, both installation costs (CAPEX) and operational costs (OPEX) in 5G network deployment will need to be reduced substantially [4]. However, to implement this new generation of technologies, there are several concerns that bring unprecedented challenges to the network operators and service providers. First, the resource management must be allocated efficiently for cellular communications, as the current available frequency spectrum has been used heavily, particularly for cellular usage [2]. The demand for additional spectrum to accommodate future user demand is difficult; therefore, alternative resource allocation solutions, utilizing the available spectrum in a shared manner, are required. Interference mitigation is another concern when dealing with a highly dense network. It is expected that small cells will be densely deployed to deliver high data rates to a massive number of users since deploying more macrocell base stations will be costly – and infrastructure-wise prohibitive [5, 6]. In the future, with the use of ultradense small cell networks (SCNs), it is expected that femtocells may be deployed with cell-center separation as small as 50 m [7]. Thus, effective interference mitigation mechanisms for heterogeneous networks (HetNets) must be designed to alleviate both cross- and co-tier interference. Next, the demand for seamless and ubiquitous communications is another crucial challenge, where users are expected to be connected anywhere and at any time [8], including in high speed vehicular environments. A survey conducted in Seoul, Korea revealed that the traffic originating from vehicular users has increased by about 8.62 times in a three-year period, and this increase is expected to be higher in the future. Another issue that arose during the 5G initial research stage is the high energy consumption caused by wireless communication technologies. Along with developing high capacity HetNets appropriate for the upcoming 5G standards, the energy consumption by all these communication technologies may reach 1700 TW/h by 2030 [9], with

50% of the relevant operating costs being energy-related [10]. The ever increasing energy consumption has triggered demand for greener networks and can be significantly mitigated by addressing the energy-related issues in network design, deployment, and operation. All the above issues pose daunting challenges and put considerable pressure on the network and service providers, while they struggle to respond to increasing demand for higher data rates, greater network capacity, and spectral efficiency, in addition to improved energy efficiency and mobility in the emerging 5G networks.

This chapter provides an overview of the requirements, important key technologies, and challenges in 5G developments. An in-depth view on the technologies that are widely used in 5G networks will be presented; including small cell densification, massive multi-input multi-output (MIMO), beamforming network, millimeter-wave (mm-wave), ubiquitous networks, and green communications. Related progress in wireless technologies, including their potential and challenges toward 5G network deployment, are also considered in this chapter.

11.2 5G Requirements

The 5G network, which was deployed in 2020, has data rates of about 10 Gbps and allows up to 1000 times greater system capacity than the 4G network [4]. Due to the proliferation of smart devices and recent advances in cellular networks, the demand for high data rates and ubiquitous networks has posed a crucial challenge to the network operators and service providers. The introduction of new handheld smart devices in different forms and a growing number of device-to-device (D2D) and machine-to-machine (M2M) applications has led to an estimated use of 26.3 billions of devices and connections during the implementation of 5G networks in 2020. Moreover, high data rate demand not only originates from indoor users but also from vehicular users [8, 11, 12]. Most of this huge traffic volume is used to support large-sized multimedia contents over faster and reliable network, for example, video streaming applications.

Fast mobility is also supported in 5G networks where lower end-to-end latency is possible, allowing the user to experience seamless and reliable connectivity by leveraging in-vehicle communications. This will provide real-time feedback especially for critical scenarios, such as in vehicular ad hoc network (VANET) communications. The 5G research also addresses energy-related issues in current and future network deployments. Energy-efficient considerations for 5G technologies can be categorized into various specific technical areas in wireless networks, as listed in [13, 14]. Large amount of energy consumed by both network infrastructure and mobile devices poses a daunting question, whether the upcoming 5G standards can really achieve green networks.

The last requirement in 5G standard is to improve user's perceived quality by optimizing the data traffic based on the type of application [15]. A reliable connectivity for uploading and downloading multimedia contents in the 4G era has provided the users with an expectation that they should be provided with better satisfaction in the next generation networks. Thus, this has led the need to seek appropriate mechanisms for dealing with different types of traffic, especially for "always on" applications[16], for example, instant social messaging and networking, and push email.

11.3 5G Cloud-Based Network Architecture

Another criterion that needs to be fulfilled by 5G networks is the ease of network setup and future modification. These important characteristics allow the industries to update the legacy of network architecture, as well as adapting the network with emerging technologies to satisfy user demand and experience. The centralized network design in 4G era failed to satisfy the latency requirements and massive data traffic and wireless connections [15], causing a significant burden on the core infrastructure. To satisfy these requirements and to support various applications, the 5G networks deployment is largely to be based on cloud-based architecture [17, 18]. In this architecture, a full virtualization of the evolved packet core (EPC) functionalities supports the network to evolve and allows it to be reconfigured and maintained easily. The cloud concept focuses on running the control plane as an application in the cloud, while simultaneously accommodating massive traffic volumes in the data plane.

According to [19], the 5G cloud-based wireless network architecture is composed of four components: 1) mobile cloud, 2) cloud-based radio access network (C-RAN), 3) reconfigurable network, and 4) big data center, as shown in Figure 11.1. Mobile clouds implement user tasks in the cloud environments, saving storage capacity, reducing device computing power, and hence prolonging the battery life [20]. In C-RAN, the baseband processing is centralized and shared among sites in a virtualized baseband unit (BBU) pool (Figure 11.2). In a conventional RAN, the capacity and coverage of a base station is limited. In order to expand network coverage, more base stations are required; however, it is difficult to deploy these large numbers of

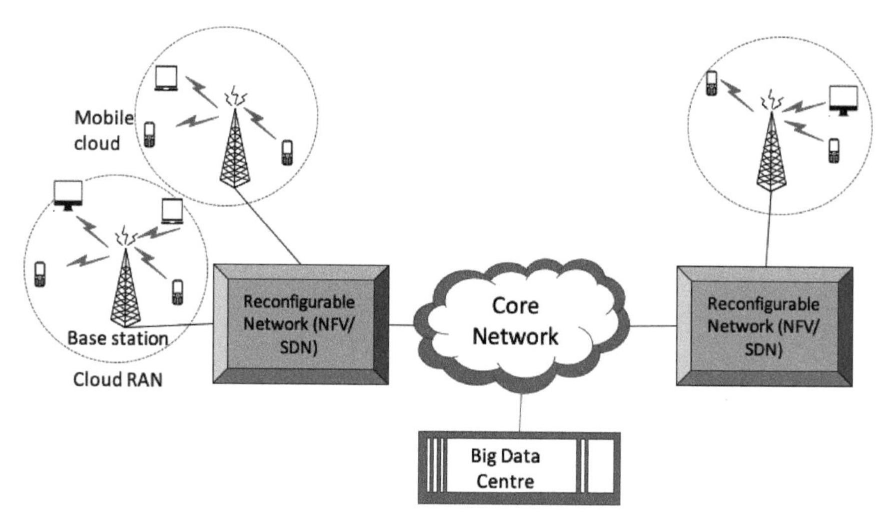

Figure 11.1 5G cloud network architecture.

base station cells due to compatibility and costing issues. Moreover, spectrum sharing between base stations is not supported whenever these base stations are in idle mode, resulting in inefficient spectrum utilization. In contrast, C-RAN significantly increases network spectral efficiency and cell edge user's throughput by addressing all the above-mentioned issues. To further increase the efficiency of C-RAN, the complexity of its network architecture and resource scheduling can be reduced by implementing a collaborative processing and an efficient scheduling within the remote radio head (RRH) clusters [19, 21].

11.4 Key Technologies

The 5G networks require a paradigm shift that includes very high carrier frequency spectra with massive bandwidths, extreme base station densities, and unprecedented numbers of antennas to support the enormous increase in the volume of traffic. There are various new technologies integrated into existing networks to achieve the effectiveness of 5G networks in achieving higher data rates and accommodating traffic demand from a very large number of users. From network densification with massive number of small cells to other latest network solutions such as mm-wave and massive MIMO, all these technologies are aimed at achieving such goals. In addition to that, network ubiquity and green communications have also been identified as

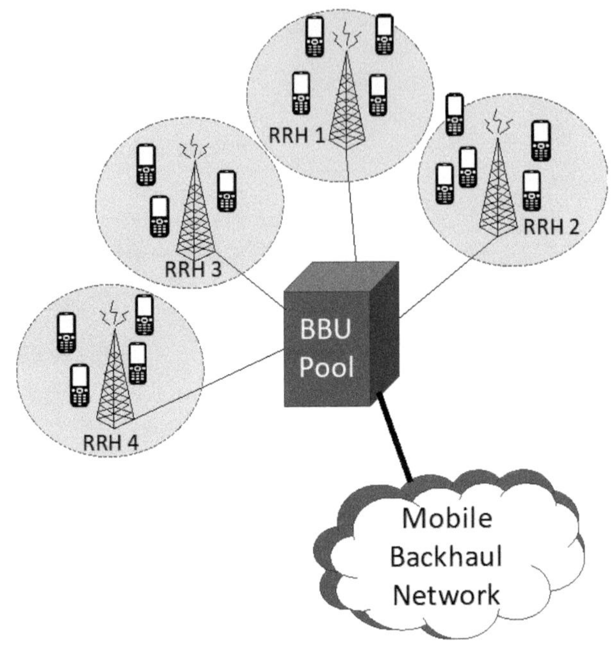

Figure 11.2 C-RAN connectivity.

important focuses in 5G, adding to its uniqueness as compared to the 4G networks.

11.4.1 Small Cell Densification

The deployment of dense small cells overlaid in the existing HetNets (Figure 11.3) has been identified as a feasible strategy to expand the network coverage and address the explosive growth of mobile data traffic. Relying solely on macrocell base stations is no longer an effective strategy, as the data volumes with associated uneven traffic distributions have increased tremendously in recent times [23]. Moreover, deploying additional macrocells would not be a viable solution due to their high installation cost and the lack of suitable sites for deployment [6]. Thus, modified networks are required in which smaller cells can be deployed within the existing macrocell area to form HetNets.

By reducing the cell size, the use of limited available bandwidth from the already scarce spectrum resources can be optimized by adopting frequency reuse[23-25]. Moreover, as these cells offer local traffic off-loading to a smaller number of users, larger portions of resources can be accommodated

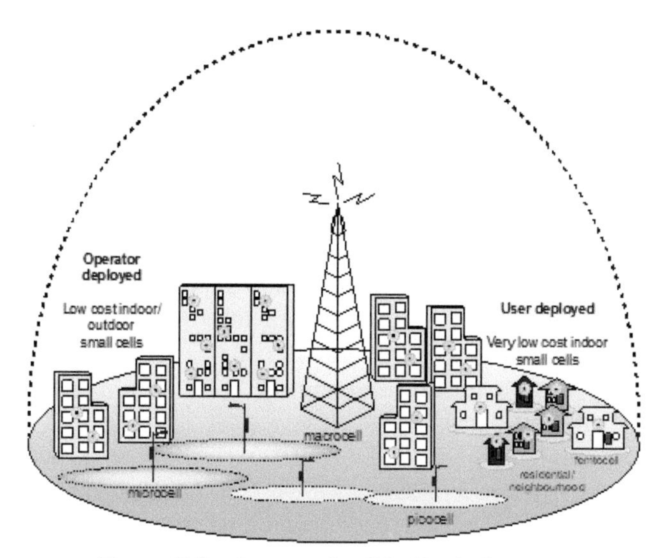

Figure 11.3 An example of HetNet deployment.

to their associated users. The shorter distance between these low-powered base stations and user devices prolongs user battery life, thus increasing both the energy efficiency and signal-to-interference plus noise ratio (SINR) due to the low loss path [25]. Although small cells (i.e. microcells, picocells, and femtocells) can facilitate high bandwidth and wireless user ubiquity, densely deployed networks introduce a new set of challenges such as backhaul connectivity, resource allocation, and energy management issues. Moreover, small cell densification also implies high inter-cell interference (ICI) among these cells due to their close proximity and arbitrary deployment.

Small cell backhaul solutions can be wired or wireless, depending on the required network coverage, installation complexity, and associated cost. Thus, providing high capacity connectivity between small cells and core network may require extensive planning before an optimal solution can be achieved. Wired connectivity is straightforward and is commonly used in small cell deployments. However, when these small cells are deployed outdoors, the backhaul connectivity becomes more complex. This raises a concern, as 75% of outdoor small cells may be backhauled by using wireless connectivity in the future [6]. Therefore, efficient backhaul solutions utilizing wireless links must be designed to address both 5G spectral and power requirements. In the 4G small cell deployment, user traffic demands are backhauled to the network operator via broadband gateways such as DSL cables. The absence of dedicated wired backhaul links can diminish the network real-time quality of services (QoS).

Indirectly, this will cause the sub-optimal small cell access point placement issue, considering that high cost is involved in connecting these small cells to an existing infrastructure. Thus, the main motivation in deploying small cells in the cell edge regions and in areas with high user density will not be realized.

Several studies have focused on determining the optimal small cell deployment while considering different performance metrics. To realize the random and flexible deployment of these small cells, optimal cell placements in an existing infrastructure must be accomplished [26]. In [27], a theoretical framework for small cell placement is designed to maximize the network spectral efficiency and mitigate co-tier interference. Work on small cell placements are also reported in [28] and [29]. In [28], an optimal solution of backhaul design for SCNs is developed by taking the network requirement constraints, such as coverage and capacity, into consideration. In contrast, the work in[29] [29] proposed small cell placement strategies to achieve a fault-tolerance network with the use of small cell self-healing features.

Despite the potential of small cells in delivering high data rates to the ever increasing number of users, another research challenge remains, particularly with respect to resource allocation to users in small cells. Small cells, specifically femtocells, are deployed by the end user with minimal intervention from network operators. Short distances between small cells cause overlapping coverage areas and hence inefficient resource management and associated interference problems. In such scenarios, inefficient resource management will result in severe interference problems. A cluster-based resource allocation scheme for femtocell-assisted macrocellular networks is proposed in [30-32]. This scheme is developed to mitigate both cross- and co-tier interference in the network. Other solutions proposed are dynamic resource allocation approach [33] and fractional frequency reuse [34, 35].

It is anticipated that small cell densification will play a critical and growing role in future wireless communications by enabling the network service providers to deliver high data rates through increased frequency reuse, enlarged coverage areas, and greater spectral and energy efficiencies.

11.4.2 Millimeter Wave

Emerging mm-wave wireless technology is a prominent solution for outdoor transmission in the 5G networks. This wireless technology relies on highly directive narrow beamwidth antennas. Such narrow mm-wave beams are what are obtained in beamforming at V band (57–66 Ghz) and E band (70–80 GHz). mm-wave operates at the higher end of frequency spectrum with wide bandwidth and has the ability to deliver higher throughput for 5G applications [36]. It also complements the microwave spectrum scarcity by offering line of sight (LoS) transmissions, which requires a clear radio path for reliable communications. Besides, its limited coverage makes mm-wave to be more immune to interference compared to the conventional microwave technology. This also allows effective frequency reuse and maximization of spectrum efficiency.

There are abundant challenges to deploying mm-wave technologies for the 5G networks. First, higher frequencies are more susceptible to propagation loss. Hence, mm-wave requires specially designed antennas for 5G purposes. In [37], a phased array mm-wave antenna was reported for large-scale applications. Other mm-wave antenna designs proposed in [38-40] achieved significant energy and spectral efficiency enhancement in 5G networks. The next challenge is to determine the channel and propagation models for mm-wave transmissions. These models are essential because the behavior of propagating signals varies with different types of shadowing material. Shadowing and multipath signal characteristics at mm-wave need also to be handled to improve the link performance. Various channel measurements for mm-wave transmissions have been conducted in both outdoor [41-43] and indoor [44-46] environments, and the results are used to estimate the environment-based performance of the transmission channel. Research work continues to investigate the feasibility of implementing mm-wave technology for moving wireless small cells, especially in vehicular environment. The last challenge is to quantify the biological implications of mm-wave signals to users in dense network deployment. As 5G will be supported by large-scale antenna arrays radiating high frequency beams, the impact on human health raises a critical concern. Thus, a clear guideline and standard for all mm-wave equipment and devices must be prepared for 5G networks, honestly addressing the science-based concerns raised.

11.4.3 Massive MIMO

Massive MIMO systems transmit and receive signals from multiple devices by focusing the radiated energy toward the intended directions. While this allows multiple users to be served, its directive beam helps to minimize intracell interference and ICI. The use of mm-wave in the 5G networks offers a larger frequency band; however, this wireless technology has shorter range coverage compared to 4G. Thus, MIMO technology is introduced to extend the range of 5G by concentrating the signal in a single, desired direction.

The relation between frequency band and the transmission range is based on the well-known equation (11.1). This equation states that the received power is inversely proportional to the square of distance between the received and transmitted antennas (see Chapter 1)

$$P_r = \frac{P_t}{4\pi R^2} \tag{11.1}$$

where P_r and P_t are the received and transmitted power, respectively, and R is the distance between transmitting and receiving antennas.

In reality, the received power is also affected by the wavelength and antenna gain. Thus, Equation (11.1) can be extended by including the wavelength (λ) and antenna gain (G). To give us the Friis equation,

$$P_r = \frac{P_t}{4\pi R^2} \left(\frac{\lambda^2}{4\pi} \right) G_r G_t \tag{11.2}$$

where G_r and G_t denote the receiver and transmitter antenna gains, respectively.

The utilization of high frequency spectrum up to 28 GHz in 5G wireless communications means that it operates at very shorter wavelengths. For instance, consider a 2 GHz 4G network and a 20 GHz 5G network. The wavelength at 2 GHz is 10 times longer than the wavelength at 20 GHz. This leads to a 100 times decrease in the received power in the 20 GHz when compared to the received power at 2 GHz. Thus, the received power in 5G will be much lower than that in the 4G communication systems.

Massive MIMO has been identified as one of the emerging technologies in 5G networks [2, 4, 47]. It provides a very large number of antenna services/antenna arrays at the base stations and at the user terminals. The term massive MIMO was used due to the collection of nearly infinite antenna arrays utilized in the base stations. Figure 11.4 illustrates the fundamental concept of massive MIMO.

There are two most important benefits of having a massive MIMO compared to a conventional MIMO [25, 48-50]. First, a massive MIMO focuses its radiation toward smaller regions with the aid of a very large number of antenna arrays. This leads to an increase in radiation energy efficiency and lower interference level. It is also said that the massive MIMO systems can increase the system capacity by a factor of 10 [51].

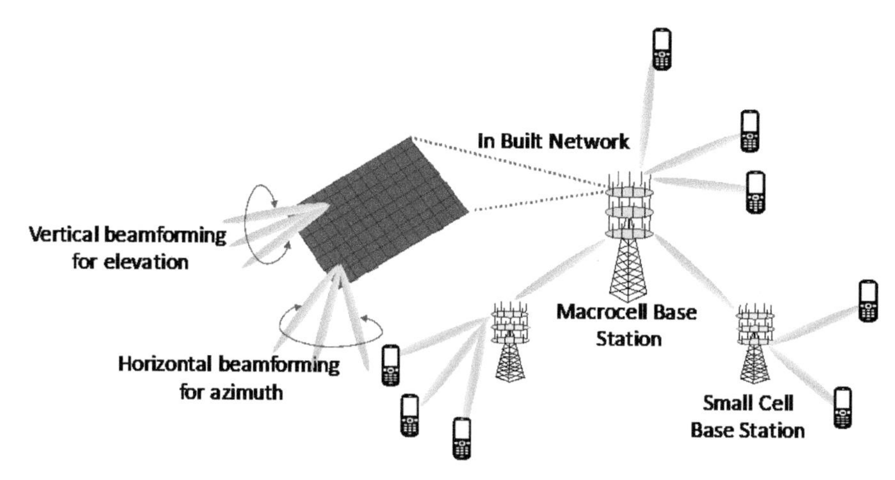

Figure 11.4 Massive MIMO and beamforming.

Another major benefit is that the massive MIMO systems can achieve significant spectral efficiency by allowing spatial multiplexing in many terminals utilizing a similar time-frequency resource. In addition to these benefits, there are also some important properties of massive MIMO that must be remembered. Massive MIMO offers green technology since the energy radiated per bit delivered from base station to user terminal can be lowered due to the highly directive beams. Moreover, it can achieve optimal performance due to its massive number of antenna arrays. It also offers flexibility in traffic offloading and network load balancing.

In massive MIMO systems, there are several unprecedented issues that need to be addressed, including complexity in digital pre-coding, modulation, and scheduling. Full digital pre-coding in massive MIMO is challenging due to its large frequency band [52]. This problem can be solved by using hybrid beamforming scheme, as proposed in [38, 53]. This scheme combines the analog beamforming and digital pre-coding to suppress the interference among non-orthogonal analog beams. Most of the massive MIMO-related works rely on mathematically tractable models [54-56]. Some of the assumptions made were unrealistic; for instance, the model for identically independent distributions [57] and the simplification of the two-dimensional spatially correlated channels [58]. Another issue that is related to the massive MIMO system is that it requires effective algorithms with a very advanced modulation technique [50] and efficient user scheduling. Due to large number of antennas used, highly correlated channel vectors operate in a frequency division duplexing (FDD) orthogonal mode, making it infeasible to perform channel estimation and sharing [59]. Because of this, the time division duplexing (TDD) mode is often chosen. There are still challenges in deploying FDD mode in massive MIMO, which include studies on covariance matrix, direction of arrival (DOA) based frequency correction, and spatiotemporal correlation [50]. Moreover, specific scheduling algorithms set for massive MIMO systems with realistic propagation channels are a challenging area.

11.4.4 Beamforming Mechanism

In 5G networks, antenna arrays with smart signal processing algorithms at the transmitter and receiver are required to increase the effectiveness of massive MIMO technology. The beam pattern can be adjusted to emphasize signals of interest and to minimize interfering signals. There are two main activities that define smart antenna arrays: 1) DOA estimation and 2) beamforming mechanism. In the former case, the direction of the signal can be estimated by using several techniques, including multiple signal classification and matrix pencil techniques. These techniques are used to find the spatial spectrum of the antenna array, and then based on the peaks of the spectrum, the signal DOA can be measured. An alternative is the electromagnetic signal processing technique reported in Chapter 7. In beamforming, a desired radiation pattern of an antenna array is created. This radiation pattern will be directed into the desired direction [60] and can be done by using finite impulse response (FIR) tapped delay line filter. The signal is directed toward the desired direction and can be done by using FIR tapped delay line filter. The signal is directed toward the desired users and other signals will act as interfering signals, as shown in Figure 11.5. The electromagnetic signal processing technique for beamforming is discussed in detail in Chapters 5 and 8. The artificial intelligence techniques are discussed in Chapter 9.

Similar to massive MIMO technology, the beamforming approach improves the signal-to-noise ratio (SNR) by using highly directional signal transmission. Moreover, by exploiting the spatial properties of the MIMO antenna, this approach also overcomes both external and internal interference that comes from a certain direction. The great importance of this technology is that it has the ability to send a "null" signal toward the interferer, allowing it to cancel the interference signal. The highly directive beam signal in this approach is stronger than an ordinary signal produced by a

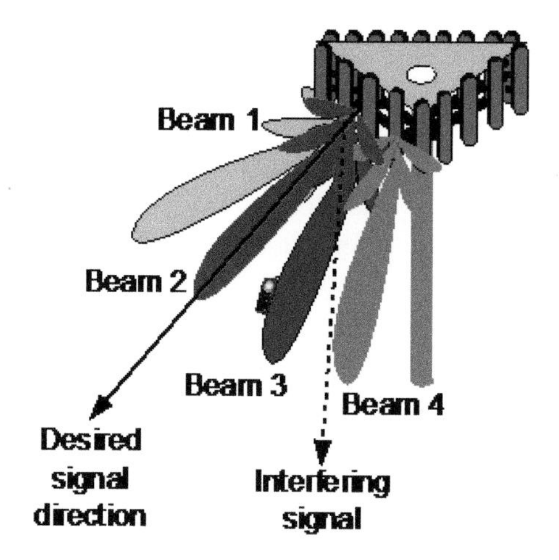

Figure 11.5 Desired and interfering signals in beamforming antennas.

regular antenna, hence much reduced wasted radio frequency (RF) energy is achieved.

Beamforming technologies can increase the efficiency of massive MIMO systems in terms of signal transmission. By having a combination of these two technologies, interference can be reduced while transmitting more information from a vast number of antennas at once. Beam directivity can be controlled by controlling beamforming weights [61]. Besides directivity, these beams also need to be appropriately selected to ensure proper alignment [47]. However, a few complications need to be resolved in beamforming technology, which includes the hidden terminal and neighbors location issues which are inherent to directional transmissions. Even though beamforming is more complex to deploy when compared to the other multiple antenna techniques, its benefits are highly desired for massive MIMO systems.

11.4.5 Ubiquitous Communications

The practical realization of the 5G to support fast and seamless mobility is through ubiquitous communications. 5G network coverage range is extended using very small cells within a micro cell [8, 62, 63]. The small cells also enable high data rates for the varied user data format to satisfy different user needs. Furthermore, the small cells enable handling of high mobility of the users [24]. However, high mobility also poses some challenges to the practical implementation of ubiquitous communications in 5G networks. Challenges include ensuring that the QoS and QoE are maintained at acceptable levels [64].

In addition to extending the coverage area in a cell, small mobile cells provide ubiquitous communications, especially for users in vehicular environments [8, 62, 63]. The mobile femtocells, which are evolved from home-based static femtocells, offer high data rates and provide seamless connectivity to a group of users in moving vehicles. These moving cells allow higher QoS and SINR by preventing high vehicle penetration loss which can be up to 25 dB [64]. While both user QoS and QoE in mobile environments can be enhanced, there are several challenges that arise along with the use of small mobile cell. The first issue is massive number of handovers as these small cells move from one serving cell to another with high vehicular speed, which can vary up to 120 km/h for cars and 360 km/h for bullet trains [65, 66]. These high speeds may cause insufficiency of time to allocate resources to these moving cells, hence degrading the network QoS. Furthermore, such small mobile cells may experience ICI as there is no frequency planning involved in developing the femtocells. Thus, it is crucial to have a dynamic resource allocation scheme, specially designed to adapt to varying speeds of small mobile cells.

Another challenge is to design wireless backhaul solutions for these small mobile cells. Unlike fixed small cells, these mobile cells are connected via wireless links (i.e. mm-wave or microwave) to the fixed cellular infrastructure. Traffic backhauling via wireless transmission may be affected by weather conditions, spectrum availability, and wireless coverage. These environmental factors affect the availability of the backhaul network. Furthermore, the cost associated with new air interface to mitigate these factors will increase the deployment cost. Such cost will increase with increase in the capacity; the 5/6G capacity is expected to be one thousand times more than the capacity of 3G/4G networks [67]. The backhaul solution will have to satisfy the coverage requirements since small cells are scattered to reach isolated users in a wider geographical area, including in the country and rural areas. In urban areas, the challenge of covering ultradense network with ultralow latency will also be a challenge to overcome [68].

A further challenge involves the allocation of resource or spectrum to each small mobile cells. As these cells move, they cause the network topology to be dynamic, hence generating a time adaptive interference set for each femtocell. The existing resource allocation solutions designed for static small cells will not effectively work in this case, as the resources need to be allocated in both time and space (spectrum) dimensions. Thus, a heuristic and adaptive resource allocation scheme is required to address this issue.

To addresses these needs to realize mobile femtocell deployment in the 5/6G networks, various efforts have been reported [8, 62, 63, 69–72]. The handover issue related to femtocell mobility was discussed in [9] and [73], while resource allocation issue for small mobile cell is discussed in [63], [70], and [74]. More investigation on the deployment of small mobile cells continues to ensure their effectiveness to deliver higher data rates and broadband services to users in vehicular environments.

11.4.6 Green Communications

Energy efficiency is another crucial aspect in all 5G communication systems, especially when wireless transmissions are involved. The increasing scale of network densification which is meant to accommodate larger network capacity has exacerbated the energy efficiency concern. Moreover, the high energy consumption in information and communication technologies (ICT) sectors alone is attracting the concern of researchers and service providers as it may reach 1700 TW/h in 2030 [9]. This demands the need for a greener network.

Along with developing high capacity advanced technologies for the 5G standards, including high network densification, overall network energy consumption must be minimized to meet both environmental and economic challenges. Close to 50% of relevant operating costs is energy-related [10]. The incorporation of green and sustainable technologies can significantly mitigate this ever increasing energy consumption for future energy saving. Much work has been done to achieve this eco-environmental goal; for instance, energy-aware base station operation which includes sleep mode [75], energy-efficient resource management [7, 54, 62], utilizing low-powered base station and devices, low energy wastage algorithms [9, 17, 76–78], and others. In [79], the energy consumed by macrocell and small cell base stations was outlined, providing valuable data for designing energy-efficient networks. Incorporation of green aspects has been included in traffic offloading in HetNets [80], causing the user traffic to be offloaded to neighboring small cells. The authors in [10], [81], and [82] proposed an energy-efficient backhaul design, which is suitable for dense SCNs. There will be more extensive research effort in the coming years to improve the energy efficiency in the 5G and future 6G communication networks.

11.5 Conclusion

The evolution of wireless communication networks, specifically in the 5/6G systems promises a manifold increase in data rates. Novel emerging techniques and technologies for this 5G and future 6G networks will have a massive impact on mobile users and on living, working, and natural environments. Various 5/6G research challenges have been presented in this chapter, along with some solutions. The crucial parts of smart antennas in the present and massive MIMO systems were highlighted. In advanced small mobile base stations, smart antennas will play a central role. The chapter has highlighted the vast amount of research challenges that are there, while reporting the advances and ideas that have already been developed and implemented. The students, engineers, and researchers working on smart antennas need to remember the broader context outlined herein, where electromagnetic signal/image processing with artificial intelligence for smart antennas are set to play a more important role compared to past signal/image processing techniques.

Acknowledgment

The authors would like to thank Universiti Malaysia Sarawak for the endless supports in providing research opportunities and encouragements to the authors. Also, they would like to thank the reviewers for their valuable efforts and comments toward improving this chapter.

References

[1] CISCO, "Cisco Visual Networking Index: Global Mobile Data Traffic Forecast Update, 2016-2021," Feb 2017.
[2] W. Cheng-Xiang et al., "Cellular architecture and key technologies for 5G wireless communication networks," IEEE Commun. Mag., vol. 52, no. 2, pp. 122-130, 2014.
[3] P. Agyapong, M. Iwamura, D. Staehle, W. Kiess, and A. Benjebbour, "Design considerations for a 5G network architecture," Communications Magazine, IEEE, vol. 52, no. 11, pp. 65-75, 2014.
[4] J. G. Andrews et al., "What Will 5G Be?," IEEE Journal on Selected Areas in Communications, vol. 32, no. 6, pp. 1065-1082, 2014.
[5] M. Coldrey, U. Engstrom, K. W. Helmersson, M. Hashemi, L. Manholm, and P. Wallentin, "Wireless backhaul in future heterogeneous networks," Ericsson Review, vol. 91, 2014.

[6] J. Hoadley and P. Maveddat, "Enabling small cell deployment with HetNet," Wireless Communications, IEEE, vol. 19, no. 2, pp. 4-5, 2012.
[7] A. Mesodiakaki, F. Adelantado, L. Alonso, and C. Verikoukis, "Energy-efficient user association in cognitive heterogeneous networks," Communications Magazine, IEEE, vol. 52, no. 7, pp. 22-29, 2014.
[8] A. S. W. Marzuki, I. Ahmad, D. Habibi, and Q. V. Phung, "Mobile small cells: Broadband access solution for public transport users," Communications Magazine, IEEE, vol. PP, no. 99, pp. 1-10, June 2017.
[9] I. Humar, G. Xiaohu, X. Lin, J. Minho, C. Min, and Z. Jing, "Rethinking energy efficiency models of cellular networks with embodied energy," Network, IEEE, vol. 25, no. 2, pp. 40-49, 2011.
[10] S. Tombaz, A. Vastberg, and J. Zander, "Energy- and cost-efficient ultra-high-capacity wireless access," Wireless Communications, IEEE, vol. 18, no. 5, pp. 18-24, 2011.
[11] A. Khan and A. Jamalipour, "Moving Relays in Heterogeneous Cellular Networks—A Coverage Performance Analysis," IEEE Transactions on Vehicular Technology, vol. 65, no. 8, pp. 6128-6135, 2016.
[12] H. Chu, P. Crist, S. Han, J. Pourbaix, and Y. Kawashima, "Funding Urban Public Transport: Case Study Compendium," in International Transport Forum Summit on Funding Transport, Leipzig, 2013.
[13] S. Zhang, Q. Wu, S. Xu, and G. Li, "Fundamental green tradeoffs: Progresses, challenges, and impacts on 5G networks," IEEE Communications Surveys & Tutorials, 2016.
[14] A. Alnoman and A. Anpalagan, "Towards the fulfillment of 5G network requirements: technologies and challenges," Telecommunication Systems, pp. 1-16, 2016.
[15] P. Ameigeiras, J. J. Ramos-Munoz, L. Schumacher, J. Prados-Garzon, J. Navarro-Ortiz, and J. M. Lopez-Soler, "Link-level access cloud architecture design based on SDN for 5G networks," IEEE Network, vol. 29, no. 2, pp. 24-31, 2015.
[16] L. Pierucci, "The quality of experience perspective toward 5G technology," IEEE Wireless Communications, vol. 22, no. 4, pp. 10-16, 2015.
[17] D. Sabella et al., "Energy Efficiency benefits of RAN-as-a-Service concept for a cloud-based 5G mobile network infrastructure," IEEE Access, vol. 2, pp. 1586-1597, 2014.

[18] K. Nagothu, B. Kelley, S. Chang, and M. Jamshidi, "Cloud systems architecture for metropolitan area based cognitive radio networks," in Systems Conference (SysCon), 2012 IEEE International, 2012, pp. 1-6: IEEE.
[19] M. Chen, Y. Zhang, L. Hu, T. Taleb, and Z. Sheng, "Cloud-based Wireless Network: Virtualized, Reconfigurable, Smart Wireless Network to Enable 5G Technologies," Mobile Networks and Applications, journal article vol. 20, no. 6, pp. 704-712, 2015.
[20] M. Shiraz, A. Gani, R. H. Khokhar, and R. Buyya, "A review on distributed application processing frameworks in smart mobile devices for mobile cloud computing," IEEE Communications Surveys & Tutorials, vol. 15, no. 3, pp. 1294-1313, 2013.
[21] J. Wu, Z. Zhang, Y. Hong, and Y. Wen, "Cloud radio access network (C-RAN): a primer," IEEE Network, vol. 29, no. 1, pp. 35-41, 2015.
[22] A. Checko et al., "Cloud RAN for mobile networks—A technology overview," IEEE Communications surveys & tutorials, vol. 17, no. 1, pp. 405-426, 2015.
[23] B. Bangerter, S. Talwar, R. Arefi, and K. Stewart, "Networks and devices for the 5G era," IEEE Commun. Mag., vol. 52, no. 2, pp. 90-96, 2014.
[24] ADRF, Bingo, COMMSCOPE, and Z. Networks, "DAS and Small Cells Evolve to Meet Today's and Tomorrow's Future Connectivity Needs," Netnet Forum, Arlington2019.
[25] M. H. Alsharif and R. Nordin, "Evolution towards fifth generation (5G) wireless networks: Current trends and challenges in the deployment of millimetre wave, massive MIMO, and small cells," Telecommunication Systems, pp. 1-21, 2016.
[26] A. H. Jafari, D. López-Pérez, H. Song, H. Claussen, L. Ho, and J. Zhang, "Small cell backhaul: challenges and prospective solutions," EURASIP Journal on Wireless Communications and Networking, vol. 2015, no. 1, p. 206, 2015.
[27] W. Guo, S. Wang, X. Chu, J. Zhang, J. Chen, and H. Song, "Automated small-cell deployment for heterogeneous cellular networks," IEEE Communications Magazine, vol. 51, no. 5, pp. 46-53, 2013.
[28] C. Ranaweera, C. Lim, A. Nirmalathas, C. Jayasundara, and E. Wong, "Cost-Optimal Placement and Backhauling of Small-Cell Networks," Lightwave Technology, Journal of, vol. 33, no. 18, pp. 3850-3857, 2015.
[29] T. Omar, Z. Abichar, A. E. Kamal, J. M. Chang, and M. A. Alnuem, "Fault-Tolerant Small Cells Locations Planning in 4G/5G Heterogeneous Wireless Networks," IEEE Transactions on Vehicular

Technology, vol. 66, no. 6, pp. 5269-5283, 2017.

[30] H. Zhang, D. Jiang, F. Li, K. Liu, H. Song, and H. Dai, "Cluster-Based Resource Allocation for Spectrum-Sharing Femtocell Networks," IEEE Access, vol. 4, pp. 8643-8656, 2016.

[31] A. Hatoum, R. Langar, N. Aitsaadi, R. Boutaba, and G. Pujolle, "Cluster-Based Resource Management in OFDMA Femtocell Networks With QoS Guarantees," IEEE Transactions on Vehicular Technology, vol. 63, no. 5, pp. 2378-2391, 2014.

[32] A. Abdelnasser, E. Hossain, and D. I. Kim, "Clustering and Resource Allocation for Dense Femtocells in a Two-Tier Cellular OFDMA Network," IEEE Transactions on Wireless Communications, vol. 13, no. 3, pp. 1628-1641, 2014.

[33] A. R. Elsherif, C. Wei-Peng, A. Ito, and D. Zhi, "Adaptive Resource Allocation for Interference Management in Small Cell Networks," IEEE Trans. Commun., vol. 63, no. 6, pp. 2107-2125, 2015.

[34] F. Jin, R. Zhang, and L. Hanzo, "Fractional Frequency Reuse Aided Twin-Layer Femtocell Networks: Analysis, Design and Optimization," Communications, IEEE Transactions on, vol. 61, no. 5, pp. 2074-2085, 2013.

[35] N. Saquib, E. Hossain, and D. I. Kim, "Fractional frequency reuse for interference management in LTE-advanced hetnets," IEEE Wireless Communications, vol. 20, no. 2, pp. 113-122, 2013.

[36] S. Scott-Hayward and E. Garcia-Palacios, "Multimedia Resource Allocation in mmWave 5G Networks," IEEE Communications Magazine, vol. 53, no. 1, January 2015 2015.

[37] W. Hong, K.-H. Baek, Y. Lee, Y. Kim, and S.-T. Ko, "Study and prototyping of practically large-scale mmWave antenna systems for 5G cellular devices," IEEE Communications Magazine, vol. 52, no. 9, pp. 63-69, 2014.

[38] S. Han, I. Chih-Lin, Z. Xu, and C. Rowell, "Large-scale antenna systems with hybrid analog and digital beamforming for millimeter wave 5G," IEEE Communications Magazine, vol. 53, no. 1, pp. 186-194, 2015.

[39] J. A. Zhang, X. Huang, V. Dyadyuk, and Y. J. Guo, "Massive hybrid antenna array for millimeter-wave cellular communications," IEEE Wireless Communications, vol. 22, no. 1, pp. 79-87, 2015.

[40] T. Y. Yang, W. Hong, and Y. Zhang, "Wideband millimeter-wave substrate integrated waveguide cavity-backed rectangular patch antenna," IEEE Antennas and Wireless Propagation Letters, vol. 13, pp. 205-208, 2014.

[41] A. Thornburg, T. Bai, and R. W. Heath, "Performance Analysis of Outdoor mmWave Ad Hoc Networks," IEEE Transactions on Signal Processing, vol. 64, no. 15, pp. 4065-4079, 2016.
[42] N. Palizban, S. Szyszkowicz, and H. Yanikomeroglu, "Automation of Millimeter Wave Network Planning for Outdoor Coverage in Dense Urban Areas Using Wall-Mounted Base Stations," IEEE Wireless Communications Letters, vol. 6, no. 2, pp. 206-209, 2017.
[43] X. Zhao, S. Li, Q. Wang, M. Wang, S. Sun, and W. Hong, "Channel Measurements, Modeling, Simulation and Validation at 32 GHz in Outdoor Microcells for 5G Radio Systems," IEEE Access, vol. 5, pp. 1062-1072, 2017.
[44] S. Sun, T. S. Rappaport, T. A. Thomas, and A. Ghosh, "A preliminary 3D mm wave indoor office channel model," in Computing, Networking and Communications (ICNC), 2015 International Conference on, 2015, pp. 26-31: IEEE
[45] X. Yin, C. Ling, and M.-D. Kim, "Experimental multipath-cluster characteristics of 28-GHz propagation channel," IEEE Access, vol. 3, pp. 3138-3150, 2015.
[46] J. Zhu, H. Wang, and W. Hong, "Large-scale fading characteristics of indoor channel at 45-GHz band," IEEE Antennas and Wireless Propagation Letters, vol. 14, pp. 735-738, 2015.
[47] M. Agiwal, A. Roy, and N. Saxena, "Next generation 5G wireless networks: A comprehensive survey," IEEE Communications Surveys & Tutorials, vol. 18, no. 3, pp. 1617-1655, 2016.
[48] W. H. Chin, Z. Fan, and R. Haines, "Emerging technologies and research challenges for 5G wireless networks," IEEE Wireless Communications, vol. 21, no. 2, pp. 106-112, 2014.
[49] V. Jungnickel et al., "The role of small cells, coordinated multipoint, and massive MIMO in 5G," IEEE Communications Magazine, vol. 52, no. 5, pp. 44-51, 2014.
[50] L. Lu, G. Y. Li, A. L. Swindlehurst, A. Ashikhmin, and R. Zhang, "An overview of massive MIMO: Benefits and challenges," IEEE Journal of Selected Topics in Signal Processing, vol. 8, no. 5, pp. 742-758, 2014.
[51] E. G. Larsson, O. Edfors, F. Tufvesson, and T. L. Marzetta, "Massive MIMO for next generation wireless systems," IEEE Communications Magazine, vol. 52, no. 2, pp. 186-195, 2014.
[52] A. Alkhateeb, J. Mo, N. Gonzalez-Prelcic, and R. W. Heath, "MIMO precoding and combining solutions for millimeter-wave systems," IEEE Communications Magazine, vol. 52, no. 12, pp. 122-131, 2014.

[53] Z. Gao, L. Dai, D. Mi, Z. Wang, M. A. Imran, and M. Z. Shakir, "MmWave massive-MIMO-based wireless backhaul for the 5G ultra-dense network," IEEE Wireless Communications, vol. 22, no. 5, pp. 13-21, 2015.

[54] E. Björnson, J. Hoydis, M. Kountouris, and M. Debbah, "Massive MIMO systems with non-ideal hardware: Energy efficiency, estimation, and capacity limits," IEEE Transactions on Information Theory, vol. 60, no. 11, pp. 7112-7139, 2014.

[55] E. Björnson, M. Matthaiou, and M. Debbah, "Massive MIMO with non-ideal arbitrary arrays: Hardware scaling laws and circuit-aware design," IEEE Transactions on Wireless Communications, vol. 14, no. 8, pp. 4353-4368, 2015.

[56] H. Insoo, S. Bongyong, and S. S. Soliman, "A holistic view on hyper-dense heterogeneous and small cell networks," Communications Magazine, IEEE, vol. 51, no. 6, pp. 20-27, 2013.

[57] T. L. Marzetta, "Noncooperative cellular wireless with unlimited numbers of base station antennas," IEEE Transactions on Wireless Communications, vol. 9, no. 11, pp. 3590-3600, 2010.

[58] Z. Jiang, A. F. Molisch, G. Caire, and Z. Niu, "Achievable rates of FDD massive MIMO systems with spatial channel correlation," IEEE Transactions on Wireless Communications, vol. 14, no. 5, pp. 2868-2882, 2015.

[59] T. E. Bogale and L. B. Le, "Massive MIMO and mmWave for 5G Wireless HetNet: Potential Benefits and Challenges," IEEE Vehicular Technology Magazine, vol. 11, no. 1, pp. 64-75, 2016.

[60] J. Liberti and T. S. Rappaport, Smart Antennas for Wireless Communications: IS-95 and Third Generation CDMA Applications. Prentice Hall, 1999.

[61] W. Roh et al., "Millimeter-wave beamforming as an enabling technology for 5G cellular communications: theoretical feasibility and prototype results," IEEE Communications Magazine, vol. 52, no. 2, pp. 106-113, 2014.

[62] F. Haider, C. Wang, B. Ai, H. Haas, and E. Hepsaydir, "Spectral-Energy Efficiency Trade-off of Cellular Systems with Mobile Femtocell Deployment," IEEE Trans. Veh. Technol., 2015.

[63] S. Jangsher and V. O. K. Li, "Resource Allocation in Moving Small Cell Network," IEEE Transactions on Wireless Communications, vol. 15, no. 7, pp. 4559-4570, 2016.

[64] E. Tanghe, W. Joseph, L. Verloock, and L. Martens, "Evaluation of Vehicle Penetration Loss at Wireless Communication Frequencies," IEEE Transactions on Vehicular Technology, vol. 57, no. 4, pp. 2036-2041, 2008.

[65] O. B. Karimi, J. Liu, and C. Wang, "Seamless Wireless Connectivity for Multimedia Services in High Speed Trains," IEEE Journal on Selected Areas in Communications, vol. 30, no. 4, pp. 729-739, 2012.

[66] E. Demir, T. Bektaş, and G. Laporte, "A comparative analysis of several vehicle emission models for road freight transportation," Transportation Research Part D: Transport and Environment, vol. 16, no. 5, pp. 347-357, 2011.

[67] M. M. Ahamed and S. Faruque, "5G Backhaul: Requirements, Challenges, and Emerging Technologies," in Broadband Communications Networks: Recent Advances and Lessons from Practice, A. Haidine, Ed.: Intech Open, 2018.

[68] D. Cohen, "What You Need to Know About 5G Wireless Backhaul ", ed: Ceragon, 2016.

[69] T. Elkourdi and O. Simeone, "Femtocell as a Relay: An Outage Analysis," IEEE Trans. Wireless Commun., vol. 10, no. 12, pp. 4204-4213, 2011.

[70] S. Jangsher and V. O. Li, "Resource allocation in cellular networks employing mobile femtocells with deterministic mobility," in IEEE Wireless Communications and Networking Conference (WCNC), 2013, pp. 819-824: IEEE.

[71] M. H. Qutqut, F. M. Al-Turjman, and H. S. Hassanein, "MFW: Mobile femtocells utilizing WiFi: A data offloading framework for cellular networks using mobile femtocells," in Communications (ICC), 2013 IEEE International Conference on, 2013, pp. 6427-6431: IEEE.

[72] S. Yutao, J. Vihriala, A. Papadogiannis, M. Sternad, Y. Wei, and T. Svensson, "Moving cells: a promising solution to boost performance for vehicular users," IEEE Commun. Mag., vol. 51, no. 6, pp. 62-68, 2013.

[73] M. Z. Chowdhury, S. H. Chae, and Y. M. Jang, "Group handover management in mobile femtocellular network deployment," in Ubiquitous and Future Networks (ICUFN), 2012 Fourth International Conference on, 2012, pp. 162-165: IEEE.

[74] S. Jangsher and V. Li, "Backhaul Resource Allocation for Existing and Newly-Arrived Moving Small Cells," IEEE Transactions on Vehicular Technology, vol. PP, no. 99, pp. 1-1, 2016.

[75] I. Ashraf, F. Boccardi, and L. Ho, “SLEEP mode techniques for small cell deployments,” Communications Magazine, IEEE, vol. 49, no. 8, pp. 72-79, 2011.

[76] F. Haider, C. Wang, B. Ai, H. Haas, and E. Hepsaydir, “Spectral-Energy Efficiency Trade-off of Cellular Systems with Mobile Femtocell Deployment,” Vehicular Technology, IEEE Transactions on, vol. PP, no. 99, pp. 1-1, 2015.

[77] A. Mesodiakaki, F. Adelantado, L. Alonso, and C. Verikoukis, “Energy-efficient user association in cognitive heterogeneous networks,” IEEE Communications Magazine, vol. 52, no. 7, pp. 22-29, 2014

[78] K. Young-Min, L. Eun-Jung, and P. Hong-Shik, “Ant Colony Optimization Based Energy Saving Routing for Energy-Efficient Networks,” Communications Letters, IEEE, vol. 15, no. 7, pp. 779-781, 2011.

[79] G. Auer et al., “How much energy is needed to run a wireless network?,” Wireless Communications, IEEE, vol. 18, no. 5, pp. 40-49, 2011.

[80] X. Chen, J. Wu, Y. Cai, H. Zhang, and T. Chen, “Energy-Efficiency Oriented Traffic Offloading in Wireless Networks: A Brief Survey and a Learning Approach for Heterogeneous Cellular Networks,” Selected Areas in Communications, IEEE Journal on, vol. 33, no. 4, pp. 627-640, 2015.

[81] S. Tombaz, S. Ki Won, and J. Zander, “On Metrics and Models for Energy-Efficient Design of Wireless Access Networks,” Wireless Communications Letters, IEEE, vol. 3, no. 6, pp. 649-652, 2014.

[82] G. Xiaohu, T. Song, H. Tao, L. Qiang, and M. Guoqiang, “Energy efficiency of small cell backhaul networks based on Gauss–Markov mobile models,” Networks, IET, vol. 4, no. 2, pp. 158-167, 2015.

12

5/6G, Smart Antennas and Coding the Algorithms: Linear ANN, Non-linear ANN, and LMS

H.M.C.J. Herath[1], H.M.G.G.J.G. Herath[1], K.M.U.I. Ranaweera[1], D.N. Uduwawala[1], and P.R.P. Hoole[2]

[1]Department of Electrical and Electronic Engineering, University of Peradeniya, Peradeniya, Sri Lanka
[2]Wessex Institute of Technology, Chilworth, Southampton, United Kingdom

Abstract

The demand for mobile communication is exponentially increasing in the 5/6G era with many novel technologies such as the Internet of Things (IoT), Industry 4.0, autonomous vehicles, smart cities, and smart healthcare. The need for better coverage, improved capacity, and higher transmission speeds is on the rise. The efficient usage of the radio frequency spectrum is one of the solutions to handle the increasing demand and technical constraints. Adaptive beamforming array antenna is a vital component of the above solution and it plays a key role in the 5G (2020) implementation and in the future 6G (2030) wireless system. This chapter presents both linear and non-linear antennas using perceptron ANN learning algorithms and its coding for adaptive beamforming. The perceptron ANN algorithm calculates the optimum weights of both linear and non-linear antenna arrays to steer the radiation pattern by directing multiple narrow beams toward the desired users and creating nulls toward interferers. The single-layer perceptron ANN is shown to give accurate beamforming with both minimum computational time and electronic memory. Four major types of activation functions are commonly used to steer the beam toward the desired direction. These are implemented in the algorithm for both linear and non-linear smart antenna arrays with different physical configurations. The MATLAB^{TM} codes for perceptron ANN algorithm to drive the smart antenna and that of the traditional Least Mean Square (LMS) algorithm driven antenna are given. The perceptron ANN driven smart antenna can perform adaptive beamforming at low computational cost while demonstrating good accuracy and fast convergence time.

Keywords: Artificial intelligence, smart antennas, beamforming, MIMO, 5G, 6G, ANN, perceptron learning.

12.1 Introduction

In recent years, developments in mobile communication technology have been observed as it evolves from 4G to fifth generation (5G) systems. Even though 4G has not been around for a very long time, because of the smartphone revolution, it is proving insufficient to handle the demands of traffic and potential applications in Internet of Things (IoT). Hence, 5G was designed with the primary goal of overcoming the limitations of 4G. From 1G to 4G wireless systems, network coverage strategies were optimized for one primary case: serving people with smartphones moving around. But with the new technologies such as IoT, it is not just smartphones anymore. It has been predicted that there will be billions of devices connected to the network. The requirement for better and reliable coverage is exponentially increasing with technology, which is developing at a rapid pace. Real-time critical services such as remote healthcare, patient monitoring, autonomous vehicles, and digital substation monitoring solely depend on the ability of the network to provide better and reliable coverage. One of the limiting factors that vendors and service providers face is the inefficient use of the wireless radiation environment. Hence, in recent years, the 5G systems have come to heavily depend on the availability and performance of intelligent or smart antennas, beamforming, and multiple input multiple output (MIMO) technologies to increase the effective usage of the radio spectrum at reduced energy.

12.1.1 Evolution of Mobile Communication System

The wireless telecommunication system has been through an exponential improvement from 1G to 5G. A new generation is named when it denotes a significant forward leap in wireless mobile technologies. The 4G wireless system works like the 3G and may be regarded as an extension of 3G but with faster internet connection, more bandwidth, and lower latency, and the 4G is about five times faster than 3G services. It used technologies such as coded orthogonal frequency division multiplexing (COFDM), MIMO array antennas, and link adaptation. There are two 4G systems. The United States developed Worldwide Interoperability for Microwave Access (WiMAX) system using orthogonal frequency-division multiplexing (OFDM), evolving from WiFi. The other 4G system is the long-term evolution (LTE) system that was developed after the WiMAX. The technologies of LTE and WiMAX are very similar. Thus, it can be stated that migrating from 3G to 4G meant a shift from low data rates for the internet to high-speed data rates for mobile video. The 5G (2021) wireless system is the fifth generation of mobile networks, a significant evolution from 4G networks. 5G has been designed to meet the very large growth in data and connectivity of the modern society. A very important advantage of 5G is the fast response time referred to as latency. Latency is the time taken for devices to respond to each other over the wireless network. 3G networks had a typical response time of 100 ms, 4G is around 30 ms, and 5G response time is said to be as low as 1 ms.

12.1.2 5G Technologies

To meet end-user requirements beyond 2020 and optimize networks upon future demands, mobile network operators (MNOs) have to find efficient solutions to enhance the Quality of Service (QoS), increase the spectrum efficiency, and maintain a healthy revenue To counter the traffic growth, build cost-efficient networks, and provide enhanced QoS to massive end-users, numerous architectures and technologies have been proposed for the 5G mobile networks. We can generally categorize them into the four categories as follows:

1) Improving the spectrum efficiency for higher data capacity by deploying advanced transmission techniques such as massive MIMO array antennas, millimeter wave (mmWave) transmission, and beamforming. Massive MIMO is an emerging technology for wireless communication systems. Featuring up to thousands of transmit/receive antennas, there is the possibility of creating extremely narrow beams for many users [6]. In a multi-user MIMO scenario, massive MIMO opens the possibility to steer many spatial streams to dozens of pieces of user equipment (UE) in the same cell, at the same frequency, and at the same time. The overall physical size of the 5G massive MIMO antennas will be

similar to 4G. However, with a higher frequency, the individual antenna element size is smaller, allowing more elements (above 100) in the same physical case.

2) Deploying small cells (picocells, femtocells, and microcells) combined with macrocells upon the existing network infrastructure. Small-cell deployments provide localized resources, filling coverage holes and maintaining service quality, allowing operators to follow traffic demands more closely, and use spectrum resources more efficiently, thus increasing network capacity. Small cells by strict definition are low-power wireless access points that operate in the licensed spectrum. They are operator-managed and provide improved cellular coverage, capacity, and applications for homes and enterprises as well as metropolitan and rural public spaces.
3) Applying emerging technologies of software defined network (SDN) and network function virtualization (NFV) to softwarize the networks. The main idea of SDN is to separate the data plane from the control plane and to introduce novel network control functionalities. Soft Air (a wireless SDN architecture) is the first comprehensive solution suite for 5G cellular systems that accelerates the innovations for both hardware forwarding infrastructure and software algorithms, enables efficient and adaptive resource sharing, achieves maximum spectrum efficiency, encourages the convergence of heterogeneous networks, and enhances energy efficiency. NFV, a complementary SDN concept, allows virtualization of hardware-related network functions before they are executed on a cloud-based infrastructure. More specifically, in conventional architectures, operators buy and install proprietary devices to implement each network function, while specialized hardware is generally very expensive but difficult to configure.
4) Optimizing and reconstructing the network architecture, specifically the RAN architecture. By bringing computation, communication, processing, and storage devices close to the edge of networks so that end-user scan accesses the data and services with low latency and high throughput. Moreover, integrate further services such as small data services as well as machine type communication (MTC) services.

12.1.3 5/6G, Health, and Environment

One of the main changes that come along with the 5G communication systems is the very dense network of antennas and transmitters. Hence, the number of higher frequency base stations and other devices will increase significantly. Therefore, the question raised is: "Will the new changes of 5G affect human health and environment negatively?"

Electromagnetic waves can be divided into two categories based on their frequency, namely ionizing and non-ionizing waves. Mid to extremely high-frequency waves such as UV rays, X-rays, and gamma rays are considered to be ionizing. These can cause cancer and other non-heating effects that may be adverse to human health. Non-ionizing radiation that operate using lower frequency waves, compared to the ionizing radiation, may still produce thermal effects or tissue heating. But at higher exposure levels, they can be harmful as well, considering the fact that thermal effects are not the only possible adverse impact of electromagnetic fields. Microwave and mmWave radiation is non-ionizing. But due to the high density of cell antennas in 5G implementation with constant exposure to mmWave radiation, it raises a variety of human health concerns and its impact on environment. Techniques like beamforming by producing sharp and focused radiation patterns targeting receiver devices can reduce the unnecessary exposure of humans to 5/6G mmWaves.

Non-ionizing radiation including mmWaves radiation is considered harmless in general due to the low power it carries, producing negligible tissue heating. But many scientists argue that a high level of pulsation of 5G can cause an issue regardless of the low power. Pulsed electromagnetic fields (EMP) may cause more biological interaction than non-pulsed electromagnetic fields. Therefore, 5G may have a negative effect on human health due to its constant abnormal pulse radiation. 5G signal pulsing may give rise to DNA damage resulting in cancer, reproductive decline, and neurodegenerative diseases. There is an ongoing debate among scientists about the adverse impact of wireless systems electromagnetic signals on health and environment.

12.1.4 Future 6G (2030) Wireless System

Although the 5G mobile communications standard is still in the early days of its development and testing, already the visioning of the 6G communication has begun (Dang *et al.*, 2020). The emphasis is to develop the 6G networks to be more human-centric compared to previous networks. Furthermore, the 6G network is set to provide high privacy and security. Many new communication techniques for the 2030s are predicted to encompass holographic calls, flying networks, teleoperated driving, and the tactile internet.

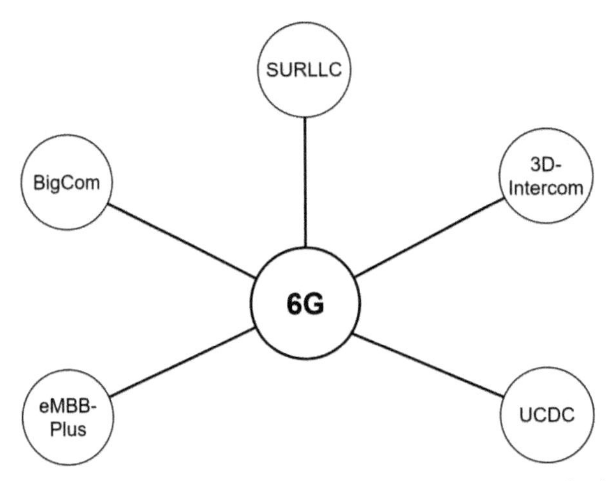

Figure 12.1 Application scenarios supported by 6G communications.

Discussions focus on practical implementations of 6G, including multiple access, air interfaces, and data centers for 6G communications. Networking patterns of 6G networks are to include cell-less architecture, decentralized resource allocation, and three-dimensional super-connectivity, as illustrated in Figure 12.1.

Targeted features of enabling communication technologies of 6G are given below.

1) Increased security and privacy: the use of higher frequencies to transmit signals is intended to improve security and privacy. Classic encryption based on RSA algorithms is being challenged by increasingly powerful computers. PHY security technologies and quantum key distribution via visible-light communications (VLC) could be the solutions to the data security challenge in the 6G network.
2) Increased affordability and full customization: 6G networks are expected to be highly affordable and fully customized. These should be two important technical indicators of 6G communications. Proposed schemes are expected to lead to higher transmission rate and reliability.
3) Less energy consumption and extended battery life: to overcome the daily battery charging requirement for most communication devices and to facilitate extensive communication services, less energy consumption and extended battery life are two targets set for 6G communications.

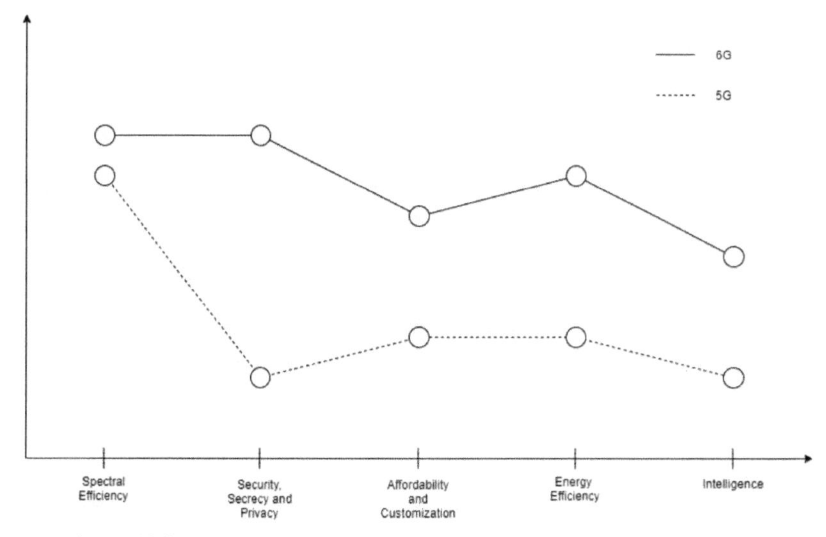

Figure 12.2 Qualitative comparison between 5G and 6G communications.

4) High intelligence: improved intelligence techniques such as operational intelligence, environmental intelligence, and service intelligence will benefit from the development of a 6G network.
5) Large bandwidth: it is proposed that the 0.1 to 10 THz frequency band be used for 6G, known as a gap band between the microwave and optical spectra.

   These futuristic 6G features are illustrated in Figure 12.2 in relation to the 5G system.

Crucial research issues relating to 6G beyond the communication technologies are as follows:

1) Dependency on basic sciences
2) Dependency on upstream industries
3) Business model and commercialization of 6G
4) Potential health and psychological issues for users
5) Social factors hindering the worldwide connectivity

12.1.5 Development of the Antenna System

The development of the antenna and antenna system is closely co-related to the increasing demand for network capacity. Omni-directional antennas are one of the simplest types of antennas which were used in the first generation wireless communication systems. They are simple dipole antennas, which radiate and receive equally in all the directions without considering the user's location. Due to this unfocused approach, signals are scattered, reaching the desired user with only a small percentage of the overall energy sent out to the environment. In an environment of multiple users, omni-directional antennas become very disadvantageous as they cannot selectively reject signals interfering with those of the user and has no spatial multi-path mitigation or equalization capabilities. Hence, omnidirectional strategies directly and adversely impact spectral efficiency, limiting frequency reuse.

Sectorized antenna systems with corner reflectors were introduced to overcome the disadvantages of omni-directional antennas. They are used to divide the cellular into sectors where each sector is managed by a directional antenna. In most modern communication systems, the base station consists of three directional antennas, each focused toward an arc of 120°. There are four and six sector antenna systems in use as well. It increases the frequency reuse by reducing the potential interferences across the cell, though increasing the complexity and weight of the mechanical structure and the number of transceivers, one for each sector.

With the exponential growth in the telecommunication industry and the increase of demand in recent years, the sectorized antennas are proving to be inefficient. Hence, the telecommunication industry started to move toward smart antenna systems. Smart antenna systems have been around since the late 1950s and were mainly used by the military. And the rapid growth of smart antennas occurred due to the development of powerful, low-cost digital signal processors, general-purpose processors, and software-based signal processing techniques.

Digital signal processing makes the antenna array smart enabling it to radiate and receive signals in an adaptive, spatially sensitive manner. Smart antennas are capable of identifying the desired user's location using the direction of arrival (DoA) estimation (see Chapter 7) and focusing the radiation pattern toward the desired user (see Chapters 5, 8, and 9) while nulling the interferences. Therefore, smart antennas have a higher signal-to-noise ratio compared to its predecessors. Smart antenna systems can be categorized into two main categories, as shown in Figure 12.3:

1) Switched beam: consists of a finite number of pre-defined beam patterns and the beams are switched according to the desired user's behavior.

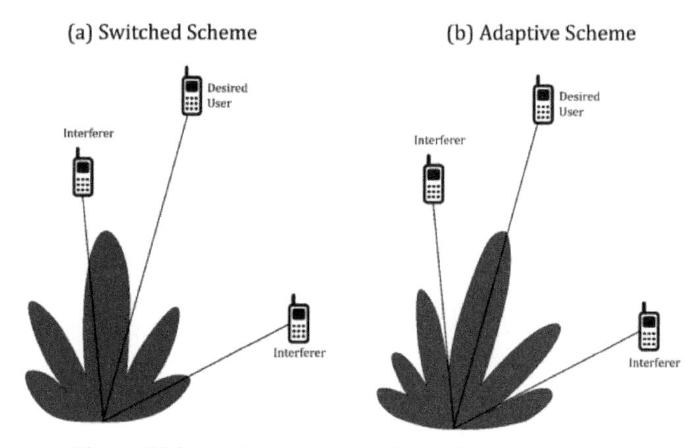

Figure 12.3 Switched beam and adaptive array schemes.

2) Adaptive array: the beam pattern is changed adaptively according to the desired user's behavior. It is more efficient than switched beam antennas since the adaptive behavior helps to avoid more interferences while achieving continuously maximum directivity toward the desired mobile user.

12.1.6 The Goals of the Smart Antenna System

The dual purpose of a smart antenna system is to augment the signal quality of the radio-based system through a more focused transmission of radio signals while enhancing capacity. The features of smart antenna systems may be summarized as follows:

1) Signal gain: inputs from multiple antennas are combined to optimize available power required to establish a given level of coverage.
2) Interference rejection: antenna pattern can be nulled toward interference sources, improving the signal-to-interference ratio of the received signals. On the reverse link or uplink, this reduces the interference seen by the base station. It also reduces the amount of interference spread in the system forward link or downlink. Such improvements in the carrier-to-interference ratio increases capacity.
3) Spatial diversity: composite information from the array is used to minimize fading and other undesirable effects of multi-path propagation.
4) Power efficiency: combines the inputs to multiple elements to optimize available processing gain in the downlink.

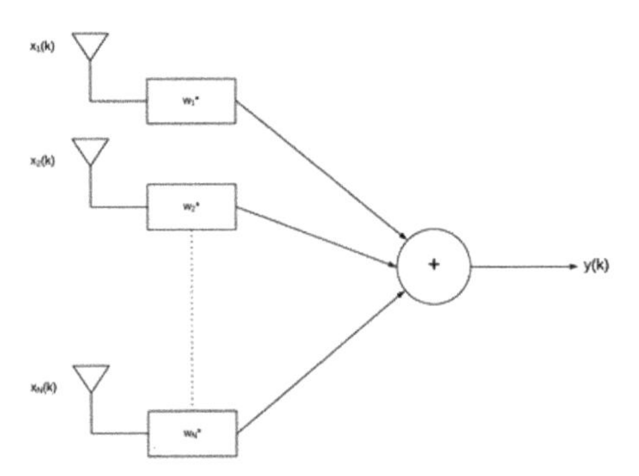

Figure 12.4 Schematic diagram of fixed weight beamformer.

12.1.7 Beamforming

Beamforming is the signal processing technique used in a smart antenna array system to form the radiation pattern by adding individual radiation pattern of each element in a way that they concentrate the energy into a narrow beam directed toward the desired user. Fixed weight beamforming is used when the DoA does not change with time. On the other hand, the adaptive algorithm continuously updates the weights according to the varying DoA. There are two kinds of antenna beamformers: the fixed weight beamformer (Figure 12.4) and the adaptive beamformer (Figure 12.5).

12.1.7.1 Fixed Weight Beamformer

A fixed weight beamformer is a smart antenna in which fixed weight is used to receive the signal arriving from a specific direction. It is called a spatially matched filter since it optimizes the signal arriving from a specific direction while attenuating signals from other directions. Fixed weight beamformers perform well given that the mobile user is stationary, but they are unable to adjust the weights once the mobile user starts moving.

12.1.7.2 Adaptive Beamformer

In the fixed weight beamforming approach, the arrival angles do not change with time; therefore, the optimum weight need not be adjusted. If the desired arrival angles change with time, it is necessary to devise an optimization scheme that operates dynamically to keep recalculating the optimum array weight. This is done using the adaptive beamforming algorithm. The task of the algorithm on a smart antenna system is to adjust the received signals

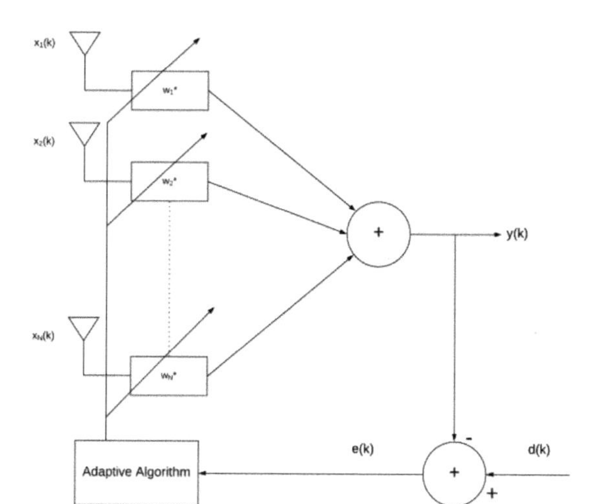

Figure 12.5 Schematic diagram of adaptive beamformer.

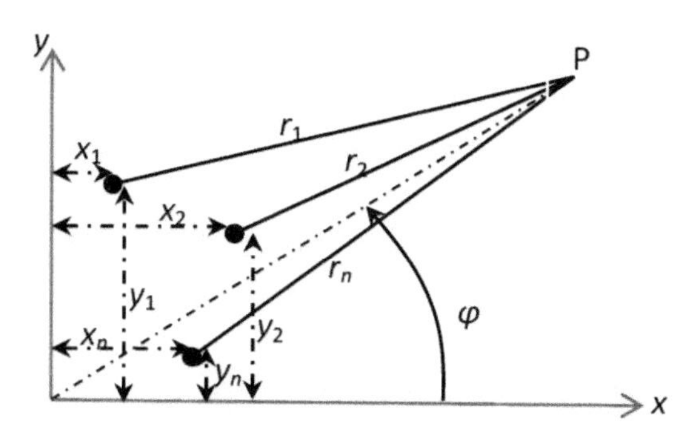

Figure 12.6 Schematic diagram of the array model [14].

so that the desired signals are effectively extracted once the signals are combined, as illustrated in Figure 12.5.

12.2 Smart antenna using ANN

In this section, we shall discuss the ability of a single neuron artificial neural network (ANN) to optimize weights of linear and non-linear arrays to obtain the desired radiation pattern of an adaptive array antenna. We shall focus on four types of beam patterns which are used to train the perceptron ANN weights and to maximize the beam in the direction of the desired user.

12.2.1 Adaptive Array Model

The electric field at the far-field observation field for the antenna element arrangement shown in Figure 12.6 can be expressed as

$$E = A_0.I_1.e^{-j\beta r_1} + A_0.I_2.e^{-j\beta r_2} + \cdots + A_0.I_n.e^{-j\beta r_n}$$

where A_0 is a constant magnitude and β is a phase constant. Expressing the electric field in terms of the weights and two-dimensional Cartesian coordinates, we get

$$E = w_1.e^{j\beta(x_1.\cos\theta + y_1.\sin\theta)} + w_2.e^{j\beta(x_2.\cos\theta + y_2.\sin\theta)} + \ldots + w_n.e^{j\beta(x_n.\cos\theta + y_n.\sin\theta)} \tag{12.1}$$

12.2.2 Single Perceptron Weight Optimization

In the perceptron ANN model, a single neuron is used with a non-linear activation function. The output is given by the activation function f

$$y = f\left(\sum_{i=1}^{N} w_i.x_i + b\right)$$

where w_i are the weights, x_i are the input signals, and b is the bias applied to the neuron. Once the output of the perceptron ANN model is calculated, the deviation of the calculated signal pattern from the desired signal pattern is calculated such that

$$\varepsilon = y_{calculated} - y_{desired}.$$

From the calculated error, weights are updated at each iteration using

$$w(k+1) = w(k) + \lambda.\varepsilon.x(k)$$

where λ is the learning rate which determines the step size of the adjustment applied to the weights at each iteration. The weight adjustment is continued until the error is below a pre-defined value or the pre-defined maximum number of iterations is reached.

12.2.3 Activation Functions

The activation function is an important part of an ANN. They basically decide whether a neuron should be activated or not. Thus, it bounds the value of the

ANN input. The activation function is a non-linear transformation that we do over the input before sending it to the next layer of neurons or finalizing it as output.

Sigmoid Activation Function: Sigmoid takes a real value as input and the output is another value between 0 and 1. It is easy to work with and has all the nice properties of activation functions: it is non-linear, continuously differentiable, monotonic function, and has a fixed output range. The advantages of the sigmoid function include non-linearity in nature. Combinations of this function are also non-linear; it will give an analog activation. Unlike the step function, it has a smooth gradient too, and it is good for a classifier and the output of the activation function is always going to be in the range (0,1) compared to (−infinity, +infinity) of a linear function. So, we have our activations bound within a range. The disadvantages of the sigmoid function include: toward either end of the sigmoid function, the output values are less responsive to changes in the input, it gives rise to the problem of "vanishing gradients," its output is not zero centered, it makes the gradient updates go too far in different directions (in the range of $0 < \text{output} < 1$), and it makes optimization harder, The sigmoid function saturates and kills gradients, and the network refuses to learn further or is drastically slow (depending on the case and until gradient computation gets hit by floating-point value limits).

The sigmoid function is defined by

$$f(z) = \frac{1}{1 + e^{-z}}.$$

Bipolar Sigmoid Activation Function. The bipolar sigmoid activation function is a variation of the sigmoid activation function where the output values are in the range of $[-1, 1]$. This activation function is a good choice to train data between 0 and 1. It is defined by

$$f(z) = \frac{1 - e^{-z}}{1 + e^{-z}}.$$

Hyperbolic Tangent Activation Function. The advantage of the hyperbolic tangent activation function is that the gradient is stronger for tanh than sigmoid (i.e. the derivatives are steeper). It is defined by

$$\tanh(z) = \frac{e^{z} - e^{-z}}{e^{z} + e^{-z}}.$$

12.3 Smart Antenna CODES: Linear/Non-linear ANN AND LMS

12.3.1 The ANN Codes: Linear and Non-Linear ANN

In this section, an explanation of the MATLAB^{TM} codes used for linear and non-linear arrays for the perceptron ANN driven smart antenna is given.

```
1   clear all;
2   close all;
3
4   %*********************** Parameters ***********************
5
6   theta_train = 0:0.01:2*pi; %training set
7   N = 5; %no of elements of the array
8   c = 3*10^8; %speed of the wave
9   f = 30*10^9; %frequency
10  lamda = c/f; %wave length
11  d = 1*lamda/2; %inter element spacing
12  k = 2*pi/lamda; %wave number
13
14  theta_i = pi/6; %interferer angle
15  theta_d = 0; %desired direction
16
17 %****************** Major Beam Types ********************
18
19  y = cos(2*(theta_train - theta_d));
20  %y = sin((pi/2)*cos(theta_train - theta_d));
21  %y = cos((pi/4)*(cos(theta_train - theta_d)-1));
22  %y = sinc(theta_train - theta_d);
23
24  %************* Weight and Bias Initialization ****************
25
26  w = ones(1, N); %initial weights
27  w = 0.05.*w;
28
29  %***************** Steering Vector ***********************
30
31  x = zeros(length(theta_train), length(N));
32
```

```
33  for t = 1:N
34      x(:,t) = exp(j*k*d*(t-(0.5*(N+1)))*cos(theta_train));
35  end
36
37
38  %************** Training Parameters ************************
39
40  lr = 0.0001; %learning rate
41  eps = 70; %no of epochs
42
43  Delta = zeros(length(eps), 1);
44  Ep = zeros(length(eps), 1);
45  delta = 1;
46  ep = 0;
47
48  %*********************** Training *************************
49
50  for i = 1:length(x)
51      delta = 1
52      ep = 0;
53      while (ep<=eps)
54          y_cal = (x(i,:)*w') ; %calculated output for the current angle
55
56          activation = tanh(y_cal); %hyperbolic activation
57          %activation = (1 - exp(-y_cal))/(1 + exp(-y_cal)); %bipolar
        sigmoid activation
58
59          delta = (y(i) - activation); %error
60
61          for j = 1:length(w)
62              w(j) = w(j) + (lr*delta*x(i,j)); %weight update
63          end
64
65          ep = ep + 1;
66          disp(ep);
67          disp(delta);
68
69      end
70      disp(delta);
```

```
71
72  end
73
74  save weights w
75  disp(w);
76 %*******************Testing ********************************
77  Y_new = abs(x*w'); %Radiation pattern using optimized weights
78  %********************** Visualization **********************
79  figure(1)
80  polarplot(theta_train, abs(y), '--k', 'LineWidth' ,2);
81  hold on
82  polarplot(theta_train, abs(Y_new), '--r', 'LineWidth' ,2,
    'LineStyle','-');
83   title(['Radiation pattern of a ',num2str(N) ' -element linear array']);
84  legend('Desired Pattern', 'Obtained Pattern');
```

In the above given code, from lines 6–15, array antenna parameters such as the number of elements, inter-element spacing, as well as the training set are specified. Angles between 0 and 2??with a step of 0.01 are taken as the training set. Lines 15 and 16 represent the interference angle and the desired mobile station MS angle, respectively. The four major beam patterns to which the perceptron ANN is trained are being specified in lines 19–22. Lines 26 and 27 display the initialization of weights, while the steering vector of the array is calculated in lines 31–35. Training parameters such as learning rate and the number of maximum iterations permitted for convergence are specified in lines 40–46.

The training of the perceptron ANN smart antenna is carried out in lines 50–74. For each iteration, the training procedure is done for all the samples in the training set and the weights are updated accordingly for each sample. The weights are saved once the error has converged. The two activation functions used in the training are specified in lines 56–57. Finally, the optimized weights are tested and the results are plotted.

The only difference in the codes for linear and non-linear cases is in the generation of the steering vector. See Chapters 5 and 9 for further details. The steering vector for a six-element non-linear array is generated with the code given below.

```
1 %*****************Steering Vector ******************************
2
3   x1 = exp(j*k*(d*sin(theta_train)))';
4   x2 = exp(j*k*((-d*sin(pi/3)*cos(theta_train))+(d*0.5*sin(theta_train))))';
5   x3 = exp(j*k*((-d*sin(pi/3)*cos(theta_train))-(d*0.5*sin(theta_train))))';
6   x4 = exp(j*k*(-d*sin(theta_train)))';
7   x5 = exp(j*k*((d*sin(pi/3)*cos(theta_train))-(d*0.5*sin(theta_train))))';
8   x6 = exp(j*k*((d*sin(pi/3)*cos(theta_train))+(d*0.5*sin(theta_train))))';
9
10  x = [x1 x2 x3 x4 x5 x6];
```

12.3.2 The Least Mean Square Code

The main difference between perceptron ANN smart antenna algorithm and the least mean square (LMS) algorithm is the frequency of the weight updates. In perceptron ANN algorithm, the angles are processed one by one and the weights are updated for each training sample. But in LMS algorithm, the angles are processed as a batch and the weights are updated once per each iteration. The following is a code for the LMS algorithm; it is for a four-element linear array:

```
1   %************************ Parameters ***********************
2
3   theta_train = 0:0.01:2*pi; %training set
4   N = 4;  %no of elements of the array
5   c = 3*10^8; %speed of the wave
6   f = 30*10^9; %frequency
7   lamda = c/f; %wave length
8   d = 1*lamda/2; %inter element spacing
9   k = 2*pi/lamda; %wave number
10
11  theta_i = pi/6; %interferer angle
12  theta_d = 0; %desired direction
13
14  %**** Radiation Patterns for Directivity increase and Interference
Suppression**********
15
16  AF_i = (k*d/2)*(cos(theta_train) - cos(theta_i))+(pi/N); %%half of the
phase deviation when the interferer angle is known
17  AF_d = (k*d/2)*(cos(theta_train) - cos(theta_d)); %half of the phase
deviation when the desired angle is known
18  %y = (1/N).*(sin(N*AF_d)./sin(AF_d)); %AF when the desired angle is
known
```

```
19  %y = (1/N).*(sin(N*AF_i)./sin(AF_i)); %AF when the interferer angle is
known
20
21%**********Major beam types ************************************
22
23  y = cos(2*(theta_train - theta_d));
24  %y = sin((pi/2)*cos(theta_train - theta_d));
25  %y = cos((pi/4)*(cos(theta_train - theta_d)-1));
26  %y = sinc(theta_train - theta_d);
27
28 %********************Weight and Bias Initialization ******************
29
30  w = ones(1, N); %initial weights
31  w = 0.05.*w;
32  b = 1; %initial bias
33
34 %***********************Steering Vector **************************
35
36  x1 = exp(j*k*(d/sqrt(2))*cos(theta_train))';
37  x2 = exp(j*k*(d/sqrt(2))*sin(theta_train))';
38  x3 = exp(-j*k*(d/sqrt(2))*cos(theta_train))';
39  x4 = exp(-j*k*(d/sqrt(2))*sin(theta_train))';
40
41  x = [x1 x2 x3 x4];
42 %***************Training Parameters ******************************
43
44  lr = 0.0001; %learning rate
45  eps = 150; %no of epochs
46
47  Delta = zeros(length(eps), 1);
48  Ep = zeros(length(eps), 1);
49  delta = 1;
50  ep = 0;
51
52  while ep$\mathrm{<}$eps
53     y_cal = x*w'; %calculated output
54     e = y'-y_cal; %error
55     w = w + lr*e'*x; %weight update
56     ep = ep + 1;
57     disp(ep)
58     disp(e)
59     disp(w)
60
61  end
```

```
62
63
64  disp(w);
65  Y_new = x*w';
66  polarplot(theta_train, abs(y), '--k', 'LineWidth' ,2);
67  hold on
68  polarplot(theta_train, abs(Y_new), '--r', 'LineWidth' ,2, 'LineStyle','-');
69
70
71  legend('Desired Pattern', 'Obtained Pattern');
```

12.4 Results and discussion

12.4.1 Linear Array Smart Antenna

The antenna beamforming code was tested for the following radiation beams specified as the desired beams to which the ANN beamformer weights have to be trained:

1) $\cos 2\left(\theta - \theta_{desired}\right)$
2) $\sin\frac{\pi}{2}\cos\left(\theta - \theta_{desired}\right)$
3) $\cos(\frac{\pi}{4}\left(\cos\left(\theta - \theta_{desired}\right) - 1\right)$
4) $sinc\left(\theta - \theta_{desired}\right)$

Selected results are given to illustrate the accuracy of the linear and non-linear beamformer. Consider the following desired antenna beam for which the ANN beamformer is to be trained:

$$\mathbf{\cos 2\left(\boldsymbol{\theta} - \boldsymbol{\theta_{desired}}\right)}.$$

It produces a radiation pattern with four major antenna radiation lobes perpendicular to each other, which can be used at a four-way road or tunnel junction. The smallest possible configuration of a linear array that is capable of producing this radiation pattern is a two-element array with an inter-element spacing of λ. From Figure 12.7, it can be seen that the radiation pattern obtained is narrower than the desired pattern in the end-fire direction and wider in the broadside direction for both tanh and bipolar sigmoid activation functions.

For a three-element array antenna, the perceptron ANN smart antenna obtained good results when the inter-element spacing is less than or equal to $\lambda/2$. In both cases when $d = \lambda/4$ and $d = \lambda/2$, the ANN generated radiation pattern amplitudes in 0° and 180° are small compared to the desired radiation pattern in these two directions. As seen in Figure 12.7, the beams become

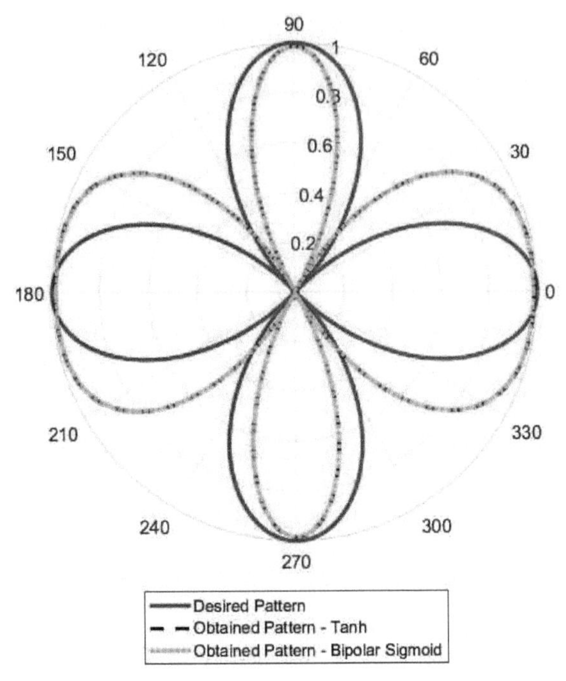

Figure 12.7 Comparison of radiation pattern between optimized beams and desired beam when the number of adaptive array elements is 2.

narrower in the broadside direction and wider in the end-fire direction when the inter-element spacing d is increased. For inter-element spacing of $d = \lambda/4$ with a five-element array, the resulting beams were similar to that with a three-element array. But when the spacing is increased to $\lambda/2$, the five-element ANN radiation pattern obtained with the tanh activation function displayed better desired characteristics than the pattern obtained using the bipolar sigmoid activation function.

Consider the desired antenna beam given by the following equation:

$$\sin\frac{\pi}{2}\cos\left(\theta - \theta_{desired}\right).$$

This radiation pattern produces two major lobes opposite to each other. The perceptron ANN smart antenna produced a beam that closely matches the desired radiation pattern using a two-element array, for both inter-element spacing of $\lambda/4$ and $\lambda/2$. As shown in Figure 12.8, the beamwidth is narrower when the inter-element spacing is $\lambda/4$ and perfect match is obtained when

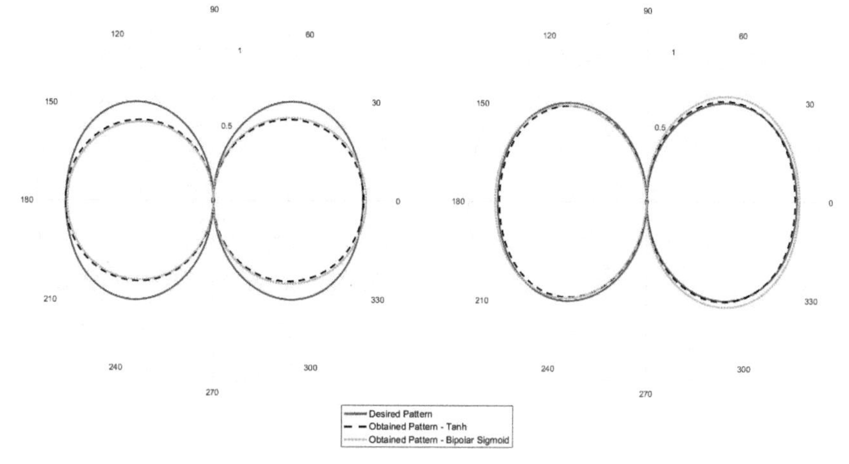

Figure 12.8 Comparison of radiation pattern between optimized beams and desired beam when the number of adaptive array elements is 2; $d = \lambda/4$ (left) and $d = \lambda/2$ (right).

the spacing d is $\lambda/2$. When the number of elements was increased to 3, good agreement was found only for an inter-element separation d of $\lambda/4$.

Third, we consider the following desired beam for which the perceptron ANN beamformer is to be trained:

$$\cos(\frac{\pi}{4}\,(\cos\left(\theta - \;\theta_{desired}\right) - 1)\;.$$

This radiation pattern can be used to cover a half-plane or single large area, as, for instance, for wireless communication from land to sea. The perceptron ANN weight optimization produced matching radiation beams for two-, three-, and five-element arrays, with an inter-element spacing of $\lambda/4$, as shown in Figure 12.9. The radiation power in the end-fire direction gets smaller due to destructive inter-element electromagnetic wave interference.

12.5 Non-Linear Array Results

12.5.1 Non-Linear Array Smart Antenna

Due to the symmetric nature of the radiation patterns obtained from linear arrays, it is impossible to generate single beam radiation patterns as well as to achieve greater adaptive behavior. Hence, the non-linear arrays are introduced to overcome the short coming of linear arrays. For further discussion on non-linear smart antennas, for instance, with the antenna elements placed on

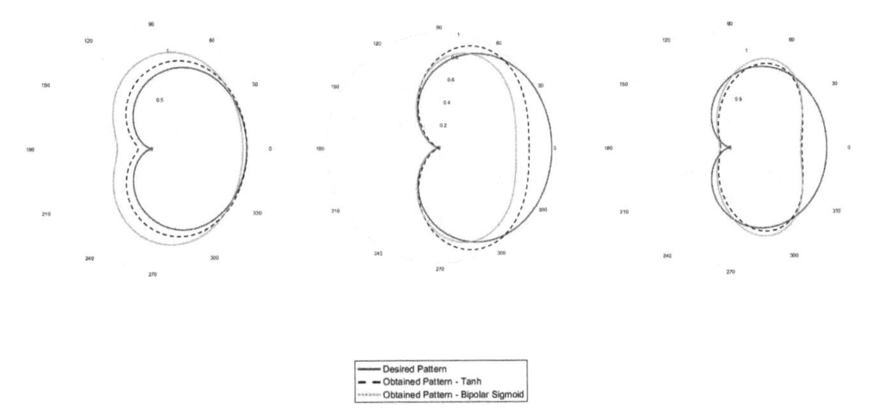

Figure 12.9 Comparison of radiation pattern between optimized beams and desired beam when $d = \lambda/4$, the numbers of adaptive array elements is 2 (left), 3 (middle), and 5 (right).

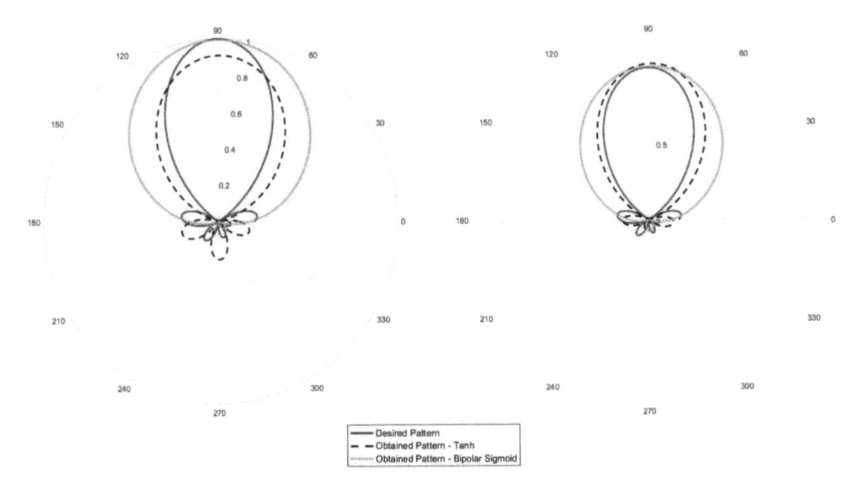

Figure 12.10 Comparison of radiation pattern between optimized beams and desired beam when the number of adaptive non-linear array elements is 4; $d = \lambda/4$ (left) and $d = \lambda/2$ (right).

the corners of a triangle for a three-element non-linear antenna, the reader is referred to Chapters 5 and 9. We consider the ANN antenna beam given by the sinc function, for which the non-linear array must be used to get the desired, single smart antenna beam with the trained weights:

$$sinc\left(\theta - \theta_{desired}\right).$$

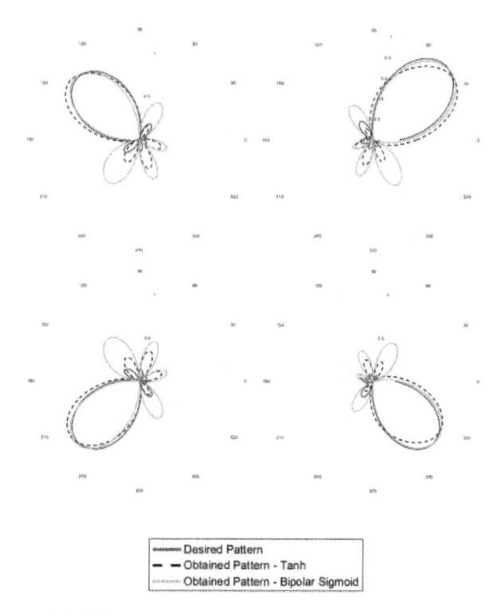

Figure 12.11 Beam rotation of six-element array.

One of the main objectives of using non-linear arrays is to obtain single beam radiation patterns which can be used to increase the directivity and reduce interferences drastically. Implementing the sinc function is a useful method to obtain a single beam pattern.

Both six- and four-element arrays displayed similar behavior when the weights are optimized using the single perceptron ANN weights, as shown in Figure 12.10. The beamwidth is larger when the bipolar sigmoid activation function is used for the array configuration with d set to $\lambda/4$. The performance can be improved by changing the inter-element spacing to $\lambda/2$. The area covered by the radiation pattern obtained with the tanh activation function is small compared to the desired pattern. Sidelobes become prominent in the pattern obtained using the bipolar sigmoid activation function.

In Figure 12.11 is shown how the single beam radiation pattern obtained from a six-element non-linear array smart antenna changes its pattern according to the desired or targeted user position when optimized with a non-linear perceptron ANN smart antenna.

Bibliography

In writing this book, the books and papers listed below were consulted. The reader should refer to them for further details or for a different perspective on the subject.

A. Abdelnasser, E. Hossain, and D. I. Kim, "Clustering and Resource Allocation for Dense Femtocells in a Two-Tier Cellular OFDMA Network," *IEEE Transactions on Wireless Communications,* vol. 13, no. 3, pp. 1628-1641, 2014.

M. Agiwal, A. Roy, and N. Saxena, "Next generation 5G wireless networks: A comprehensive survey," *IEEE Communications Surveys & Tutorials,* vol. 18, no. 3, pp. 1617-1655, 2016.

P. Agyapong, M. Iwamura, D. Staehle, W. Kiess, and A. Benjebbour, "Design considerations for a 5G network architecture," *Communications Magazine, IEEE,* vol. 52, no. 11, pp. 65-75, 2014.

M. M. Ahamed and S. Faruque, "5G Backhaul: Requirements, Challenges, and Emerging Technologies," in *Broadband Communications Networks: Recent Advances and Lessons from Practice*, A. Haidine, Ed.: Intech Open, 2018.

A. Alkhateeb, J. Mo, N. Gonzalez-Prelcic, and R. W. Heath, "MIMO precoding and combining solutions for millimeter-wave systems," *IEEE Communications Magazine,* vol. 52, no. 12, pp. 122-131, 2014.

E. Ali, Ismael M., Nordin R. and Nor F. A."Beamforming techniques for massive MIMO systems in 5G:overview, classification, and trends for future research,"Frontier of Information Technology & Electronic Engineering, Vol. 18, Issue 6, pp. 753, 2017

A. Alnoman and A. Anpalagan, "Towards the fulfillment of 5G network requirements: technologies and challenges," *Telecommunication Systems,* pp. 1-16, 2016.

M. H. Alsharif and R. Nordin, "Evolution towards fifth generation (5G) wireless networks: Current trends and challenges in the deployment of millimetre wave, massive MIMO, and small cells," *Telecommunication Systems,* pp. 1-21, 2016.

I. Ashraf, F. Boccardi, and L. Ho, "SLEEP mode techniques for small cell deployments," *Communications Magazine, IEEE,* vol. 49, no. 8, pp. 72-79, 2011.

P. Ameigeiras, J. J. Ramos-Munoz, L. Schumacher, J. Prados-Garzon, J. Navarro-Ortiz, and J. M. Lopez-Soler, "Link-level access cloud architecture design based on SDN for 5G networks," *IEEE Network,* vol. 29, no. 2, pp. 24-31, 2015.

C.R. Anderson, and Rappaport T.S, "In-building wideband partition loss measurements at 2.5 and 60 GHz," IEEE Trans. Wireless Communication, Vol. 3, No. 3, pp.922-928, 2004.

J. G. Andrews *et al.*, "What Will 5G Be?," *IEEE Journal on Selected Areas in Communications,* vol. 32, no. 6, pp. 1065-1082, 2014.

G. Auer *et al.*, "How much energy is needed to run a wireless network?," *Wireless Communications, IEEE,* vol. 18, no. 5, pp. 40-49, 2011.

M.D. Austin, Velocity adaptive handoff algorithms for microcellular systems, *IEEE Trans. Vehicular Technology*, **43**, 549–560, 1994.

B. Bangerter, S. Talwar, R. Arefi, and K. Stewart, "Networks and devices for the 5G era," *IEEE Commun. Mag.,* vol. 52, no. 2, pp. 90-96, 2014.

E. Ben-Dor, Rappaport. T.S, Qiao. Y, and Lauffenburger. S.J, "Millimeter-wave 60 GHz outdoor and vehicle AOA propagation measurements using a broadband channel sounder," in Proc. IEEE Global Telecomm. Conf. (Globecom), 2011, pp. 1-6.

J. Bae, Choi Y.S, Kim J.S, and Chung M.Y, "Architecture and performance evaluation of mmWave based 5G mobile communication systems," in Proceedings of International Conference on Information and Communication Technology Convergence (ICTC), 2014, pp. 847-851

R. Cichheti, A. Faraoane, D. Caretelli, M. Simeoni, Wideband, Multiband, Tunabe and Smart Antenna Systems, for UWB and Mobile Wireless Systems, Int J of Antennas and Propagation, Hindawi, 2017.

C.A. Balanis, *Antenna Theory*, 2nd edn, John Wiley, New York, 1997.

C.A. Balanis,. and L.M. Claypool, *Panayiotis: Introduction to Smart Antennas (Synthesis Lectures on Antennas).* 2007.

M. R. Bhalla, and Anand Vardhan Bhalla. "Generations of mobile wireless technology: A survey." International Journal of Computer Applications. Vol. 5, No. 4, 2010

ADRF, Bingo, COMMSCOPE, and Z. Networks, "DAS and Small Cells Evolve to Meet Today's and Tomorrow's Future Connectivity Needs," Netnet Forum, Arlington2019.

E. Björnson, J. Hoydis, M. Kountouris, and M. Debbah, "Massive MIMO systems with non-ideal hardware: Energy efficiency, estimation, and capacity limits," *IEEE Transactions on Information Theory,* vol. 60, no. 11, pp. 7112-7139, 2014.

E. Björnson, M. Matthaiou, and M. Debbah, "Massive MIMO with non-ideal arbitrary arrays: Hardware scaling laws and circuit-aware design," *IEEE Transactions on Wireless Communications,* vol. 14, no. 8, pp. 4353-4368, 2015.

D. W. Bliss, Keith W. Forsythe, and Amanda M. chan "MIMO Wireless Communication", Lincoln Labt. Journal, Vol. 15, No. 1, 2005

F. Boccardi, Heath R.W, Lozano A, Marzetta T.L, and Popovski P, "Five disruptive technology directions for 5G," IEEE communications Magazine, Vol. 52, No. 2, pp. 74-80, 2014.

S.K. Bodhe, Hogade B.G. and Shailesh D. Nandgaonkar, "Beamforming Techniques for Smart Antennas using Rectangular Array Structure," *International Journal of Electrical and Computer Engineering (IJECE)*, vol.4, no.2, pp. 257-264, 2014.

T. E. Bogale and L. B. Le, "Massive MIMO and mmWave for 5G Wireless HetNet: Potential Benefits and Challenges," *IEEE Vehicular Technology Magazine,* vol. 11, no. 1, pp. 64-75, 2016.

W.G. Carrara, R.S. Goodman and A. Majewski, *Spotlight Synthetic Aperture Radar Signal Processing Algorithms*, Artech House, Boston, 1995.

P. Cardieri and Rappaport T.S, "Application of narrow-beam antennas and fractional loading factor in cellular communication systems," IEEE Trans. Veh. Technol., Vol. 50, No. 2, pp. 106-113, 2014.

L. Castedo, An adaptive beamforming technique based on cyclostationary signal properties, *IEEE Trans. Signal Processing*, **43**, 1637–1650, 1995.

Chan Teck Chi and Ngo Ken Hui, *An Alternative Method of Image Compression and Restoration*, School of Electrical and Electronic Engineering, Nanyang Technological University, Singapore, Report, March 1999.

A. Checko *et al.*, "Cloud RAN for mobile networks—A technology overview," *IEEE Communications surveys & tutorials,* vol. 17, no. 1, pp. 405-426, 2015.

C.N. Chen and D.I. Hoult, *Biomedical Magnetic Resonance Technology*, Adam-Hilger, Bristol, 1989.

S. Chen and Zhao J., "The requirements, challenges, and technologies for 5G of terrestrial mobile telecommunications," IEEE Communications Magazine, Vol. 52, No. 5, pp. 36-43, 2014.

N. Chen., Kadoch M. and Rong B." SDN Controlled mmWave Massive MIMO Hybrid Precoding for 5G

Heterogeneous Mobile Systems,"Hindawi Publishing Corp., 2016, pp.4

M. Chen, Y. Zhang, L. Hu, T. Taleb, and Z. Sheng, "Cloud-based Wireless Network: Virtualized, Reconfigurable, Smart Wireless Network to Enable 5G Technologies," *Mobile Networks and Applications,* journal article vol. 20, no. 6, pp. 704-712, 2015

X. Chen, J. Wu, Y. Cai, H. Zhang, and T. Chen, "Energy-Efficiency Oriented Traffic Offloading in Wireless Networks: A Brief Survey and a Learning Approach for Heterogeneous Cellular Networks," *Selected Areas in Communications, IEEE Journal on,* vol. 33, no. 4, pp. 627-640, 2015.

W. Cheng, Zhang. X, and Zhang. H, "RTS/FCTS mechanism based full-duplex MAC protocol for wireless networks," in Proc. Global Commun. Conf. (GLOBECOM), 2013, pp. 5017-5022.

W. Cheng-Xiang *et al.*, "Cellular architecture and key technologies for 5G wireless communication networks," *IEEE Commun. Mag.,* vol. 52, no. 2, pp. 122-130, 2014.

W. H. Chin, Z. Fan, and R. Haines, "Emerging technologies and research challenges for 5G wireless networks," *IEEE Wireless Communications,* vol. 21, no. 2, pp. 106-112, 2014.

S. Choi, Design of an adaptive antenna array for tracking the source of maximum power and its application to CDMA mobile communications, *IEEE Trans. Antennas Propagation*, **45**, 1393–1404, 1997.

R.R. Choudhury, Yang. X, Ramanathan. R, and Vaidya. N.H, "On designing MAC protocols for wireless networks using directional antennas," IEEE Trans. Mobile Comput., Vol. 5, No. 5, pp. 477-491, 2006

M. Z. Chowdhury, S. H. Chae, and Y. M. Jang, "Group handover management in mobile femtocellular network deployment," in *Ubiquitous and Future Networks (ICUFN), 2012 Fourth International Conference on*, 2012, pp. 162-165: IEEE.

H. Chu, P. Crist, S. Han, J. Pourbaix, and Y. Kawashima, "Funding Urban Public Transport: Case Study Compendium," in *International Transport Forum Summit on Funding Transport*, Leipzig, 2013.

A.C. Cirik, Rong. Y, and Hua. Y, "Achievable rates of full-duplex MIMO radios in fast fading channels with imperfect channel estimation," IEEE Trans. Signal Process., Vol 62, No. 15, pp. 3874-3866, 2014.

CISCO, "Cisco Visual Networking Index: Global Mobile Data Traffic Forecast Update, 2016-2021," Feb 2017.

D. Cohen, "What You Need to Know About 5G Wireless Backhaul ", ed: Ceragon, 2016.

M. Coldrey, U. Engstrom, K. W. Helmersson, M. Hashemi, L. Manholm, and P. Wallentin, "Wireless backhaul in future heterogeneous networks," *Ericsson Review,* vol. 91, 2014.

S. Collonge, Zhharia G, and Zein G.E, "Influence of the human activity on wide-band characteristics of the 60 GHz indoor radio channel," Vol. 3, No. 6, pp.2396-2406, 2004.

Comparison of speed 2G to 4G.Adapted from"3G VS 4G Technology". Retrieved from https://technicgang.com/3g-vs-4g-technology-what-is-the-difference-between-3g-and-4g/amp/

R.T. Compton, *Adaptive Antennas: Concepts and Applications*, Prentice-Hall, Englewood Cliffs, 1988.

H. Cox, R. Zeskind and M.M. Owen, Robust adaptive beamforming, *IEEE Trans. Acoustics*, **35**, 1987.

S. Dang, O. Amin, B. Shihada and M-S, Alouini, What should 6G be?, Nature Electronics, vol. 3, 3030, pp 20-29.

D.J. Daniels, *Surface Penetrating Radar*, Institution of Electrical Engineers, London, 1996.

S.J.P. Dekleva,. Shim, Upkar Varshney, and Geoffrey Knoerzer. "Evolution and emerging issues in mobile wireless networks." Communications of the ACM 50, no. 6 (2007): 38-43.

E. Demir, T. Bektaş, and G. Laporte, "A comparative analysis of several vehicle emission models for road freight transportation," *Transportation Research Part D: Transport and Environment,* vol. 16, no. 5, pp. 347-357, 2011.

Y. Deville, A neural network implementation of complex activation function for digital VLSI neural networks, Microelectronics Journal, vol. 24, pp. 259-262.

S. Drabowitch, A. Papiernik, H. Griffiths, J. Encinas and B.L. Smith, *Modern Antennas*, Chapman and Hall, London, 1998.

A.A. El-Sherif., Mohamed. A, " Joint routing and resource allocation for delay minimization in cognitive radio based mesh networks," IEEE Trans. Wireless Commun., Vol. 13, No. 1, pp. 186-197, 2014

A.H. El Zooghby, Christodoulou, C.G. and Georgiopoulos, M. "Performance of radial basis function network for direction of arrival estimation with antenna arrays," *IEEE Transactions on Antennas and Propagations*, vol. 45, pp.1611-1617, 1997.

A.H. El Zooghby, Christodoulou, C.G. and Georgiopoulos, M. "Neural Network-Based Adaptive Beamforming for One and Two- Dimensional Antenna Array," *IEEE Transactions on Antennas and Propagations*, vol. 46, pp.1891-1893, 1998.

T. Elkourdi and O. Simeone, "Femtocell as a Relay: An Outage Analysis," *IEEE Trans. Wireless Commun.,* vol. 10, no. 12, pp. 4204-4213, 2011.

A. R. Elsherif, C. Wei-Peng, A. Ito, and D. Zhi, "Adaptive Resource Allocation for Interference Management in Small Cell Networks," *IEEE Trans. Commun.,* vol. 63, no. 6, pp. 2107-2125, 2015.

T.C. Farrar and E.D. Becker, *Pulse and Fourier Transform NMR*, Academic Press, London, 1971.

Z. Feng, and Zhang Z, "Dynamic spatial channel assignment for smart antenna," Wireless Pers. Commun. Vol. 11, No. 1, pp. 79-87, 1998.

R.P. Feynman, R.O. Leighton and M. Sands, *Lectures on Physics*, Vol. 2, Addison-Wesley, London, 1964.

P.J. Fitch, *Synthetic Aperture Radar*, Springer-Verlag, Berlin, 1988.

"5G and EMF explained." Retrieved from https//emfexplained.info/?ID=25916

5G-FD MIMO."Retrieved 5G and Large MIMO, from https://www.sharetechnote.com/html/5G/5G_MassiveMIMO_FD_MIMO.html

A.S. Gangal,. Kalra, P. K. and. Chauhan, D. S., Performance Evaluation of Complex Valued Neural Networks Using Various Error Functions, World Academy of Science, Engineering and Technology, 2007

Z. Gao, L. Dai, D. Mi, Z. Wang, M. A. Imran, and M. Z. Shakir, "MmWave massive-MIMO-based wireless backhaul for the 5G ultra-dense network," *IEEE Wireless Communications,* vol. 22, no. 5, pp. 13-21, 2015.

G.M. Georgiou,and Koutsougeras, C., Complex domain back propagation, IEEE Transaction On Circuits and Systems – II: Analog and Digital Signal Processing, Vol.39, No. 5. pp. 330–334, 1992.

Goh Seow Hee and Tan Pek Hua, *A Signal Processor for Tracking and Imaging a Landing Aircraft*, School of Electrical and Electronic Engineering, Nanyang Technological University, Singapore, Report, April 1998.

M.X. Gong, Akhmetov. D, Want. R, and Mao. S, "Multi-user operation in mmWave wireless networks," in Proc. IEEE Int. Conf. Commun., 2011, pp.1-6.

S. Goyal, Liu. P, Panwar. S.S, DiFazio. R.A, Yang. R, and Bala. E, "Full-duplex cellular systems: Will doubling interference prevent doubling capacity?" IEEE Commun. Mag., Vol. 53, No. 5, pp. 121-127, 2015

F. Gross, Smart Antennas with MATLAB^{TM}, McGraw Hill, 2nd Ed, 2015

X Gu, X-H Peng and G C Zhang, 2006. "MIMO systems for broadband wireless Communications", BT Technology Journal, Vol 24 No 2, April 2006.

N. Gustavo, Radio Link Performance of Third Generation (3G) Technologies of Wireless Networks" Faculty of Virginia Polytechnic Institute and State University, chp.2, pp.3, 2002

W. Guo, S. Wang, X. Chu, J. Zhang, J. Chen, and H. Song, "Automated small-cell deployment for heterogeneous cellular networks," *IEEE Communications Magazine,* vol. 51, no. 5, pp. 46-53, 2013.

G. Gür, Multimedia transmission over networks fundamentals and challenges,"Bogazici University, pp.717

F. Haider, C. Wang, B. Ai, H. Haas, and E. Hepsaydir, "Spectral-Energy Efficiency Trade-off of Cellular Systems with Mobile Femtocell Deployment," *IEEE Trans. Veh. Technol.,* 2015.

F. Haider, C. Wang, B. Ai, H. Haas, and E. Hepsaydir, "Spectral-Energy Efficiency Trade-off of Cellular Systems with Mobile Femtocell Deployment," *Vehicular Technology, IEEE Transactions on,* vol. PP, no. 99, pp. 1-1, 2015.

R. Hansch and Olaf Hellwich, "Classification of Polarimetric SAR data by Complex Valued Neural Networks," proceedings of ISPRS workshop 2009, vol. 38, Hannover, Germany, 2009.

Hamid,Jalab, A., Rabha and Ibrahim, W. , "New activation functions for complex-valued neural network", *International Journal of Physical Sciences*, vol. 6(7), pp 1766-1772, 2011.

S. Han, I. Chih-Lin, Z. Xu, and C. Rowell, "Large-scale antenna systems with hybrid analog and digital beamforming for millimeter wave 5G," *IEEE Communications Magazine,* vol. 53, no. 1, pp. 186-194, 2015.

A. Hatoum, R. Langar, N. Aitsaadi, R. Boutaba, and G. Pujolle, "Cluster-Based Resource Management in OFDMA Femtocell Networks With <roman>QoS</roman> Guarantees," *IEEE Transactions on Vehicular Technology,* vol. 63, no. 5, pp. 2378-2391, 2014.

G.C. Hess, *Land-Mobile Radio Engineering*, Artech House, Boston, 1993.

Y. Ho and M.A. Ingran, Design of partially adaptive arrays using the singular value decomposition, *IEEE Trans. Antennas Propagation*, **45**, 1997.

J. Hoadley and P. Maveddat, “Enabling small cell deployment with HetNet,” *Wireless Communications, IEEE,* vol. 19, no. 2, pp. 4-5, 2012.

R. Hohno and H. Imai, Combination of an adaptive array antenna and a cancellor of interference for DS-SSMA system, *IEEE Journal Comms*, **8**, 1990.

W. Hong, K.-H. Baek, Y. Lee, Y. Kim, and S.-T. Ko, “Study and prototyping of practically large-scale mmWave antenna systems for 5G cellular devices,” *IEEE Communications Magazine,* vol. 52, no. 9, pp. 63-69, 2014.

P.R.P. Hoole (Editor), *Electromagnetic Imaging in Science and Medicine: With Wavelet Applications*, WIT Computational Mechanics Publishers, Southampton, 2000.

P.R.P.Hoole, Smart Antennas and Signal processing for Communication, Biomedical and Radar Systems, WIT Press, UK, 2001.

S.R.H. Hoole, Computer Aided Analysis and Design of Electromagnetic Devices, Elsevier, New York, 1987.

S.R.H. Hoole (Editor), *Finite Elements, Electromagnetics and Design*, Elsevier, New York, 1995.

S.R.H. Hoole and P.R.P. Hoole, *A Modern Short Course in Engineering Electromagnetics*, Oxford University Press, New York, 1996.

P R P Hoole, K. Pirapaharan and S R H Hoole , *Engineering Electromagnetics Handbook*, WIT Press, UK, 2013.

P.R.P. Hoole, S. T. Ong, A. Balasuriya, S.R.H. Hoole, Shore to ship Steerable Electromagnetic Beam System based Ship Communication and Navigation, Proc. The 28th International Review of Progress in Applied Computational Electromagnetics, 6 pages, Michigan, USA, April 2012

P.R.P. Hoole, L.M. Abdul Rahim, S.R. H. Hoole, An Electromagnetic Signal Processor for Beam-forming a Wireless Mobile Station: Strengthening the Desired Signal and Nulling Main Interference, Proc. The 28th International Review of Progress in Applied Computational Electromagnetics, 5 pages, Michigan, USA, April 2012.

P.R.P. Hoole, K. Pirapaharan, and S. R. H. Hoole' An Electromagnetic Field Based Signal Processor for Mobile Communication Position-Velocity Estimation and Digital Beam-forming: An Overview, Journal of Japan Society of Applied Electromagnetics and Mechanics, Japan, Vol, 19, Nov 2011, pp. S33-S36

N. Hosein, and Hadi. B, "Sparse code multiple access," in Proc. IEEE 24th Int. Symp. Pers. Indoor Mobile Radio Commun. (PIMRC), 2013, pp. 332-336.

I. Humar, G. Xiaohu, X. Lin, J. Minho, C. Min, and Z. Jing, "Rethinking energy efficiency models of cellular networks with embodied energy," *Network, IEEE,* vol. 25, no. 2, pp. 40-49, 2011.

H. Insoo, S. Bongyong, and S. S. Soliman, "A holistic view on hyper-dense heterogeneous and small cell networks," *Communications Magazine, IEEE,* vol. 51, no. 6, pp. 20-27, 2013.

A. H. Jafari, D. López-Pérez, H. Song, H. Claussen, L. Ho, and J. Zhang, "Small cell backhaul: challenges and prospective solutions," *EURASIP Journal on Wireless Communications and Networking,* vol. 2015, no. 1, p. 206, 2015.

W.C. Jakes, A comparison of specific space diversity techniques for reduction of fast fading in UHF mobile radio systems, *IEEE Trans. Vehicular Technology*, **20**, 81–92, 1971.

S. Jangsher and V. O. K. Li, "Resource Allocation in Moving Small Cell Network," *IEEE Transactions on Wireless Communications,* vol. 15, no. 7, pp. 4559-4570, 2016.

S. Jangsher and V. O. Li, "Resource allocation in cellular networks employing mobile femtocells with deterministic mobility," in *IEEE Wireless Communications and Networking Conference (WCNC)*, 2013, pp. 819-824: IEEE.

S. Jangsher and V. Li, "Backhaul Resource Allocation for Existing and Newly-Arrived Moving Small Cells," *IEEE Transactions on Vehicular Technology,* vol. PP, no. 99, pp. 1-1, 2016.

F. Jean-Luc, Diane Titz, Fabian Ferrero, Cyril Luxey, Eric Dekneuvel, and Jacquemod, G. "Phased Array Antenna Controlled by Neural Network FPGA," Loughborough Antennas and Propagation Conference, Loughborough, UK, pp. 1-5, 2011.

S.S. Jeng, Experimental evaluation of smart antenna system performance for wireless communications, *IEEE Trans. Antennas Propagation*, **46**, 749–757, 1998.

Z. Jiang, A. F. Molisch, G. Caire, and Z. Niu, "Achievable rates of FDD massive MIMO systems with spatial channel correlation," *IEEE Transactions on Wireless Communications,* vol. 14, no. 5, pp. 2868-2882, 2015.

F. Jin, R. Zhang, and L. Hanzo, "Fractional Frequency Reuse Aided Twin-Layer Femtocell Networks: Analysis, Design and Optimization,"

Communications, IEEE Transactions on, vol. 61, no. 5, pp. 2074-2085, 2013.

E.A. Kadir., Shamsuddin S. M., Rahman T. A., Ismail A. S,"Big Data Network Architecture andMonitoring Use Wireless 5G Technology," Int. J. Advance Soft Compu. Appl, Vol. 7, No. 1, March 2015, pp.9

. O. B. Karimi, J. Liu, and C. Wang, "Seamless Wireless Connectivity for Multimedia Services in High Speed Trains," *IEEE Journal on Selected Areas in Communications,* vol. 30, no. 4, pp. 729-739, 2012.

A. Khan and A. Jamalipour, "Moving Relays in Heterogeneous Cellular Networks—A Coverage Performance Analysis," *IEEE Transactions on Vehicular Technology,* vol. 65, no. 8, pp. 6128-6135, 2016.

F. Khan, Pi Z, and Rajagopal S, "Millimeter-wave mobile broadband with large scale spatial processing for 5G mobile communication," in Proc. 50th Annual Allerton Conf. Commun. Control Comput. 2012, pp. 1517-1523.

M.S. Kim. and Guest, C. C. "Modification of Back-propagation for complex-valued-signal processing in frequency domain,"*International Joint Conference on Neural Networks*, (San Diego, CA), pp. III-27 – III-31, June 1990.

T. Kim, and Adali, T., Fully Complex Back-propagation for Constant Envelop Signal Processing, in Proceedings of IEEE Workshop on Neural Networks for Sig. Proc., Sydney, pp. 231–240, Dec. 2000.

J. Kim, and Kim. I.G, "Distributed antenna system-based millimeter-wave mobile broadband communication system for high speed trains," in Proc. IEEE Int. Conf. ICT Convergence, 2013, pp. 218-222.

T. Korakis, Jakllari G, and Tassiulas L, "A MAC protocol for full exploitation of directional antennas in ad-hoc wireless networks," in Proceedings of AMC international Symposium on Mobile Ad Hoc networks and computing, 2003, pp.97-108.

J.D. Kraus, *Antennas*, McGraw-Hill, New York, 1988.

E. Krestal (Editor), Imaging Systems for Medical Diagnostics, Siemens, 1990.

H.N. Kritikos and D.L. Jaggard, *Recent Advances in Electromagnetic Theory*, Springer, New York, 1990.

W.C.Y. Lee, Mobile Communication Engineering Theory and Applications, McGraw-Hill, New York, 1998.

J.C. Liberti and T.S. Rappaport, *Smart Antennas for Wireless Communications*, Prentice-Hall, Englewood Cliffs, 1999.

T.M. Lillesand and R.W. Kiefler, *Remote Sensing and Image Interpretation*, 3rd edn, John Wiley, New York, 1994.

J. Litva and T.K.Y. Lo, *Digital Beamforming in Wireless Communications*, Artech House, Boston, 1996.

Chun Loo and Norman Second, Computer model for fading channel with application to digital transmission, *IEEE Trans. Vehicular Technology*, **40** (4), 700–707, 1991.

M.T. Ma, Theory and Application of Antenna Arrays, John Wiley, New York, 1974.

V. J. MacDonald., "The Cellular Concept," Bell Systems Technical Journal, Vol. 58, No. 1, pp. 15-43, 1979.

A. S. W. Marzuki, I. Ahmad, D. Habibi, and Q. V. Phung, "Mobile small cells: Broadband access solution for public transport users," *Communications Magazine, IEEE,* vol. PP, no. 99, pp. 1-10, June 2017.

Massive MIMO." Retrieved from https://techblog.comsoc.org/tag/massive-mimo/ [63]

Y. Mehmood, Afzal. W, Ahmad. F, Younas. U, Rashid. I, and Mehmod. I, "Large scaled multi-user MIMO system so called massive MIMO systems for future wireless communication networks," in Proc. Int. Conf. Autom. Comput., 2013, pp. 1-4.

A Mehrotra, *GSM System Engineering*, Artech House, Boston, 1997.

A. Mesodiakaki, F. Adelantado, L. Alonso, and C. Verikoukis, "Energy-efficient user association in cognitive heterogeneous networks," *IEEE Communications Magazine,* vol. 52, no. 7, pp. 22-29, 2014.

R.A. Monzingo and T.W. Miller, *Introduction to Adaptive Arrays*, John Wiley, New York, 1980.

H. Mott, Antennas for Radar and Communications: A Polarimetric Approach, John Wiley, New York, 1992.

T.S. Naveendra and P.R.P. Hoole, Near field computation for medical imaging, *IEEE Trans. Magnetics*, **35**, 1999.

T.S. Naveendra and P.R.P. Hoole, A generalized finite sized dipole model for radar and medical imaging. Part I: Near field formulation for radar imaging, in J.A. Kong (Editor), *Progress in Electromagnetics Research*, PIER 24, pp. 201–225, 1999.

T.S. Naveendra and P.R.P. Hoole, A generalized finite sized dipole model for radar and medical imaging. Part II: Near field formulation for magnetic resonance imaging, in J.A. Kong (Editor), *Progress in Electromagnetics Research*, PIER 24, pp. 227–256, 1999.

T.S. Naveendra, *Near Electromagnetic Fields in Medical and Radar Imaging*, School of Electrical and Electronic Engineering, Nanyang Technological University, Singapore, Thesis, 1999.

T. Nguyen.,"Small Cell Networks and the Evolution of 5G (Part 1)," 2017. Retrieved from https://www.qorvo.com/design-hub/blod/small-cell-netwroks-and-the-evolution-of-5g.

Ng Joo Seng and Seow Chee Kiat, *Near Range Communication and Radar Network for Traffic Guidance and Control of Airport Surface*, School of Electrical and Electronic Engineering, Nanyang Technological University, Singapore, Report, 1998.

D. Noble, "The History of Land-Mobile Radio Communications," IEEE Vehicular Technology Transactions, pp. 1406-1416, 1962.

M. Jiang, and Hanzo. L, "Multiuser MIMO-OFDM for next-generation wireless systems," Proc. IEEE, Vol. 95, No.7, pp. 1430-1469, 2007.

V. Jungnickel *et al.*, "The role of small cells, coordinated multipoint, and massive MIMO in 5G," *IEEE Communications Magazine,* vol. 52, no. 5, pp. 44-51, 2014.

Oh Kok Leong and Ng Kim Chong, *Beamforming with Position and Velocity Estimation in Cellular Communication*, School of Electrical and Electronic Engineering, Nanyang Technological University, Singapore, Report, April 1999.

M.S. Kim. and Guest, C. C., Modification of back propagation for complex-valued signal processing in frequency domain, Int. Joint Conference on Neural Networks, San Diego, vol. 3, pp. 27-31, June 1990.

J. Kim, Lee. H.W, and Chong. S, "Virtual cell beamforming in cooperative networks," IEEE Journal of Selected Areas in Communications," Vol. 32, No. 6, pp. 1126-1138, 2014.

Mamta, Abhishek Roy and Navrati Saxana, "Next Generation 5G Wireless Networks: A comprehensive Survey," IEEE Communications Surveys & Tutorials, Vol. 18, No. 3, pp. 1617-1655, 2016

E. G. Larsson, O. Edfors, F. Tufvesson, and T. L. Marzetta, "Massive MIMO for next generation wireless systems," *IEEE Communications Magazine,* vol. 52, no. 2, pp. 186-195, 2014.

J. Liberti and T. S. Rappaport, *Smart Antennas for Wireless Communications: IS-95 and Third Generation CDMA Applications*. Prentice Hall, 1999.

L. Lu, G. Y. Li, A. L. Swindlehurst, A. Ashikhmin, and R. Zhang, "An overview of massive MIMO: Benefits and challenges," *IEEE Journal of Selected Topics in Signal Processing,* vol. 8, no. 5, pp. 742-758, 2014.

Lwin Maw Abdul Rahim[1], Hikma Shabani[2], Al-Khalid Hj Othman[2], Norhuzaimin Julai[2], Ade Syaheda Wani Marzuki[2], P.R.P. Hoole[2] and S. R. H. Hoole[3], Linear Smart Antenna Configurations for a Transceiver in a Multisignal Environment, Journal of Telecommunication, Electronic and Computer Engineering (JTEC), 2017

T. Maksymyuk, *Deep Learning Based Massive MIMO Beamforming for 5G Mobile Network.* in *2018 IEEE 4th International Symposium on Wireless Systems within the International Conferences on Intelligent Data Acquisition and Advanced Computing Systems (IDAACS-SWS).* 2018..

T. L. Marzetta, "Noncooperative cellular wireless with unlimited numbers of base station antennas," *IEEE Transactions on Wireless Communications,* vol. 9, no. 11, pp. 3590-3600, 2010

Widrow, B., McCool, J. and Ball, M., The Complex LMS algorithm, Proceedings of the IEEE, 1975.

Howard E. Michel, Abdual Ahad, S. Awwal and David Rancour, "Artificial Neural Networks using Complex number and Phase encoded weights – Electronic and Optical Implementations," *International joint conference on Neural Networks*, 2006.

M. Mavromoustarkis, G. Mastorakis, J. Batalla (Editors), Internet of Things (IoT) in 5G Mobile Technologies, Springer, 2016.

K. Nagothu, B. Kelley, S. Chang, and M. Jamshidi, "Cloud systems architecture for metropolitan area based cognitive radio networks," in *Systems Conference (SysCon), 2012 IEEE International*, 2012, pp. 1-6: IEEE.

A.F. Naguib, A. Paulraj and T. Kailath, Capacity improvement with base station antenna arrays in cellular CDMA, *IEEE Trans. Vehicular Technology*, **43**, 1994.

M. Nasimi., Hashim F. and Kyun C.N.,"Characterizing energy efficiency for heterogeneous cellular networks,"IEEE Student Conf. on Research and Dev.,pp. 200, 2012

P. Niroopan, and Chung. Y.H, "A user-spread interleave division multiple access system," Int. J. Adv. Res. Comput. Commun. Eng., Vol. 1, No. 10, pp. 837-841, 2012.

Osseiran, J.F. Monserrat, P. Marsch (Eds), 5G Mobile and Wireless Communications Technology, Cambridge University Press 2016

T. Omar, Z. Abichar, A. E. Kamal, J. M. Chang, and M. A. Alnuem, "Fault-Tolerant Small Cells Locations Planning in 4G/5G

Heterogeneous Wireless Networks," *IEEE Transactions on Vehicular Technology,* vol. 66, no. 6, pp. 5269-5283, 2017.

N. Palizban, S. Szyszkowicz, and H. Yanikomeroglu, "Automation of Millimeter Wave Network Planning for Outdoor Coverage in Dense Urban Areas Using Wall-Mounted Base Stations," *IEEE Wireless Communications Letters,* vol. 6, no. 2, pp. 206-209, 2017.

J.D. Parsons, M. Henze, P.A. Ratliff and M.J. Withers, Diversity techniques for mobile radio reception, *IEEE Trans. Vehicular Technology*, **25**, 75–84, 1976.

A.J. Paulraj and C.B. Papadias, Space–time processing for wireless communications, *IEEE Signal Processing Magazine*, pp. 33–41, November 1997.

Z. Pi and Khan F, "System design and network architecture for a millimeter-wave mobile broadband (MMB) systems," in Proc. IEEE Sarnoff Symposium, 2011, pp.1-6.

Z. Pi, and Khan F, "An introduction to millimeter-wave mobile broadband systems," IEEE Communications Magazine, Vol. 49, No. 6, pp. 101-107, 2011

L. Pierucci, "The quality of experience perspective toward 5G technology," *IEEE Wireless Communications,* vol. 22, no. 4, pp. 10-16, 2015.

K. Pirapaharan, Kunsei. H, Senthilkumar. K.S, Hoole. P.R.P, and Hoole. S.R.H, "A Single Beam Smart Antenna for Wireless Communication in Highly Reflective and Narrow Environment," in Proceedings of International Symposium on Fundamentals of Electrical Engineering (ISFEE), 2016, pp. 1-5.

K. Pirapaharan, Herman Kunsei, K.S. Senthilkumar, P.R.P Hoole, S.R.H. Hoole, A Single Beam Smart Antenna for Wireless Communication in Highly Reflective and Narrow Environment, Proc. International Symposium on Fundamentals of Engineering, IEEE Xplore Digital Library. 2017

K.Pirapaharan, P.R.P. Hoole, Norhuzaimin Julai, Al-Khalid Hj Othman, Ade S W Marzuki, K.S. Senthilkumar, , and S.R. H. Hoole, , Polygonal Dipole Placements for Efficient, Rotatable, Single Beam Smart Antennas in 5G Aerospace and Ground Wireless Systems Journal of Telecommunication, Electronic and Computer Engineering (JTEC), 2017

A.Prashanth, Investigation on complex variable based back propagation algorithm and applications, Ph.D. Thesis, IIT, Kanpur, India, March 2003.

J. Qiao, Shen. X, Mark. Q, Shen. Y, He. Y, and Lei. L, "Enabling device-to-device communications in millimeter-wave 5G cellular networks," IEEE Commun. Mag., Vol. 3, No. 1, pp. 209-215, 2015.

M. H. Qutqut, F. M. Al-Turjman, and H. S. Hassanein, "MFW: Mobile femtocells utilizing WiFi: A data offloading framework for cellular networks using mobile emtocells," in *Communications (ICC), 2013 IEEE International Conference on*, 2013, p. 6427-6431: IEEE.

V. Rajinder, "A Comprehensive Survey of the Wireless Generations," International Journal of Research in Computer Applications and Robotics, Vol.3, No. 9, pp. 5-51, 2015.

C. Ranaweera, C. Lim, A. Nirmalathas, C. Jayasundara, and E. Wong, "Cost-Optimal Placement and Backhauling of Small-Cell Networks," *Lightwave Technology, Journal of,* vol. 33, no. 18, pp. 3850-3857, 2015.

L.H. Randy, Phase-only adaptive nulling with a genetic algorithm, *IEEE Trans. Antennas Propagation*, **45**, 1997.

T.S. Rappaport, Wireless communications: principles and practice. Vol. 2. New Jersey: Prentice Hall PTR, 1996

T.S. Rappaport, Smart Antennas, Adaptive Arrays, Algorithms, and Wireless Position Location: Selected Readings, IEEE Press, London, 1998.

T.S. Rappaport T.S., "Millimeter wave mobile communications for 5G cellular: It will work!," IEEE Access, Vol.1, pp. 335-345, 2013.

Rappaport T.S, Gutierrez F, Ben-Dor E, Murdock N.J, Qiao Y, and Tamir J.I, "Broadband millimeter wave propagation measurements and models using adaptive beam antennas for outdoor urban cellular communications," IEEE Transactions on Antennas and Propagation, Vol. 61, No. 4, pp. 1850-1859, 2013.

T.S. Rappaport, Roh.W and Cheun K., "Wireless engineers long considered high frequencies worthless for cellular systems. They couldn't be more wrong," IEEE Spectrum, Vol.51, No. 9, pp. 34-58, 2014.

B. Razavi, *RF Microelectronics*, Prentice-Hall, Englewood Cliffs, 1998.

W. Roh *et al.*, "Millimeter-wave beamforming as an enabling technology for 5G cellular communications: theoretical feasibility and prototype results," *IEEE Communications Magazine,* vol. 52, no. 2, pp. 106-113, 2014

J. Rodriguez, *Fundamentals of 5G mobile networks*. 2015: John Wiley & Sons

E. Roubine and J.C. Bolomey, *Antennas*, Vols 1 and 2, North Oxford Academic, 1987.

W. Roh., "Millimeter-wave beamforming as an enabling technology for 5G cellular communications: Theoretical feasibility and prototype results," IEEE communication Magazine, Vol. 52, No. 2, pp. 106-113, 2014.

A.W. Rudge, K. Milne, A.D. Olver and P. Knight, *The Handbook of Antenna Design*, Peter Peregrinus, London, 1986.

D. Sabella *et al.*, "Energy Efficiency benefits of RAN-as-a-Service concept for a cloud-based 5G mobile network infrastructure," *IEEE Access,* vol. 2, pp. 1586-1597, 2014.

Y. M. Tsang, Poon. A.S.Y, and Addepalli. S, "Coding the beams: Improving beamforming training in mmwave communication systems," in Proc. IEEE Global Telecomm. Conf. (Globecom), 2011, pp. 1-6

S. Tombaz, A. Vastberg, and J. Zander, "Energy- and cost-efficient ultra-high-capacity wireless access," *Wireless Communications, IEEE,* vol. 18, no. 5, pp. 18-24, 2011.

N. Saquib, E. Hossain, and D. I. Kim, "Fractional frequency reuse for interference management in LTE-advanced hetnets," *IEEE Wireless Communications,* vol. 20, no. 2, pp. 113-122, 2013.

S. Scott-Hayward and E. Garcia-Palacios, "Multimedia Resource Allocation in mmWave 5G Networks," *IEEE Communications Magazine,* vol. 53, no. 1, January 2015 2015.

G. Sebastiani and P. Barone, Mathematical Principles of Basic Magnetic Resonance Imaging in Medicine, *J. Signal Processing*, 1991.

K.S. Senthilkumar, K.Pirapaharan, G.A.Lakshmanan, P.R.P Hoole, S.R.H. Hoole, Accuracy of Perceptron Based Beamforming for Embedded Smart and MIMO Antennas, Proc. International Symposium on Fundamentals of Engineering, IEEE Xplore Digital Library, Jan 2017,

K.S. Senthilkumar, Lorothy Singkang, P.R.P Hoole, Norhuzaimin Julai, S. Ang, Shafrida Sahrani, Kismet Anak Hong Ping, K. Pirapaharan, S.R.H. Hoole, A Review of a Single Neuron Weight Optimization Model for Adaptive Beam Forming Journal of Telecommunication, Electronic and Computer Engineering (JTEC), 2017

K.S. Senthilkumar, K. Pirapaharan, Norhuzaimin Julai, P.R.P Hoole, Al-Khalid Hj Othman,, R. Harikrishnan, S.R.H.Hoole, Perceptron ANN Control of Array sensors and transmitters with different activation functions for 5G wireless systems, Proc. IEEE Int Conf on Signal Processing and Communication, India, July 2017

M. Shiraz, A. Gani, R. H. Khokhar, and R. Buyya, "A review on distributed application processing frameworks in smart mobile devices for mobile cloud computing," *IEEE Communications Surveys & Tutorials,* vol. 15, no. 3, pp. 1294-1313, 2013.

P. Sibi., Jones S. A. and Siddarth, P., Analysis of different activation functions using back propagation neural networks. Journal of Theoretical and Applied Information Technology, Vol. 47, pp. 1264-1268, 2013

H. Silverman, Complex Variables, Houghton, Newark, USA, 1975.

M. Soumekh, *Fourier Array Imaging*, Prentice-Hall, Englewood Cliffs, 1994.

H.L. Southall, Simmers, J.A. and O'Donnell, T.H. "Direction finding in phased arrays with a neural network beamformer," *IEEE Transactions on Antennas and Propagations*, vol. 43, pp.1369-1374, 1995.

G.W. Stimson, *Introduction to Airborne Radar*, Hughes Aircraft Co., El Sugundo, California, 1983.

G.L. Stuber, *Mobile Wireless Communications*, Kluwer, Dordrecht, 1996.

W.L. Stutzman and G.A. Thiele, *Antenna Theory and Design*, John Wiley, New York, 1998.

S. Sun, T. S. Rappaport, T. A. Thomas, and A. Ghosh, "A preliminary 3D mm wave indoor office channel model," in *Computing, Networking and Communications (ICNC), 2015 International Conference on*, 2015, pp. 26-31: IEEE.

N. Suttisinthong, Seewirote, B., Ngaopitakkul, A. and Pothisarn, C., Selection of Proper Activation Functions in Back-propagation Neural Network algorithm for Single-Circuit Transmission Line, Proceedings of the International Multi Conference of Engineers and Computer Scientist, Vol II, IMECS 2014, March 12 - 14, 2014.

S. Taira, M. Tanka and S. Ohmori, High gain airborne antenna for satellite communications, *IEEE Trans. Aerospace and Electronic Systems*, **27** (2), 354–359, 1991.

S. Talwar, Choudhury. D, Dimou. K, Aryafar. E, Bangerter. B, and Stewart. K, "Enabling technologies and architectures for 5G wireless," in Proc. MTT-S Int. Microw. Symp. (IMS), 2014, pp. 1-4.

K. Taehwan kim, Tulay Adali, "Fully complex multi-layer perception network for non-linear signal processing," *Journal of VLCI signal processing*, vol. 32, pp. 29-43, 2002.

E. Tanghe, W. Joseph, L. Verloock, and L. Martens, "Evaluation of Vehicle Penetration Loss at Wireless Communication Frequencies," *IEEE Transactions on Vehicular Technology,* vol. 57, no. 4, pp. 2036-2041, 2008.

A. Thornburg, T. Bai, and R. W. Heath, “Performance Analysis of Outdoor mmWave Ad Hoc Networks,” *IEEE Transactions on Signal Processing,* vol. 64, no. 15, pp. 4065-4079, 2016.

S. Tombaz, S. Ki Won, and J. Zander, “On Metrics and Models for Energy-Efficient Design of Wireless Access Networks,” *Wireless Communications Letters, IEEE,* vol. 3, no. 6, pp. 649-652, 2014.

A. Uncini, Vecci, L., Campolucci, P. and Piazza, F., Complex Valued Neural Networks with Adaptive Spline Activation Functions, IEEE Trans. on Signal Processing, vol. 47, no. 2, 1999.

R. Vijayan, A model for analyzing handoff algorithms, *IEEE Trans. Vehicular Technology*, **42**, 351–356, 1993.

E.W. Vook, Ghosh A, and Thomas, T.A, “MIMO and beamforming solutions for 5G technology,” in Proc. IEEE Microwave Symposium (IMS), 2014, pp. 1-4.

X. Wang, P.R.P. Hoole and E. Gunawan, An electromagnetic-time delay method for determining the positions and velocities of mobile stations in GSM network, in J.A. Kong (Editor), *Progress in Electromagnetics Research*, PIER 23, pp. 165–186, 1999.

Z. Wang, Li H, Wang H, and Ci S, “Probability weighted based spectral resources allocation algorithm in HetNet under Cloud-RAN architecture,” in Proceedings of International Conference on Communication China Workshop, 2013, pp. 88-92.

N.Y. Wang, Agathoklis, P. and Antoniou, A. “A new DOA Estimation Technique Based on Sub array Beam forming,” *IEEE Transactions on Signal Processing,* vol. 54, no.9, pp. 3279-3290, 2006.

W. Wharton, S. Metcalfe and G.C. Platts, *Broadcast Transmission Engineering Practice*, Focal Press, UK, 1992.

B. Widrow and S. Stearns, *Adaptive Signal Processing*, Prentice-Hall, Englewood Cliffs, 1985.

T. Wu, Rappaport T.S, Collins C.M, “Safe for generation to come: Consideration of safety for millimeter waves in wireless communications,” IEEE Microwave Magazine, Vol. 16, No. 2, pp. 65-84, 2015.

J. Wu, Z. Zhang, Y. Hong, and Y. Wen, “Cloud radio access network (C-RAN): a primer,” *IEEE Network,* vol. 29, no. 1, pp. 35-41, 2015.

G. Wunder. G, “5G NOW: Non-orthogonal, asynchronous waveforms for future mobile applications,” IEEE Commun. Mag., Vol. 52, No.2, pp. 97-105, 2014.

P. Xia, Yong. S.K, Oh. J, and Ngo. C, "Multi-stage iterative antenna training for millimeter wave communications," in Proc. IEEE Global Telecomm. Conf. (Globecom), 2008, pp. 1-6.

P.. Xia, Yong. S.K, Oh. J, and Ngo. C, "A practical SDMA protocol for 60 GHz millimeter wave communications," in Proc. 42nd Asilomar Conf. Signals Syst. Comput., 2008, pp. 2019-2023.

G. Xiaohu, T. Song, H. Tao, L. Qiang, and M. Guoqiang, "Energy efficiency of small cell backhaul networks based on Gauss–Markov mobile models," *Networks, IET,* vol. 4, no. 2, pp. 158-167, 2015

W. L. Yang, "Ergodic and Outage Capacity of Narrowband MIMO Gaussian Channels", Department of Electrical and Computer Engineering, The University of British Columbia, Vancouver, British Columbia.

T. Y. Yang, W. Hong, and Y. Zhang, "Wideband millimeter-wave substrate integrated waveguide cavity-backed rectangular patch antenna," *IEEE Antennas and Wireless Propagation Letters,* vol. 13, pp. 205-208, 2014.

X. Yin, C. Ling, and M.-D. Kim, "Experimental multipath-cluster characteristics of 28-GHz propagation channel," *IEEE Access,* vol. 3, pp. 3138-3150, 2015.

K. Young-Min, L. Eun-Jung, and P. Hong-Shik, "Ant Colony Optimization Based Energy Saving Routing for Energy-Efficient Networks," *Communications Letters, IEEE,* vol. 15, no. 7, pp. 779-781, 2011.

W. R. Young "Advanced Mobile Phone System: Introduction, Background, and Objectives," Bell System Technical Journal, Vol. 58, No. 1, p.7, 1979.

S.J. Yu and J.H. Lee, Design of two-dimensional rectangular array beamformers with partial adaptivity, *IEEE Trans. Antennas Propagation,* **45** (1), 1997.

S. Yutao, J. Vihriala, A. Papadogiannis, M. Sternad, Y. Wei, and T. Svensson, "Moving cells: a promising solution to boost performance for vehicular users," *IEEE Commun. Mag.,* vol. 51, no. 6, pp. 62-68, 2013.

A. Zakrzewska, Ruepp. S, and Berger. M.S, "Towards converged 5G mobile networks-challenges and current trends," in Proc. IEEE ITU Kaleidoscope Acad. Conf. Living Converged World Impossible Without Stand.?, 2014, pp.39-45.

Y. Zeng, Zhang. R, and Chen. N, "Electromagnetic lens-focusing antenna enabled massive MIMO: Performance improvement and cost reduction," IEEE Journal of Selected Areas in Communications," Vol. 32, No. 6, pp. 1194-1206, 2014.

J. A. Zhang, X. Huang, V. Dyadyuk, and Y. J. Guo, "Massive hybrid antenna array for millimeter-wave cellular communications," *IEEE Wireless Communications,* vol. 22, no. 1, pp. 79-87, 2015.

S. Zhang, Q. Wu, S. Xu, and G. Li, "Fundamental green tradeoffs: Progresses, challenges, and impacts on 5G networks," *IEEE Communications Surveys & Tutorials,* 2016.

H. Zhang, D. Jiang, F. Li, K. Liu, H. Song, and H. Dai, "Cluster-Based Resource Allocation for Spectrum-Sharing Femtocell Networks," *IEEE Access,* vol. 4, pp. 8643-8656, 2016.

X. Zhao, S. Li, Q. Wang, M. Wang, S. Sun, and W. Hong, "Channel Measurements, Modeling, Simulation and Validation at 32 GHz in Outdoor Microcells for 5G Radio Systems," *IEEE Access,* vol. 5, pp. 1062-1072, 2017.

G. Zheng, "Joint beamforming optimization and power control for full-duplex MIMO two-way relay channel," IEEE Trans. Signal Process., Vol. 63, No. 3, pp. 555-566, 2015.

J. Zhu, H. Wang, and W. Hong, "Large-scale fading characteristics of indoor channel at 45-GHz band," *IEEE Antennas and Wireless Propagation Letters,* vol. 14, pp. 735-738, 2015.

A. El Zooghby, Smart Antenna Engineering, Artech House, 2005

Index

About the Author

Paul R.P. Hoole completed his first electrical and electroninc engineering degree to postgraduate degrees in the United Kingdom. He holds an M.Sc degree in Electrical Engineering with a Mark of Distinction from the University of London and an MSc degree in Plasma Science from University of Oxford. His engineering doctorate, the D.Phil. degree, is from the University of Oxford. In his engineering career he has spent time in Singapore, Papua New Guinea, USA, Malaysia and Sri Lanka. His recent publications and research interests have been in the areas of advanced wireless systems, electromagnetic signal processing, lightning-aircraft electrodynamics and wireless systems based supervisory control and data acquisition network for renewable energy electric power generation and drone imaging in remote villages and systems. Prof. Hoole has authored several papers and books in engineering. His books *Electromagnetic Imaging in Science and Medicine* and *Electromagnetics Engineering Handbook* (with K. Pirapaharan and S.R.H. Hoole), were published by WIT Press, UK.